RESOLVE AND FORTITUDE

RESOLVE AND FORTITUDE

Microsoft's "SECRET POWER BROKER" breaks his silence

JOACHIM KEMPIN

Copyright © 2012 by Joachim Kempin.

Library of Congress Control Number: 2012919149
ISBN: Hardcover   978-1-4797-3201-2
               Softcover   978-1-4797-3200-5
               Ebook   978-1-4797-3202-9

All rights reserved. No part of this book may be reproduced or transmitted in any form or by any means, electronic or mechanical, including photocopying, recording, or by any information storage and retrieval system, without permission in writing from the copyright owner.

Indexed by Reginald Raymund Caturza
Copyedited and reviewed by Fatimah Imam

This book was printed in the United States of America.

To order additional copies of this book, contact:
Xlibris Corporation
1-888-795-4274
www.Xlibris.com
Orders@Xlibris.com

TABLE OF CONTENTS

PREFACE IX

GETTING MY FEET WET... 1

THE BUSINESS.. 9
    Insight the power nexus 9
    Cleaning house 15
    Renewing customer focus.. 23
    Dead on arrival.. 29

FLYING HIGH 35
    The world belongs to me... 35
    Historical customer visits 39
    Getting out of bondage 47

TURBULENCE.. 51
    Étude in *f-moll* 51
    A serious wake-up call 55
    Crisis in the far east 59
    Pirates of the world 61
    Falling behind.. 65
    AIMing 71
    Assignments and threats 75

DISRUPTIONS 79
    Ups and downs 79
    Front line partnership 85
    Rescuing little critters... 89
    Monopoly accusations 93
    Patent horrors 103

The Saga of Big Blue 107
- A sweetheart deal 107
- In the shadow of Chicago 111
- Alliance round two.. 115
- Trust and verify... 119

Start me up 123
- Microsoft nearly missed it 123
- Hurdles before launch 127
- On stage and beyond...131

Business "as usual"... 139
- Of opportunities and threats 139
- Run by committee 147
- Compaq going astray...151
- A strongheaded response...153
- Internet explorer to win..157
- Prelude in the Senate..159
- Another alliance attempt 161

In the shadow of the feds 167
- The only man Bill Gates may fear...................................167
- Back in Jackson's court 171
- Internal politics 179

The business must go on... 183
- Windows 98, finally 183
- Software minutemen..185
- Price wars 193
- Piracy in china..195
- Nobody is immune to failure 199
- Forging ahead 209

Justice on the block 213
- Trial setup and strategies 213
- Witness parade 221
- In the lion's cage 229

Betrayal of justice 241
- Rebuttal...241
- Monopoly power..245
- Verdict 255

My last hurrah 257
- A new CEO 257
- Xbox 263
- Late challenges 267
- Searching for a successor 273
- Appellate court response 275
- Parting 281

A state of disorder 289
- What *l'audace*? 289
- PC down spiral 305
- A cool renaissance to be reckoned with! 309
- A last Word 323

Appendix: How Microsoft got her stripes 327
- And how I earned mine 345

Acknowledgments 349

Glossary 351

References 371

Index 373

Preface

"Join us and run Microsoft's German subsidiary!"

The trip from Paris to Seattle had not been wasted. They wanted me! About time somebody recognized this German guy can do more than manage Apple's small European software-marketing team. It took Bill Gates, co-founder and CEO of Microsoft, the better. The only thing left was to say *"Yes*, for the right amount" and take the challenge head-on.

Onward I went, and more than thirty years later, let me now share with you what happened after I sealed the deal. It did not stop with running Microsoft Germany successfully. Headquarters called on me a second time and offered me a promotion, and off I went to emerge into Microsoft's power nexus from where I managed—for over fourteen years—her dealings with personal computer (PC) manufacturers.[1] In doing so, I achieved senior vice president status long before I left the company in 2002 to retire happily.

Being trusted with that job was the unforeseen highlight of my career. The primary objective of the group I managed was to sell operating systems—the software code which makes computers tick—to PC manufacturers. The most important software we offered, Windows, defines Microsoft's heart and soul. Its success made the company the number one juggernaut in the information technology universe. When this book will be published, at version 8, Windows will have experienced much iteration since I had left. It still dominates the PC world, but Microsoft has struggled to get a foothold in the tablet and smartphone market with its derivatives—a humbling experience for a company that once reigned so supreme! Antitrust concerns, internal turf wars, and certain unwillingness to forcefully defeat competitors or boldly conquer new

---

1   We called them OEMs, original equipment manufacturers.

grounds are the main reasons for failing to keep the once-uncatchable, seeming-frontrunner position.

Aptly, my story converges on how the company fought and won operating system wars—how major triumphs and breakthroughs were achieved and how my group contributed to foment Microsoft's lead. As long as Microsoft was the hardworking underdog aiming for the coveted number one spot in the information technology industry, the press and the public showered her with admiration. Eventually, her exorbitant success bred jealousy and spite, echoing volleys of gunfire from competitors and regulators alike—tremendously changing the public perception of the company.

My group's actions and the inner machinations of Microsoft landed crosshairs dead center in the middle of two major Department of Justice headline trials. Intimately involved, I sweat through all of them. Not necessarily what I had signed up to when I said *yes* to Germany.

A journalist once wrote that like an "enforcer," I "wielded the pricing sword" for Microsoft's operating systems. Others called me "Microsoft's secret power broker." Not needing monikers, I kept them guessing—a game I loved to play. Then, the public had no right to know! Yet the time has come to reveal what went on behind the scenes and lastly unearth the bold-faced truth and intimate details I lived through and contributed to. The turmoil, the challenges, the victories, and the defeats. Occurrences I undeniably and irrefutably put my mark on as I cherished every moment of being part of the by-now historic PC revolution.

Let's warm up in Germany and follow the path my group took to win the operating system wars. Let the voyage continue with a critical review of Microsoft's antitrust trial. I invite you to examine from a key witness's perspective what economic theories and legal tactics our pursuers harnessed to ultimately get the company condemned and partially regulated. Following this, I will issue a report card for the company's current management and investigate how her lead got eroded by computing paradigm shifts, emerging competitors, and the failure to respond timely and vigorously. After examining the impact of Windows version 8 in regard to serving an increasingly mobile and socially active computing crowd and its potential to reverse Microsoft's fortune, I will finish by challenging Microsofts' board and leadership team to restructure the company. I firmly believe that drastic changes are needed so Microsoft can conjure some future magic and impress the investment community anew.

An educational, revelatory, and exhilarating trip is waiting. Illuminated and put into perspective by my thirty years of work experience in the information technology industry. My views being controversial, disagreements will be abounding. Let the debate begin!

Joachim Kempin, October 26, 2012, Seattle WA, USA

PS: For readers less familiar with some of the technology mentioned in the book, the state of the personal computer market from the late 1970s to the mid-1980s, or my biography, please study the appendix in more detail.

Getting my feet wet

"Sacrifice for the cause and you will prove your worth!"

Words spoken by my new boss, Scott Oki, still ring in my ears! That third-generation Japanese American preached what he practiced. Highly disciplined, energetic, and intelligent, Scott had gained the trust of Microsoft's founder, Bill Gates, through hard work and superior management skills. He shared Bill's dream of dominating the personal computer software industry and made sure the crew he managed worked feverishly to make this vision come true. High ethical work standards made him a tough boss. He communicated his intents unambiguously and made clear that only success would earn you a longer leach. Scott—he kept his word. As the German subsidiary grew from zero to $15 million during his three years as my boss, he entrusted me to coordinate the retail aspect of our business for all of Europe.

I maintain the utmost respect for Scott and regretted his departure in '86 to take over Microsoft's ailing US retail business. All country managers chipped in and gave him a symbolic farewell present: a glass replica of Caesar's thumb. No doubt he had us under his, but performing well, I never felt uncomfortable. We learned from each other. His appreciation of my performance later caused him to again influence my career by recommending me to take over the OEM sales and marketing group.

Life as subsidiary manager was quite different from my earlier work assignments. The buck stopped with me. Microsoft (MS) was a company on the frontiers of technology, sizzling with ambition, innovative passion, and the white heat urgency of the desire to succeed. My job was to create a crew who shared the company's goals passionately and to find a way to let our results stand out. The mother ship created winning products; the subsidiary established effective distribution systems and worked the local PR channels.

Geographically I was responsible for doing commerce in Germany, Austria, Switzerland, and all Eastern-bloc countries. Our retail business targeted Apple and IBM PC users. We reached them through distributors in merchant outlets. The products we sold in that fashion were PC programming languages, application software, and a few PC hardware add-ons. To promote these, we used creative print advertising and means of public relations.

The most exciting part of our business was dealing with local PC manufacturers, conducting what consisted of selling MS Disk Operating System software (MS-DOS) and early versions of MS Windows to local PC manufacturers. Most of them simply purchased DOS licenses from us. Others, like Siemens, had piled on behind Windows early and enthusiastically. In Triumph-Adler and Siemens we acquired two customers for our Xenix operating system—an MS proprietary UNIX[2] version.

One aspect helping us win was our ability to produce localized product versions well ahead of competition. In the spreadsheet category, having a German version of Multiplan available propelled us right past our competitors Lotus and VisiCalc. Soon our sales in Germany surpassed the nearly unrivaled US product leader Lotus 1-2-3, resulting in the early dismissal of Lotus's first subsidiary manager. In the word processor category, it took a bit longer to beat the entrenched WordStar competition, but eventually we succeeded there as well. A German version of MS-DOS gave our main operating system (OS) competitor Digital Research Inc. (DRI) a huge headache while awarding us the edge in dominating the local OEM market.

We fought hard for these gains, pursuing our missions vigorously and with immense focus. I submerged in and was so inspired by MS's culture that I made the company my second home. My family accused me of being married to her; somehow I was and it paid off. Germany was soon MS's largest subsidiary in Europe, generating 10 percent of corporate revenue. Our success created some jealousy and rivalry. My French colleague Bernard Verges could not stomach our premier position and, in a more or less friendly way, challenged our sales results every month. With a larger market potential and a dedicated team on my side, he never surpassed us. Thank you, old crew!

Having zero experience running a subsidiary and considering the daunting challenges we faced, I was forced to learn quickly on the job. To accomplish our goals, I encouraged employees to experiment boldly when exploring new opportunities. I solicited unorthodox ideas and solutions, and after

---

2 An operating system originally developed and licensed by AT&T.

careful critique, we courageously applied them to the contests at hand. It was a golden and compelling time to work for this entrepreneurial enterprise! We all enjoyed the abundance of freedom we had serving customers. If bad decisions were made, people just learned from their mistakes! As long as the offense was inspired by an honest objective to improve the company's performance and services, forgiveness trumped.

I experienced this when we started selling directly to large enterprises. Assuming that bypassing our distribution partners would cause retaliation and bad blood, my boss was not supportive and recommended against it. Convinced we had to try, we did it anyway! Three months later, our success made all other European subsidiaries jump on the bandwagon. Taboos existed only to be broken—I never got a reprimand. I intuitively realized that my boss, like myself, wanted empowered and impassioned employees who perceived their jobs as vital missions and pursued them with unrelenting energy. In the true spirit of the early American pioneers, instead of winning the West, we set out to dominate the information technology (IT) world.

Lessons learned during my German military service and in former management training sessions helped me master my new assignment. As I observed how the company was run at the corporate level, an unmistakable metric of parallels surfaced. Her leadership was definitely based on the principle of *Auftragstaktik*—a mission-oriented command philosophy that establishes a leadership style[3] based on general guidance as opposed to prescriptive supervision. It endorses soldiers' (or, in this case, employees') initiatives and encourages independent decision making at all levels. Yes, I know Germany lost two world wars, so on the face of it, her military practices can't be superior. Yet the core elements of this empowerment philosophy are easily applied not only to the military but to reengineer an enterprise as well.[4]

The premise of this mission-oriented command philosophy has its roots in the early nineteenth century after the Prussian army experienced painful losses during the Napoleonic wars. Her generals were forced to analyze the reasons behind the multiple defeats, starting by questioning the petrified command structure of the army hierarchy they had created and lived under. They were shocked by what they found. Despite nearly exponential increases in battle complexity and the number of troops deployed, commanders were leading, in centuries-old custom, from behind. At a time when the scope of engagements made all-encompassing real-time observations no longer feasible, they had to rely on an ancient messenger system—on foot or on horseback—to obtain information needed to direct and deploy their troops.

---

3   *Führungsstil* in German.
4   See http://ezinearticles.com/?expert=Robert_G._Ogilvie.

With this system being unreliable and lacking speed, commanders could therefore neither maintain sufficient situational awareness nor react fast enough to changing battlefield conditions.

To make things worse, fighting men—per army doctrine of that time—weren't empowered to deviate from previously giving orders. You risked your life if you did not follow them unabatedly. When battleground conditions changed and communication failed, soldiers consequently dared not to react except to bolt for their lives. Judging the engineering of their carefully crafted battle strategies as Prussian thorough, the investigators deducted that, by allowing situation-conscious subcommanders not enough autonomy to operate independently, failure, from cumbersome rigidity, was a built in inevitability.

Over the next fifty years, the elite of the Prussian generals—Scharnhorst, Gneisenau, Moltke the Elder and Clausewitz—set out to elevate command behavior to the demands of modern combat. In peacetime, when their armies were bound by a system of strict discipline, management was effective. In wartime when leadership beyond was required, the system broke down. In a 1990 *Harvard Review* article, John P. Kotter correctly points out that management deals primarily with complexity, while leadership first and foremost addresses change. The Prussian army, with its deeply layered structure and proven regulations, was highly capable at dealing with the former. Generals, like a lot of corporate executives, excelled in complex managerial tasks such as planning, budgeting, organizing, staffing, and controlling cost. Yet most were ill equipped to align and inspire people toward a common goal and trust them enough to adjust independently in the heat of a battle without taking their eyes off the desired objective.

To change such a deeply entrenched command system was a formidable task. Nobody wanted to endanger the basic disciplinary structure an army (or a corporation) requires to function. Firm belief in a religiously fixed, hierarchically command discipline—working well in peacetime—lay squarely in the way of modernization. Even for these powerful Prussian figures, mastering the desired behavioral makeover presented a daunting balancing act. They never let go! Two generations went by before their work paid off and laid the foundation for winning the Austro-Prussian War of 1866 and the Franco-German War in 1870–71. Later their doctrine enabled the successful Blitzkrieg campaigns during the early phase of WWII.

Only in the late twentieth century, following the collapse of the Soviet Union, did the Anglo-American army officially, yet hesitantly, adopt an Auftragstaktik-like leadership style. With the marine corps being ahead of other branches in the US military as it historically relied on the commander's intent as a guiding principle for carrying out orders, specifying the desired result while leaving open the method of execution. Nevertheless, exemplary, forward-looking generals—Patton for instance—already studied and applied the concept extensively during WWII, contributing both to his success and to the controversy of his legend.

Moltke characterized this revolutionary Prussian command doctrine best: "An order shall never include anything a commander can do by himself." Establishing a mission-oriented leadership style means leaving the tactical details of achieving an objective to lower levels in the chain of command, thus ensuring flexibility and rewarding initiatives. As long as the commander's original intent is unwaveringly maintained and defined objectives are relentlessly pursued, no permission is required when adapting to shifting battleground (or competitive) situations.

This was exactly how I judged MS's early operating style. If you could not license an operating system at list price and a competitive threat loomed on the horizon, sweetening the pot needed zero permission and would not be punished but rewarded. Business first, pride later!

This style of command management is only successful when all goals are precisely formulated so the intent of the one in charge is unmistakably understood by all subordinates. The commander then has to extend trust to all levels in the chain and believe in the determined execution capabilities of his or her troops. This requires considerable training, confidence in the leadership, and the ability to trust each other. A collective willingness to cooperate, with zero tolerance for conformists or careerist self-protectionism, follows. This is much easier to accomplish with a smaller group than with armies of individuals employed in large enterprises!

Upon looking at one of my old computer programs, a professor of mine once implored, "Why do you always have to do things differently than other people?" He saw it otherwise; I considered his question a compliment. As long as my version performed better than the one formerly designed—by him—he should have congratulated me. I understood the aim of his assignment perfectly, and had given my best to accomplish the task at hand. This was the spirit I wanted people working for me to operate in! If, in the process, they superseded my own achievements, all the better!

The final element in this context is the vital importance of speed for an enterprise. As on the battlefield, competitive encounters demand swift decision making and therefore should not be judged later on accuracy alone. This is not to say that I appreciate hasty conclusions and careless responses. Yet delaying decisions does not win the day.

I soon recognized that managing the still-young tech start-up with such a mind-set was no easy task. The company had just over four hundred employees and the average age was merely 23.5 years. Being over 40 years of age, I was considered a handicap. Energy and stamina was desired; experience counted less. Any lack of it was made up for with enthusiasm and seventy- to eighty-hour workweeks with no consideration of overtime pay!

What else impressed me? No resting on your laurels! If moss can't gather on rolling stones, imagine meteors. This management philosophy created not only competitive products, but also ever-successful teams generously rewarding breakthrough contributions, resolve, and fortitude. As a result, we all passionately and tirelessly strived for excellence following the motto drilled into us: "Only best products and services will eventually be victorious!" Best for consumers. Best in industry. Giving up was not an option in management's parlance. Contentedness with success or status quo? Unacceptable behavior. With the industry moving at lightning speed, falling behind produced fatal results, and even small lapses severely endangered our hard-earned position.

As on a battlefield, the intense work environment demanded up-to-date information. To facilitate this, MS installed a proprietary messaging system, instantly converting the company into an e-mail-addicted community. At first the German subsidiary was not connected—left out. We tried to keep up via phone, fax, and old-fashioned telegrams (!) but quickly became frustrated suffering unacceptable response times. "E-mail or die!" Only after we got hooked up via an expensive private overseas network did we feel like real Microsofties. This breakout advantage of nearly religious e-mail usage so early in the trajectory of organizational modernism contributed immensely to MS's success. At once I made it my personal habit to respond to every mail within twenty-four hours even while on the road. The ripple effect of communication momentum proved to be an incalculably powerful tool for me personally.

E-mail addiction was paired with the less-welcome principle of "bad news travels the fastest." Bill was the one who had personally instilled his disgust for holding back unfavorable information into everyone. Delaying good news meant corks would be popped later; delaying bad news was a crime, hindering speedy remedies! People learned quickly, and mostly the hard way.

There were failures and flaws to be sure. Projects were abandoned or delayed, and concepts flopped. But after all the dust settled in those early days, the company had charted an astonishingly successful course over IT territories never before conquered and into places never dreamed of, evolving MS rapidly into one of the most fascinating and successful companies in the history of modern commerce.

Even with all the freedom I possessed running Germany, I sometimes felt myself being dragged along by the overwhelming gravitational power of the host planet. After four years, I longed to be closer to the power nexus. A chance to run all of Europe had come and gone, and a VP position to run the application group had not materialized. Lotus tried to lure me away with an attractive offer, MS fortunately countered, and I stayed put. Out of the blue, Scott Oki, my ex-boss, contacted me: "The VP in charge of the US OEM group is retiring. Would you be interested?"

MS's intriguing OEM business provided huge leverage for the company and was her most profitable one. Running that group demanded well-honed management skills and excellent technical knowledge of computer and OS technologies, as well as strong sales and negotiation skills. Confident I could measure up, I applied and moved on.

THE BUSINESS

INSIGHT THE POWER NEXUS

"We are vulnerable and can disappear in the blink of an eye!"

This cautionary mantra expressed by MS's top management still resonates. In the eyes of our top echelons, beating Lotus, WordStar, and DRI in Germany, while impressive, did not matter as much as I thought it would. For them, only relentless pursuit of excellence guaranteed our future.

As I arrived in Seattle for my new job, Bill's philosophy and the company's progressive execution to evolve PC technology with OEMs and independent software vendors (ISVs)—the ones who write the key applications programs needed to make a computing platform a success—was in full swing. This was my chance to have a huge impact on the growth prospects of MS, her OEM customers, and the industry in general. Trusting me with the critical assignment of running the OEM group felt good, but I knew there was a staggering amount of work ahead. I had to prove myself all over again in this new job—no laurels to rest on!

My first encounter was with my new boss Jon Shirley, a former exec from Tandy/RadioShack who had led her PC division successfully. Jon was a class act manager with razor-sharp business instincts. A solid and astute, always-probing-for-just-the-right-answer executive of the highest credentials! For MS's tennis-shoes-and-jeans culture, he was a bit too buttoned-up but made up for it with experience and being personal and inviting. When you talked to him, he expected you to be well prepared. And if you weren't, he didn't hesitate administering a nice dressing-down, employing his dry but stinging humor. Jon dwelled in a world of numbers and details. His favorite inspection area during a subsidiary visit was the warehouse. Every subsidiary had one in the days before MS's logistics for Europe got streamlined in Ireland. His walk-throughs were as legendary as his method of spotting excess inventory. If you were discovered, he would demand an immediate explanation of what you planned to do about it. With Jon around, you always had to be on your toes!

In our first one-on-one encounter, he immediately expressed the need for a thorough analysis and evaluation of my new crew's performance and every single individual it contained. I liked his straightforward management style and well-meant guidance. He was direct, tough, and intense but never prescriptive.

I had experienced this personally back in Germany when our US manufacturing operations had let us hang out to dry by not supplying products in a timely fashion—for several months. Instead of accepting the unacceptable, I decided to produce them locally, only to be caught by Jon during his next warehouse inspection. Right away he spotted minute variances in our local package designs—a touch too small and a slight color deviation. Yes, the man was all detail and always in the trenches. I got an earful and expected to be fired on the spot. Not the way Jon operated. Back in the United States, after investigating the incident, he read our manufacturing VP the riot act. I was forgiven, and he ordered me to obtain his approval before another need to replenish our supplies locally ever arose.

In doing so, Jon neatly endorsed a principle I firmly believed in. Do not punish well-intending actions, in this case prompted by the singular purpose of achieving healthy quarterly revenue numbers. Therefore, he tolerated the violation of a long-standing policy once and replaced it with a more sensible one, setting a memorable example.

In our unforgettable first meeting, he encouraged me to consult and discuss any OEM related issues directly with Bill Gates. Since founding the company in 1975 with Paul Allen, he had stayed close to MS's OEM customers. Selling BASIC[5] to these clients had started the company, and in his heart, he sometimes still saw himself as chief of OEM sales. Unfettered access to Bill! I was flattered. I only hoped the arrangement wouldn't cause conflicts in the chain of command or lead to end-running manipulations. These two guys had wildly different qualities. Jon was a details fiend with an office too neat to be comfortable—typical for an operations guy. The eccentric Bill was some kind of a clutter man, all awhirl in concept, brainstorms, and books to the ceiling kept neat by the enduring efforts of his admin. My original suspicion luckily turned out to be unfounded. Having two bosses offering different perspectives helped me to function in my new position with the best degree of knowledge and insight possible.

---

5   A programming language called Beginner's All-purpose Symbolic Instruction Code.

Let's meet Bill next, Jon's opposite twin—in looks and in interests. Nonassuming with a strong hint of boyish nerdiness, he couldn't care less about his underwhelming first impression. He sported outsized glasses and began rocking metronomically in his chair the second the discussion turned technically challenging. Prescient and brilliant, Bill had an uncanny, almost Einsteinian, grasp of the universe of technology—its dimension, velocity, and business implications. He certainly knew numbers, regularly shaping them into context more meaningful than the most prepared managers when probing the depths of their intellects with challenging questions. Projections. Deductions. Logarithmic inquiries. Modern information technology was where he passionately excelled. If he accepted your opinion in the world of tech, you had gained not only his interest but also his enduring respect. More distant than Jon, he could often been spotted floating in a cloud overhead. Never to be underestimated! Leaping from topic to topic, his mind always racing in overdrive, he was less predictable. By comparison, Jon, the methodical, always reached the synthesis of any issue he pursued; Bill got there if the people in attendance stimulated his intellect.

At age thirty-two, he was already solemnized by the surrounding aura of a visionary guru status. Complementing this was a perceived precociously shrewd business sagacity that inspired some to loathe and others to envy him. Steering the company successfully for twelve years had allowed him to forge close and lasting relationships with top leaders on the frontiers of the PC transformation. He had personally coined the company's mantra: "A PC in every home and on every desk." For my colleagues and me, by extension, the slogan meant "With a lot of MS software utilized on them." Suspecting competitors found it aggressive and arrogant. For the company's employees, it portrayed an unrelenting commitment to transform the nascent industry into an IT superturbine.

We all believed in this transformation. In my first exchange with Bill, we therefore focused on the roles OEMs were playing in this bigger picture and what he expected from me in my new role. He opened up by saying, "I consider nurturing and fomenting the OEM customer base a key building block for our aspirations." He then stated that MS software innovations were impelled by this customer base and their component suppliers and that this lockstep was what created a tsunami force of revolutionary information appliances. He then lectured me that neither of us could prosper without independent software vendors (ISVs). As I knew from my work in Apple, they delivered the applications that enriched these information appliances, like icing on the cake. As such, they fashioned the cutting edge needed to transform PC-based information technology into an unbeatable dynamo. "Powerless without them" sounded quite humble coming from him.

Our discussion turned even more invigorating, inspiring, and sharply informative when he briefed me on MS's evolving product strategies centered on the importance of making Windows a success and the challenge IBM posed in this regard. He further outlined his interest in the high-tech endeavors OEMs were engaged in and suggested we meet frequently to share freshly gathered intelligence. Hardware-driving software and vice versa ensured the insight I provided would keep him apprised of key novelties. His challenge to me: "Your new job is not just a sales and marketing job. I expect you to provide me with relevant information and strategic recommendations so I can make better decisions for the company." The primary reason Jon Shirley had extended me unencumbered access was evident. Obviously, Bill wanted unfiltered reports and an opportunity for sharing his assessments of events directly with the person running the most profitable and strategic group in the company to date. I felt honored to receive such a refreshingly new level of trust and responsibility. And I made myself a promise to answer his calling by pairing the highest level of performance with perfect execution!

Completing MS's executive trio was Bill's closest friend, then in charge of R & D for all operating systems and today's CEO of MS, Steve Ballmer. He made up for any shortfall through sheer voltage and a personality inimitably his own. I heard people say that he looked like a linebacker with quarterback ambitions. Eventually he made that grade! Burly and nearly bald, he bristled with crosscurrents of energy. When raising his voice and booming with enthusiasm, he became easily carried away. People who did not know him well were scared or intimidated by some of his outbursts. Nevertheless, Steve remains the most intuitive people manager I have ever met. He presented himself as a confident, eternal optimist but played the compulsory doom-and-gloom card brilliantly—effectually offering you a contrary, chilling dimension. He had worn many different hats in the young company's history. Bill's "fixer" was one. Supposedly ever loyal, he had straightened out the thorniest predicaments. Holding degrees from Harvard in mathematics and economics, this chest-thumping showman was another numbers guy at heart. Numbers, finally, were what grounded him. I quickly learned how to use the numerical to reach him. The clear absolute language of numbers and their certainty and involuntary warmth had the effect of calming Steve and of speaking directly to his careening sensibilities.

Over the years, I had gotten acquainted with our exec trio through frequently scheduled review meetings. An ironclad practice that kept the company on track and focused and top management in the loop. Formal and well-structured, these meetings were complemented by casual gatherings where our top guys informed us about sea changes they'd spotted shaking up the industry. Bill's updates offered a wealth of insight into PC manufacturers' affairs, their upcoming product announcements, and the horizons of hardware component manufacturers. They also caught up with MS software competitors, their expected directions and potential strategies. Bill was uncannily well-informed, and his alchemistic industry foresights nearly always came through. He used these gatherings as an effective method to foster trust and ignite confidence in MS's destiny. Participants left pumped up, filled in, and utterly convinced. Leadership at its best! His only shortfall was a virtue called patience. I was in good company.

Steve kept us abreast of operating system developments, the latest feature changes, embarrassing delays, and joint development activities with IBM. He considered himself the guardian of that company's loyalty, ensuring her support for at least MS-DOS and OS/2—a new operating system jointly developed with IBM to replace DOS and MS Windows. What surprised me the most was that the trio viewed IBM simultaneously as a vitally important collaborator and a potentially deadly cobra! In the mid '80s, IBM was the largest software company in the world. The bulk of her software was meant for use on her mainframe[6] platforms. Nothing prevented her from challenging us with alternative PC software. MS was loaded with tons of talent brimming with ambitions but was tiny and with rather scant resources compared to IBM. The top guys rightfully worried.

Another competitor we watched and examined closely was Apple Computer. As she went from her earlier platforms to the Lisa workstation and the most innovative Macintosh (Mac), Bill admired her. Nevertheless, she constituted another formidable competitor well ahead in operating system (OS) design. A true threat to the IBM PC platform and potentially MS's livelihood! We nonetheless supported the MAC with advanced office productivity applications such as MS Excel and MS Word, an experience that helped us when we later adapted them for MS Windows.

No one in MS ever quite understood why Apple didn't license her Mac OS to other manufacturers. Back in '84/'85, Bill suggested this several times to Steve Jobs, one of Apple's founders, who was in charge of Mac development. If he had acted upon that recommendation, he would have radically altered the PC landscape, MS's future, and my career.[7] Jobs disagreed with Bill because he wanted end-to-end control over all aspects of Apple's products and despised what made MS so successful—freely licensing OSs. After Jobs got ousted in 1985, the new team did not pursue this unique opportunity either and stayed on Jobs's course until Mike Spindler, my ex-boss, took over as Apple's CEO in '93. Lucky us!

To keep the troops alert, Bill and Steve were fond of depicting an atmosphere

---

6   Large computer systems used for commercial and scientific computing tasks.
7   The book *Steve Jobs* published in 2011 sheds some light on this.

of chilling paranoia, of predators waiting to devour us. Monsters in the darkness, a competitor du jour—IBM, Lotus, WordStar, DRI, any of them. Their warnings instilled far more than merely in-depth comprehension of the company's goals and objectives. They made certain we stayed on our toes, putting competitors on notice while vigilantly attempting to predict their next moves. But the wildfires of fear and obsession sputtered for me at times. The rants became a bit theatric, a match in the wind. The German-born mathematician would have preferred more realistic and analytical evaluations, and many of my colleagues shared my belief. To inspire us to jump in the trenches and blaze away as if our lives depended on it, the top guys led us to believe an army of dangerous Huns was always advancing over the horizon, fully intent on annihilating us. Competing for our execs meant keeping enemies constantly in sight—the favorite tactic from their portfolio of persuasion. The only one! Fearful explicitness was their form of communicating competitive dangers to all employees, painting competitors in vivid black and white; shades of gray did not exist. Did this gladiator goading induce ever-greater aggression and anticompetitive behavior among the foot soldiers? I might never know.

Cleaning House

"We have an unfair advantage, and we intend to keep it that way."

This is how VP Jim Harris, the predecessor in my new job, described how he had run the business. I was appalled and wondered what else was wrong with him or my new group. By the spring of '87 MS had gained a dominant position in the OS market for IBM PCs, and I didn't believe a market leader needed to express such arrogance when dealing with clients. Yet Jim, the entrenched pit bull—as some called him—insisted.

This intense sales executive had joined MS from Intel in '83 and, through grit, charm, and exhaustive effort, developed a healthy $50 million OEM business for the company. However, my first meeting with him surprised me. Not only had he cleaned out his desk, but he also indicated his desire to stay at home for the next three months, preparing for his move to a new destination. Sure, he would take my calls and make introductions to key customers, but otherwise, he would be hands off. I was tickled by his confidence. Retreating so quickly nevertheless felt strange!

Getting acquainted with my new colleagues and charges and fully coming to terms with policies and procedures of the OEM branch was next on my agenda. A journey chocked with jolting revelations! What I discovered absolutely rocked me. The OEM group consisted of roughly thirty-five salespeople along with five logistic personnel plus several administrative assistants contributing largely by selling MS-DOS and a much smaller number of MS Windows units. The sales team was divided into four groups. Three were serving the larger customers. The fourth was dedicated to a staggering number of smaller ones. There were no marketing people whatsoever, and no written policies and procedures could be found.

By far the most extraordinary black hole in the OEM universe, salespeople were prohibited from possessing the official price guideline (PGL). Jim had designated the maintenance of this top secret document to one of his lieutenants. No one else had direct access. Management feared the covert document could fall into customer hands. Not having faith in the sales teams relegated them to lowly messenger roles. My first thought: *No trust, no power, no customer respect, change ASAP.* My management style differed sharply when directing people. Right then I realized a little grassroots empowerment in good old German military tradition would go a long way with my new crew!

To test my conclusion, I began dropping into offices to get to know my new cohorts, showing interest for special customers' projects, their plans, and high-profile challenges. My new teammates, not overly accustomed to management addressing them directly and soliciting advice, initially hesitated to respond. What was going on here? Languishing on this, I rapidly recognized that gazing into history served no purpose. I needed to urgently move on, introduce my kind of trust-generating management style. Auftragstaktik to the rescue!

Attached to the group were several attorneys responsible for developing standard contracts for broad-stroke clientele and unique versions negotiated mostly with large customers. A knowledgeable and reliable controller handled revenue reporting and billing. The group overall was roughly the same size as the one I'd left behind in Germany but was raking in twice the proceeds.

An evident impediment to progress led me to attack the PGL matter first. In Europe we had carefully fashioned and put into motion a single one for all countries. Jeremy Butler, who was now VP in charge of all non-US business, was pleased when I approached him, proposing unified worldwide pricing. In confidence he told me Jim Harris had been adamant about maintaining his sovereignty and staunchly resisted constructing a one-size-fits-all global list. My pact with Jeremy thoroughly pissed off the current PGL keeper, inciting hostile arguments. I glossed over them, and being the spreadsheet expert I was, I single-handedly engineered the new PGL from then on with MS Excel. Both Jon and Bill were pleased with our decision. A first and important step in treating all customers around the world equally! Unified contracts were next.

I perceived contracts as a necessary evil to conduct business, as I preferred handshakes. In German, a contract is called *ein Vertrag*. It originates from the verb *vertragen*, which I prefer to translate into "to live in harmony" instead of "to agree." Thus, agreeing in the moment gets superseded in spirit by expecting to live in harmony forever. If attorneys would compose agreements with that intent in mind, prosperity and peace would undoubtedly be fostered. Let's see how MS postulated her intent to live in harmony with her OEM customers.

Until '83, MS licensed MS-DOS on a flat-fee basis to OEMs. After paying it—typically around $50,000—OEMs could preinstall a copy of MS-DOS on every PC they cared to produce but were restricted from distributing MS-DOS stand-alone. Only IBM was exempted from that general rule.

All licenses were version specific. Product versions were identified by a primary version number, and two additional numbers behind a decimal point reserved for smaller upgrades. Licensees received the lesser ones for free, enabling them to keep their PCs up-to-date. Adding major new features to an OS was indicated by changing the number to the left of the decimal point. For these, a new license was needed and a new flat fee became due.

Even with several hundred OEM customers, MS could hardly turn a profit using this business model since, as a rule, two to three years normally expired between major releases. As MS-DOS matured and grew in size, development costs increased exponentially. MS was left with few choices for making that business profitable. Bumping up the flat-fee price was one, charging for smaller upgrades the other. Instead she chose to ask her customers to pay a royalty for every OS unit they shipped. Introducing the change gradually by navigating the sea of expiring agreements and new product versions, the company eventually converted all customers. From then on she received a smoother and vastly enhanced revenue stream, making this segment of her business finally profitable. DRI, her toughest competitor, had always licensed her OSs on a per-unit basis; therefore, no competitive disadvantage was experienced.

OEMs were offered two choices under the new scheme: The first option resulted in a pay as you go accord—a "per-copy license" as we called it. This type of agreement gave OEMs maximum flexibility in deciding which PC would be shipped with or without our OS. The second option required OEMs to install it on all PCs with the same model name. We labeled such an accord a "per-system license." This reduced royalty rates by $1 per PC. The reduction was justifiable because the per-system commitment made had the potential to produce a more dependable and larger revenue stream.

Extra volume discounts were made available for firmer purchases. To earn them necessitated OEMs to make an unwavering annual "minimum commitment"—a typical and well-accepted practice in most industries. Customers reported their unit shipments quarterly, and based on their stated volume and royalty rates, MS invoiced them. To remain in good standing, they typically had to pay their bills within thirty days.

As accountable and enduring business entities, our customers typically planned their PC models twelve to twenty-four months ahead. Estimating PC sales even for such a period was a challenge—more art than science. Business assumptions had to take several factors into account: economic uncertainties, demand fluctuations, technology advances, and the competitive landscape. Important tactical decisions such as timing product launches, setting attractive price points, and determining appealing configurations influenced the outcome as profoundly. Most OEMs understood the risks in forecasting future sales and were circumspect in doing so. Incorrect targets based on reckless assumptions resulted in inventory that was difficult to sell or, worse, forced a company out of business. Most of our customers, unenthused about overcommitting, played safe by projecting realistic sales objectives.

Entering into an annual guarantee with us became pretty straightforward. We probed customers' assumptions, and if they were deemed too ambitious, we wanted them turned down. Our sales reps were not incentivized, as we never paid any commissions. Encouraging customers to enter into the impossible was deemed counterproductive and was discouraged all the way up the chain of command. Some overly ambitious CEOs nevertheless insisted on achieving a lower per-unit price, hoping that committing the company to a larger unit target might dial up long-needed sales mojo. Unrealistic or not, leaders do take risks, and asking them to scale back the lofty underlying assumptions is not always an easy task. If all attempts to haze a customer back to a more sensible commitment level failed or—worse case—a viable competitive threat surfaced, winning the business superseded lingering objections, and an unrealistic deal was concluded. Shame on all of us!

When bitter realism caught up with pipe dream–like projections most customers saw the signs on the wall and renegotiated their payment horizons in due time, paying higher per-unit prices. Too often, however, an accord ran out unrevised with customers expecting nothing less than mercy. Nobody wanted to kiss overpayments good-bye by just writing them off. The argument we heard most: "Yes, we have prepaid for licenses, but we have not consumed them either. Therefore, payments made for nonconsumed goods should be paid back or credited." Their voluntarily signed agreements expressed the opposite, and they cavalierly forgot that the extra discounts they had received were in exchange for paying us for the contracted units.

Finding ourselves in legal arguments with customers, though, was an unwholesome predicament that almost always headed for troubled waters. Deep inside, I was sympathetic to their claims. In all honesty, we had not earned the money, and we certainly desired to continue to grow a healthy business relationship. But the overpaid money in its entirety wasn't simply heading back their way. Most sinners understood our lucid and logical position when we offered to accrue a reduced amount for future purchases. I blamed my sales reps and their managers as much for these painful predicaments as the irresponsible customers—aren't they always right? With a little common sense and oversight, most of these train wrecks could have been avoided.

Negotiating settlements of this nature was undoubtedly one of the least appreciated jobs. The simple fact of customers not making sales projections, often right from the beginning, cast a dark shadow and had the potential to turn into an ugly fight. Our cold-blooded competitors accused us of setting the bar artificially high for minimum commitments to deliberately lock them out. Far from the truth as explained above.

If OEMs had bought too many CPUs[8] from Intel, would she have taken excess back and returned the money? Not on your life. They would probably have wound up on the black market. Intel, like MS, prohibited OEMs from selling products without having them installed inside a PC. The difference between Intel's CPUs and the prepurchased OSs: our customers bought a license and no physical goods. Licenses could be used up later. CPUs, on the other hand, had a shorter shelf life, and their value deflated over time. How else to view this? Business, with all its quirks, permutations, and natural checks and balances, has thrived and evolved with vigor in the real world since the Cro-Magnon man swapped spears for skins. Once I'd traded my best six spears for your oxhide, I wasn't giving you the hide back because you had only used three of them. You'd thought you needed all six getting the mastodon. Now we should trade them back, with me receiving what—a worthless half hide?

Condemning our minimum commitment practices, competitors further claimed that OEMs only continued licensing OSs from us from fear of losing their overpayments. Customers had better reasons to stay with us. They and their clients obviously liked and appreciated our products. Staying loyal, they put their trust in a vendor viewed as having marketplace longevity through steadily increasing product values. Their bet was on the forward-looking standard setter!

After licensing a competitive OS, OEMs returned to us for a reason. They believed in our vision and leadership, willingly accepting higher royalties. You will always pay extra for the leading brand. If one wishes to drive a Porsche, one pays for a Porsche; if a Chevy, then a Chevy. Sharply different

---

8   Central processor unit, electronic component capable of executing the binary instructions for a computer system

experiences dictated by volition, demand, and quality! Our customers committed along with us.

Last but not least, let me mention MS policy for recognizing OEM revenues. Our accountants always accrued overpayments. They never contemplated recognizing them fully until OEMs reported actual shipments. Therefore, my group never received any credit for the accrued sums. In other words, there was no incentive for my sales group or me personally to encourage customers to inflate purchasing guarantees.

The new PGL was completed just in time for the 1987/88 financial planning exercise. I was astonished to see my controller, Tim Beard, doing most of the work instead of the individual sales reps. He struggled with a dearth of data, doing his best to generate a valid bottom-up, OEM-by-OEM, and system-by-system forecast. This concept had worked nicely in Germany, enhancing planning acumen especially when the data was derived with customers' input. Afterward, management compared the final numbers with easily obtainable top-down market estimates from independent analysts. No one in my new group had been trained to apply these simple safeguards. Eventually, we finished the budget on time, but it contained an uncomfortable degree of ambiguity. Luckily, my first fiscal year ended with us achieving the projected numbers. Comfortable only with exceeding forecasts—a personal disappointment. Perfecting the flawed process was made a high priority.

Having concluded I had a dysfunctional management team, I replaced its members within the next year—one by one. Meanwhile, my outstanding controller, the evolving management team, and our experienced attorneys volunteered to take a stab at scripting a basic operating procedure. Together we bore down with inspired intensity, reviewing long- and short-term team goals, and basic day-to-day ops principles engineered to provide a solid framework for ensuring success. Eventually, a new ops manual was given to every salesperson, including standard licensing contracts and an up-to-date price guideline. My salespeople finally had the basic tools to excel in their profession and act with confidence.

To make my own office functional, I asked our facility group to fit me out with a desk containing no drawers. Drawers encourage putting things off. I preferred a couple of larger filing cabinets in my admin's office and a conference table with chairs in my own. I made a point of sitting down with employees at my conference table instead of staying at my desk. Lowering the employee-boss barrier made meetings less intimidating and more productive. On top of my desk, documents I was currently working on were in plain sight for their perusal. No reason to conceal the tasks at hand.

Every time I observed managers with chronically empty desks, I couldn't help but wonder how hard they were really working. Perhaps I'm making too much of my personal habits, but the top performers in my group never had a clean desk even when on vacation.

Renewing customer focus

"Know your customers and you know your business."

This is a principle my predecessor seemed to have neglected during the last year of his reign. Instead, he and his team spent excessive time in never-ending weekly review meetings. So I cancelled them. Assuming my instructions were crisp and precise, anything beyond monthly reviews felt like babysitting and endangered the empowerment principle I believed in.

Emphasizing customer understanding and encouraging visits made the account teams redefine account strategies with the goal of enriching client relationships and expanding our business. For mission control, I insisted on inviting top management to join us, soliciting exec input before we plotted our next steps. Bill and Jon especially liked to participate and offer encouragement, fanning the flames of progress and future engagements. They contributed fabulously and were appreciative of our degree of preparedness. Both had considerable history with and insight into most of our key customers and didn't hesitate to challenge the account teams. Imagine the awkward moments of embarrassment when one of our execs displayed a deeper familiarity with a client's account than the presenting sales rep! Serving an all-for-good and enlightening purpose, I learned a ton while observing the painful and often rib-tickling entanglements in good humor.

These synergizing and invigorating powwows served as confidence boosters bolstering our sales reps. No longer doubting the mission, they strengthened sales abilities and injected an exhilarating pioneering spirit when serving a customer-driven world with renewed purpose. As I had anticipated, encouragement and backup from top management produced astonishing results by slowly but surely turning our SRs into confident and empowered customer advocates. They savored the results!

When visiting US customers for the first six months, I had to play catch-up.

So I required the account teams to provide a thorough review of each and every outstanding issue beforehand. Jim Harris helped me get the ball rolling. Once he believed I had a good handle on things, he graciously stepped aside, allowing me to move seamlessly into existing relationships. A handful of these early meetings were emblematic and memorable.

I remember visiting the Michels duo from SCO[9] with Jim. We were licensing Xenix—MS's version of UNIX—to them, which no longer fit into our portfolio. During my first visit, we began preparing them for a technology transfer, allowing MS to concentrate primarily on MS-DOS, Windows, and OS/2.

SCO was founded in '79 by a colorful father-and-son team, Larry and Doug. They were a couple of fun-loving good old tech warriors of an earlier stripe, given to deep number-diving, raw humor, and multiple entrees. Generously proportioned chowhounds who did business for real and for fun! We met them at a boisterous fish-fry eatery on Santa Monica Pier and didn't get around to discussing serious business issues until their second round of plates arrived.

"Well, men," said Jim H., "now that you've met Joachim, you know exactly who's going to be pampering you even more than me from now on. If that's possible!"

"No," snickered Doug around a mouthful of fried squid, "it isn't."

Larry winked at me, chewing bawdily, and gave a thumbs-up with one hand, waving the waitress over with the other. Jim just sat smiling contentedly. You could see it wasn't his first rodeo with these guys. He enjoyed it deeply, grinning at me, contributing to the growing sense of certainty that the new relationships were going to be enjoyably successful.

Another notable visit with Jim was with execs of Phoenix Computers. They were eager for an agreement to sell MS-DOS together with their BIOS[10] chips (containing a piece of software enabling the CPU of an IBM PC to perform basic input and output operations). Phoenix sold her version loaded into so-called read-only memory (ROM) chips, ready to be inserted onto a PC motherboard socket. In 1981 she had bought Seattle Computer Products (SCP)—the old Tim Paterson company. The same-self firm MS had acquired QDOS[11] from—our starting point for creating MS-DOS. In her by-now-ancient contract, SCP had retained the right of distributing her own

---

9   Santa Cruz Operation
10 Basic Input/Output System
11 Quick and Dirty Operating System

hardware with MS-DOS, and the new owners had decided to exercise this long-forgotten distribution privilege.

As much as Jim and I resented having Phoenix as a second supplier for MS-DOS, we had come to negotiate the desired deal in good faith. Priced at around $5, Phoenix's BIOS chip hardly qualified as hardware, as outlined in the old agreement. It therefore needed careful pricing considerations. To avoid potential conflicts and overlaps, we proposed they target system builders,[12] which we served poorly, for such a combo pack. There was no shortage of these entrepreneurial garage-band front-runners who, for us to be thorough in an energized low-end marketplace, had to be reckoned with.

If Phoenix could become a bona fide distributor for this market segment with a track record of denying us revenue through piracy, she would be of great help. Her management team fell in at once, but enthusiasm waned when we revealed our royalty expectations. Our potential partner was already disappointed by our initial pricing proposal when Jim Harris, adding insult to injury, dropped the derogative bombshell I headlined. I was deeply embarrassed. The Phoenix guys, visibly upset, shared my sentiment. I couldn't blame them. Up to then, we'd had have a lively, constructive, and amicable meeting. Why pee in the pool? Jim was wise and diplomatic enough to excuse himself at once, and with me at my best behavior, we managed to arrive at a balanced agreement.

By then Compaq had become MS's most important OEM customer. The company occupied the lofty position of most innovative PC manufacturer, along with being a close ally and the world's number two PC manufacturer. Bill had befriended her chairman and main investor Ben Rosen, remained close to her CEO Rod Canion, and enjoyed the demanding intellectual exchanges with her R & D SVP Gary Stimac. Steve's buddy was her sales and marketing VP Mike Swavely. MS's early engagement with Compaq had helped the start-up in '81 to a quick breakthrough. By now she was firmly established, fast growing and profitable, and respected throughout the industry by friends and foes alike. Product-development organizations in both our companies continued to work closely with each other, ensuring new Compaq PCs would function smoothly with all MS software. To IBM's dismay, our partnership resulted in Compaq staying well ahead of the innovation curve. Unfortunately, our mutual working relationship was focused solely on MS-DOS. Compaq had neither licensed nor actively promoted Windows. My personal mission to change!

The challenge of moving Compaq to Windows grew into a long-term struggle

---

12   Very small PC manufacturers

despite our continuing and mutually inspiring bond. The introduction of version 2.03 at the end of '87 supporting Intel's 80386 CPU promised to change this as we succeeded to persuade her execs to license it for one model. Against our advice, Compaq's product management skimped on its memory configuration to strike an attractive price point. The result was predictable: running in super slo-mo, Compaq's Windows PCs failed to shine. Compaq blamed it on bloated code; we cited missing memory. From then on, convincing Compaq of the glories and marketing miracles of Windows amounted to a herculean task. I had my work cut out!

Our business relationship with Digital Equipment Corporation (DEC), one of my ex-employers, was another complex one. She had entered the PC business relatively late. Ken Olsen, her cofounder and CEO, couldn't quite bring himself to believe PCs would ever gain traction in the real world. The respected, wise old oracle was famous for quoting, "There is no reason for any individual to have a computer in his home." Oh boy!

I met him personally for the first time in a sales training class in '74 in Maynard, Massachusetts, DEC's headquarters. He was a leader and an engineer by heart and a business executive by accident. Success had never gotten into his head. Whenever permitting, he made room in his schedule to attend the last day of a new-hire sales training class. His visit with me in attendance was advertised as a luncheon meeting. To the participants' great disappointment, lunch was never served nor paid for by the company. In his excessive stinginess, Ken brought his own, normally a sandwich prepared for him by his wife. He expected his sales personnel to do the same or buy theirs in the company's cafeteria. "As a good salesperson, you have to be underpaid" was the phrase always omitted during Ken's visits. The incident reminds me of the slogan for one of the cooking channels: "Stay hungry." Underpaid was his way of expressing that hungry salespeople will perform better—I subscribe to that.

When he reluctantly allowed Digital Equipment Corporation (DEC) to enter the PC fray, the company did so by licensing the needed OS from DRI. Her selection made no friends in MS and created a myriad of incompatibilities for her customers. DEC paid a tall price by not choosing the better partner and by not making her PCs true IBM PC clones. As losses mounted, DEC finally switched to MS-DOS. Living successfully in a proprietary world, her hardware, even in round two, was still not engineered for 100 percent IBM PC compatibility. Trying to be different and adding proprietary touches did not work. Failing twice was difficult to accept for the leading minicomputer vendor. When DEC eventually fell in line, she found herself far behind and had overpaid MS several millions in royalties.

Cleaning up the mess fell squarely in my lap when visiting DEC in '87 for the first time in my new capacity. I saw an urgent need to alter her payment schedule to stop her from bleeding cash. Not present when she entered the PC business, I carried no baggage from her fling with DRI and was best qualified to reset her license without prejudice. After two false starts, Ken Olsen had put VP John Rose in charge of her PC business. John was a rational and friendly executive who openly acknowledged the deep hole DEC had dug herself into. Honesty is the best foundation for a successful negotiation, and it did not take long to arrive at an agreeable solution.

One company I visited relatively late was Packard Bell (PB). Her founder and CEO, Beny Alagem, an ex-Israeli tank commander, was as bold and aggressive in business as he had been in the tank battles of the '73 Yom Kippur War. He was a brilliant, quick-on-his-feet sort of guy, blessed with an abundance of wit and humor. A talented and shrewd negotiator who sold PCs exclusively through retail. When I first met him, he was expanding into Europe, for which I offered him valuable insight. He had been reluctant to buy Windows—the reason for my visit. Laser focused on the supercompetitive and price-conscious consumer market, the resulting increase in costs had hindered him to commit.

The man was notorious for obtaining the best possible prices for all PC components; software royalties were no exceptions. Every meeting or phone call with Beny always ended with him requesting a better deal. He worked on rock-bottom margins, making pennies count. My visit did make a small difference as we agreed to a Windows trial for a number of higher-end systems. A surprise for his competitors who never gave Beny the credit he deserved, even when his company grew beyond wildest expectations. In the meantime, I enjoyed a constructive, harmonious business relationship with a man who never stopped calling me Jochen.

Observing Beny in action, I recognized his adopted management principles were similar to mine. The successful Israeli defense forces had adopted the principle of Auftragstaktik to guide its troops early on. Watching Beny at work proved he had had taken his military training lessons to heart as he crisply defined his motives, methods, and goals and issued clear instructions. He never lost sight of his aim. His refreshing openness and honesty was no doubt the reason the two of us hit it off at once. Respect naturally followed.

Visiting numerous customers set an example for my troops, increased my understanding of their business perspectives, and last but not least, affected my continuing restructuring effort. It motivated me to refine and tune our policies and to match personalities between customers and sales reps. As a result, I chartered Jim Cecil, a consultant, to independently conduct annual customer surveys to improve customer satisfaction. Increased marketing campaigns with OEMs were the other most obvious consequence. My now-intimate understanding of the galloping US PC market enabled me to produce an accurate business forecast for my second fiscal year. As the economy picked up, we beat our budget handsomely. I had my bragging rights back!

As I was the new kid on the block, the press had initially been eager to talk to me. As manager of the German sub, I had given interviews freely and initially without much thought. Journalists covering our local activities and the mother ship were critical but open and fair. Once in a while, an interviewer was a bit sneaky, making me increasingly wary over time, planting the first seeds of deep circumspection in my perception of the media.

What I came to experience in the United States was considerably worse. Most journalists were driven by self-serving agendas, often pretty hostile ones—based on speculation, seldom on facts. In my first couple of interviews, reporters put unrecognizable words in my mouth, with no relation to the verbatim. My position, with intimate proximity to the highest ranking MS and customer executives, provided plenty of knowledge of goings-on behind the curtains. I became hypervigilant about disseminating or confirming confidential information, on or off the record. I soon concluded the press was using me to suit a narrow, high-handed agenda of sensationalism and lowbrow controversy. I couldn't quarrel with publishers wanting to sell papers, but misinforming their readers was a tactic I wanted no part of. I consider myself a pretty straightforward guy, and it irked me to watch my words become twisted into journalistic knots.

Further agreeing to be interviewed by sensationalists could simply lead to a world of hurt with my loyal customers and my trusting superiors. Disputing or denying parts of an interview, once published, came off poorly and served only to enhance the misguided story. By then the cat was out of the bag. I therefore began declining interviews in the United States, leaving the thankless chore to others. Understandably, my snubbing did not create accolades. I continued giving interviews in other parts of the world where reporting was less slanted and self-serving. In retaliation, the US press called me all manners of names and unearthed any gossip they could find. All the mud hurling bothered me little. I had a terrific and important job, and I fortunately didn't need the distracting glare of the limelight. Less enabled me to succeed where it counted.

Dead on arrival

"I believe OS/2 is destined to be the most important OS . . . of all times."

Who would not have shared this opinion after reading what MS's visionary chairman expressed in the programming guide for OS/2 published in '87? The business world and trade press certainly bought the story as advertised. Early disagreements and lingering tensions between IBM and MS about its specifications and development schedule had by now been overcome. Despite the thorniness of the past, a release date had been set for the fall.

This release date made my OEM customers in the summer of '87 hurry to jump on the bandwagon! Under no circumstances would they risk missing the transition or allow IBM the opportunity to establish another lead. As agreed upon with IBM, OEMs could license OS/2 from either company, presenting my group with a potential challenge. Yet our initial fears were soon proven groundless. Licensing from the number one competitor was just not in the cards, and to my delight, we obtained that business almost exclusively.

Lacking an effective licensing department was probably one factor driving OEMs to us. The way IBM's contracts or prices were contrived might have been the other, but with my customers never disclosing any details. In that respect, I was blinded. We drew up our pricing guides without ever talking to IBM and experienced no resistance. A nearly feverous atmosphere incited a perfect storm of a sign-up frenzy and, unbeknownst to us, a world of severe headaches in the not-too-distant future.

One of the companies eager to deploy OS/2 was Hewlett-Packard (HP). Jon Shirley accompanied me on my first visit to Palo Alto, California, meeting her execs. The game plan was to conduct a relationship review, introduce me as the new VP, and then address an OS/2 issue HP had previously brought up. HP produced and marketed her own PCs while operating—with the exception of her printer division—like a slow-moving utility company. Her bread and butter was selling mainframe like minicomputers and terminals, running a profitable software business, and developing her consulting practice. As with IBM, MS judged her cooperation with healthy skepticism. The reasons can be found in her DRI-based terminal business and her nearly religious UNIX engagements. The overall review went reasonably well, and we agreed to follow up with her printer division for a cooperative alignment.

Convening separately with HP's PC division, we veered quickly into a vicious contract dispute. HP was one of the companies still holding on to an old MS-DOS flat-fee agreement. According to her contract interpretations, it included OS/2. Jon considered this a ruse. Having familiarized myself with the old contract language, I couldn't find any reason supporting HP's claim. Her hairsplitting attorneys' adamancy was based on language unearthed by reading an ill-worded press announcement positioning OS/2 as the successor to MS-DOS and Windows. Forgetting that Windows was not part of DOS per se and HP's unwillingness to acknowledge the substantial differences between OS/2 and MS-DOS made their arguments look like an irrational ploy. During the meeting, Jon, the ever the inspiring leader, assigned me the task of resolving the dispute. Empowerment at its best without yielding the driver seat!

Given the steely insistence of holding the line, my initial go-around with HP's predisposed negotiation team was a rough one. Over the next couple of months, we eventually arrived at what I considered a reasonably balanced resolution. Signing the new agreement at last, her team still bore reservations. My role in settling the conflict was not considered an acceptable victory for HP, in particular by her lawyers wanting to run the show. Worse yet, my interaction left bruised executive egos behind, and being internally branded a tough cookie strained my relationship with HP from the beginning.

Though we did initiate a printer software development project with HP. Her printer division was run by Richard Belluzzo, an eventual MS president. Being the dominant PC printer supplier, failing to team up with HP would have been futile. In return for her development participation, HP received most favorable licensing terms. For us to develop and license a printer OS was an exciting and inspiring project. The idea of achieving a leading position in yet another relatively high-volume market segment drove our ambitions to believe we could create a second OEM leg. Off we went to new shores, targeting the printer divisions of Nippon Electric Company (NEC), Sharp, Canon, Epson, and IBM.

As we concluded the HP deal, OS/2 was just reaching the market. To ship on time, a compromise had been agreed upon. Version 1.0 would be released without a GUI.[13] A potentially dire handicap and disadvantage compared to operating a Windows PC or a Mac. The two partners considered this low risk assuming the glaring deficiency would not hamper the targeted business community's interest. Accustomed to character-based applications, this customer class then cared less about the benefits of operating a computer with point-and-click mouse and graphically powered windows technology. Hard to imagine today!

The trade press positioned OS/2 as successor to the DOS/Windows combo. Using her clout with enterprise customers, IBM promoted it as the most advanced OS for business PCs. We had spent the long, hot summer closing deals with all OEMs interested in selling PCs to this community. Embarrassingly and disappointingly, early reviews gave it a tepid reception. Having only a text-based user shell had disenchanted few. Missing the mark to fully support Intel 80386–powered PCs, the ones the business community preferred to deploy, was deemed a catastrophe. It effectively cut potential system performance in half. Windows and MS-DOS fully supported it. Why bother with OS/2?

Soon thereafter, real trouble started. OEMs selling PCs to the business community had made huge commitments to us. We had priced OS/2 at approximately $60 per copy, and OEMs had pledged over $60 million in the form of firm guarantees. The fear of IBM gaining an advantage had caused a stampede of commitments. The lukewarm reviews made its demand plummet. Potential buyers' opinion of the software bombed, tanked, and crashed. Suddenly, there was no longer a sizable market; even IBM was struggling. Enterprise customers decided to wait for the next version. Yet the payment clock had started ticking. I logged a crescendo of complaints. Hoping the crisis would be resolved sooner than later, I decided to sit tight until a more promising next version arrived.

I readily admit juxtaposing OS/2 versus DOS and Windows was not an easy task. In contrast, IBM positioned it to the IT world as the mother of all OSs. The bet was on! Her covert goal was to ultimately wrestle the lucrative business away from us. To accomplish this, she demanded that the next version comply with IBM's newly concocted, hyperambitious, mainframe-encompassing software architecture. The crux a preponderance of Windows programs would have to be altered to run in the latest OS/2 environment violating the holy grail of preserving software compatibility. MS battled this tooth and nail but was forced to give in—all of a sudden facing a long and painful waltz to IBM's tunes.

---

13   Graphical user interface

Our top trio immediately stopped tooting OS/2's horn enthusiastically and avoided calling it a replacement for DOS/Windows. A hurricane of inside opinions regarding the value of continuing the joint development surfaced and swirled. Among MS executives, there was a growing fear of not being able to determine the company's destiny much longer. Would we have to follow IBM's lead or, worse, live forever beneath the pinstriped, belabored, and unforgiving rule of her increasingly ironfisted dictatorship?

Anyone with a pulse and a newspaper knew. I got my updates from SVP Paul Maritz, a lanky Rhodesian educated in computer sciences at the Universities of Natal and Cape Town. Calm, deliberate, sardonic, and self-effacing, Paul offered welcome counterpoints to Bill's and Steve's frothiness. Part of his brilliance was his penchant for thinking things through and taking his time to do so. Where Bill and Steve would hip-shoot you an opinion in thirty seconds, Paul might take a day for better judgment. I liked working with him; he was no fear monger, always keeping his cool.

He dealt with IBM's development groups. Steve handled the continuing and politically delicate negotiations with her line management, like IBM's SVP James Cannavino, an outspoken and prickly executive. Mistrusting the relationship, MS had semicovertly begun to improve Windows's performance. I was given a preliminary demo and was left much impressed. This was going to be fun to sell!

Plugging along and not considered the most talented group, our OS/2 team displayed far less enthusiasm. The stubbornly rigid IBM bureaucracy imposing archaic development and testing rules left a mark on our programmers. Containing the disagreements and rising conflict was tough. Once in a while, a disgruntled developer from either company called a journalist, spilling the anonymous beans. The whole ordeal became a bit of a public secret. With a confidentiality agreement in place, talking to the press could have meant dismissal. Were these leaks planted? I really didn't know, but rumors about the tensions were swirling every which way.

My customers heard about them, grew skeptical, and observed how fast OS/2 became IBM-centric, creating an unfair disadvantage for non-IBM peripheral sales. Supporting such a trend was absolutely not in their best interests. In talking to independent software developers, they found that applications for OS/2 were not their priority. If not paid by IBM, they finished their Windows versions first. The momentum was changing rapidly.

At the beginning of '88, a decision had to be made to chart avenues beyond OS/2. I soon learned that MS's product management—encouraged by the progress made—was accelerating Windows. Meanwhile, Dave Cutler, a top-notch Michigan-born software architect from Digital Equipment Corporation, had been hired. Not fond of OS/2's internal design, he concocted a new architecture, later called MS Windows NT.[14] Its emergence guaranteed a head-on collision with IBM.

Tinkering with OS/2, accelerating Windows, and brainstorming about NT's future was a high-risk gamble at a time when Steve Jobs was secretly approaching IBM's top echelons to convince them to bet on NeXTSTEP as the future OS supplier for IBM PCs. He had developed this UNIX-based OS after his ouster from Apple at NeXT, a company he founded to make superior PCs for the education market. Jobs's motivation to approach IBM was twofold. First, he wanted to get even with Bill, who had refused to write applications for his new baby. Secondly, he desperately needed money to keep his fledgling company alive. IBM turned him down, and when Bill and Steve heard about this from James Cannavino, they made sure IBM had no second thoughts about this opportunity.

As head of a sales team, I was struggling to clarify why my customers should bet any longer on OS/2. The for-now-fictional Windows NT was internally already positioned to succeed it. NT was being designed with a robust, secure, and portable architecture in mind. Its plan called for adding sophisticated networking and mainframe connectivity for enterprise customers, exceeding OS/2's capabilities. Selling four OSs in parallel challenged our positioning knack, confusing OEMs, end users, and foremost the backbone of our success: ISVs. There were rumors the nebulous NT would be named OS/2 version 3.0. For my taste, MS was waffling instead of communicating compelling reasons to stay on course with Windows.

Even after version 1.1, OS/2 sales hardly improved, except for IBM's, who continued to basically give it away. Having decided to wait before addressing the overpayments quagmire, I had to act. Our relationships were being severely strained, particularly with OEMs who had signed long-term contracts. My disbelief in the success of OS/2 ultimately influenced me to stop my customers' bleeding of cash by scaling down their unrealistic commitments. We both had grievously erred by trusting the unproven to succeed.

IBM never gave up on OS/2, and when version 1.2 arrived, her drive to increase its popularity eventually affected Windows sales. With its improved version delayed further, Bill and I were getting greatly concerned. The pendulum was swinging slowly but surely in favor of OS/2! One of the key reasons: neither Lotus nor WordPerfect was willing to move their applications to Windows. With commercial customers depending on them,

---

14 New Technology

IBM had a valid selling point. I suspected that IBM gave both companies monetary incentives to keep them away from Windows. Bill made several calls to Lotus execs, beseeching them to alter course. They never wavered. No one ever expected that betting on the wrong horse would one day cost Lotus her independence.

It got more confusing. To cover up the simmering conflict, IBM and MS announced an extension of their partnership, broadening it to MS-DOS. The weird and politically motivated move made IBM responsible for its next version. It contained nearly four hundred bugs when it was released. IBM had flunked the test. Unable to shift the blame, MS was embarrassed and set out to correct its deficiencies. Some insiders accused IBM of sabotage. The reality was different: we were resource bound. The renewed focus on Windows, the still-manpower-sucking OS/2, and the ambitious NT plans had left us with no resources to spare to take care of our bread-and-butter product. In addition, top developers excited by up-to-the-minute Windows technologies were no longer overeager to work on good old MS-DOS. My customers lost confidence and started looking toward DRI. A hit for my business was a foregone conclusion.

Not wanting this to happen, I started selling non-OS products to receptive OEMs like MS Money, MS Encarta, and MS Works. These applications had consumer appeal, and it did not take long to find interested OEM customers with focus on retail. But I knew the only things that would truly get us out of crisis was an improved version of DOS, the release of Windows 3.0, and a further weakening of OS/2 demand.

Flying High

The world belongs to me

At the tail end of '88, MS's board requested my presence to analyze the OS/2 situation and the future dimensions of the OEM business. I must have passed muster because soon thereafter, the company went through reorganization, and I was promoted to run the OEM business worldwide. What a career move! Jeremy Butler, who had been my second boss in MS when managing Germany, was my new superior, reporting to Jon. Being moved down mattered little because I maintained my direct access to Bill, so I went to see the world!

I had shared my sales strategies with the international team in the past, but the regional guys, loving their sovereignty, did not always follow them. Therefore, policies differed by region at a time when our customers were establishing beachheads in foreign countries and expected equal treatment. To get a grip around this challenge, I centralized the OEM organization, something that ran counter to MS's general management philosophy. The change allowed me to enforce pricing discipline, explore opportunities simultaneously and without delay, respond to a crisis faster, and stamp out local favoritism. I appointed US-based managers—and not all of them United States nationals—to take command of each of the different regions in the world.

OEM personnel working in local MS offices around the world still reported into their country teams. As such, they were part of a matrix organization with the caveat of taking OEM business directives only from headquarters. Being commanders in the field while remaining closest to their customers, they were welcome to accept tactical inputs from their country teams as long as this did not endanger the overall aim of our missions. This is squarely along the lines of how Auftragstaktik principles work best. Initiatives desired and rewarded!

People working in the newly formed OEM division appreciated the clarity provided by the new structure. Several GMs and Area VPs still regarded the newly centralized organization with skepticism, yet business benefitted from streamlined communications and timely executions. Over time, the nonbelievers came around. The new org allowed me at last to add a small nucleus of marketing people to the team. It eliminated an Achilles's heel and created a vital and effective weapon for the division, keenly focused and decisively driven to increase share and defeat competitors.

With the new organization in place, I addressed some lingering customer issues. Customers were complaining that reporting royalties on a per-system basis was too burdensome, time-consuming, and costly. The sheer number of new systems rolled out had exploded, and they recommended bunching the ones carrying the same central processor together instead of listing every model separately. My key managers agreed in principle but warned me that losing these details would make our forecasting less precise. I acknowledged their concern but sided with the customers. From then on, we allowed them to license and report on a "per processor" basis.

Next, customers were asking for extended contract terms. Demanding these beyond our customary two-year term meant long-term price guarantees for them. If I had run a PC manufacturer, I'd have been challenged to commit beyond two years, however tempting. For a number of customers, price stability trumped the uncertainty of turbocharged market transformations. So we gave them the option to lock-in prices for up to five years.

Revising our licensing options presented a golden opportunity to reduce the number of pages in our contracts. I loathed the ever-ballooning page count, and so did my customers. So I challenged my attorneys, and they astoundingly reduced them by 40 percent. Encouraged, I made it a habit throughout my tenure to keep agreements as tight as possible, saving negotiation time and legal expense for all involved.

To discuss next year's objectives and to personally get to know my top managers, I invited them to attend a conference in the USA. For the first time, they all sat as one team in one room and listened to the self-same messages in unity. I opened the event with a key note, which I started by playing a video clip containing Patton's monologue from the movie named after him. Watching George C. Scott reenact this passage, calling the Germans Huns and listening to the gory descriptions of how to kill Krauts had its effect. The audience loved it and took notice of this German fellow who had the guts to use such a drastic animation as introduction. After I finished, they all understood why I had opened with Patton's monologue. A brutal fight was looming with DRI. That company wasn't taking prisoners, and our job was to be smarter and win against them all over the world. Our key weapon as I postulated and demanded then: significantly improve customer relationships. The expected result: increased Windows sales and zero losses to DRI and beyond!

The message got across to my sales reps, who found the newly created per-processor variety easiest to sell. Customers liked its reporting convenience and the longer terms we now offered. Consequently, my new team secured a ton of new business as it gave DRI a run for her money. This prompted her to march—hat in hand—to the Department of Justice screaming for intervention. Unable to beat us with winning products, she and other competitors lobbied the Feds to regulate us. Proving once again what the Prussian general Clausewitz once advocated that the "defeated feel their setbacks more than the victors enjoy their triumphs." The Feds lent an ear, and by the end of '89, the Federal Trade Commission launched a secret investigation. OEM practices were not the only issues of their interest. As its commissioners began scrutinizing our dealings, a few of our joint OS/2 announcements with IBM smelled of collusions. Eventually, the latter was dropped, and the simple customer accommodations we had extended were given foremost attention. Unbeknownst to me, dark thunderclouds loomed on the horizon.

The following year, I made all OEM personnel attend the gathering and utilized it annually from then on to propagate next year's business objectives. The event became legendary and instrumental in making my group wickedly effective. Mixing business instructions with team sports, fun, and recreational activities fostered a notably energized sales and marketing force. People talked for years about our adventures in the wilds of nature and the great team spirit they engendered. Creating such a driven and high-spirited crew was extremely rewarding. Someone used the nom de plume Marines for the OEM group. I was visibly proud and felt privileged to be their commandant!

Historical Customer Visits

A visit to Europe opened my eyes to DRI's successful counters. My first stop was to the UK to meet with London-born Alan Sugar, founder of Amstrad, which had roots in the consumer electronics business. Performing as a highly efficient copycat, his company caused disruptions in the retail landscape by undercutting brand-name pricing through reduced production costs. Her entry into the PC business came in the fall of '86 with her legendary PC model 1512. Aggressively priced with a low-end CPU, it cost roughly half of what comparable PCs were going for.

Amstrad's OS vendor of choice was DRI, instantly catching the eyes of Bill Gates, who was adamant about gaining a piece of Amstrad's business. Winning any deal with such a cost-conscious supplier meant major pricing concessions. At first, Amstrad did not budge. Unexpected help arrived from a German consumer electronic company named Schneider. Their management's ultimatum: if Amstrad wanted to be the supplier of Schneider PCs, it had to deliver them with MS-DOS. This ultimatum prompted Scott Oki to make a cold call and cut a shocking rock-bottom deal with Alan. Since Schneider was located in Germany, he gave me, then the German country manager, a courtesy call. Being under Bill's gun had translated into lowering his pants in order to win. My request: "Keep the deal quiet." To my utmost surprise, Alan Sugar kept his word. After all these years, thank you!

Over the next couple of years, Amstrad captured nearly 20 percent market share in Europe. As I prepared for my visit, the luster was fading from Sugar's company. Struggling with quality issues and reputation, he was considering returning to his consumer roots. I had been briefed to expect a flat-out arrogant and unfriendly CEO. His played-up animosity was used to pry up a better deal and was reserved for suppliers who did not dance to his tune. When we met, I found a depressed and not-exactly-overloquacious executive. Rather than work hard to regain his reputation and get his PC business back on track, his overall interest had wandered off. This did not prevent him from attempting to wheedle lower prices out of me and threaten me with going back to DRI.

In Sir Alan's mind, software was an unnecessary evil. It just added cost and questionable value to his PCs. Fortunately, he was no longer speaking from a position of strength. I certainly didn't relish a direct confrontation with him, but I nevertheless told him straight-out that another sweetheart deal was most likely not forthcoming. At that time, we achieved somewhat of a truce, but after our current contract expired, I expected renewed fireworks and pricing pressure. I was frankly delighted when the short meeting concluded.

The meetings in Germany were civilized and friendly. Siemens, the gigantic industrial engineering conglomerate, was a loyal MS customer. I knew most of her executives from my days in Germany. The only time we truly disappointed them was in '86 after they licensed Xenix together with Windows, Word, and Excel tailored for this OS. This was a unique deal personally approved by Bill and Steve, which was never repeated anywhere in the world! Before I left Germany, MS changed direction, and we had to undo the agreement. The relationship setback was severe, but we managed to keep Siemens close.

The leading German PC manufacturer by volume was a much smaller company called Vobis. Her PCs, mostly manufactured in Germany, were sold through company-run PC stores. While mainly focused on the German market, Vobis had expanded into other European countries. This included opening a store in Paris on the famous Avenue des Champs-Élysées. Talk about an expensive ego!

The marketing campaigns Vobis concocted to increase showroom traffic were legendary. Her energetic CEO and co-owner, Theo Lieven, deserves credit for their success. Easily approachable, Theo was nevertheless a supertough and shrewd negotiator and quite a piece of work. In private he was an accomplished concert piano player and would later go on to found a piano institute from a Lake Como palazzo in Italy. The local press liked his outspoken and eccentric style. MS's business with his company was dismal. He was DRI's best customer and sold some PCs without any OS. MS Windows had wound up in his stores as a stand-alone product, but there was no chance to conclude an OEM deal for it. Our first meeting was uneventful beyond receiving yet another request for cheaper MS-DOS prices as a prerequisite to gain his business. He wanted me to slash our price to meet DRI's offer. Not needing another Amstrad deal, I said "Thanks but no thanks" and departed.

My next conference with Manfred Schmitt from Escom was more productive. He had started a kind of Vobis clone company by opening his own PC stores. He had flirted with DRI a couple of times but was reconsidering his decision to better differentiate from Vobis and contemplated bundling Windows. We established a good rapport, but I came away with an uneasy sense of his integrity. The rivalry between the two companies and their vying owners nevertheless offered up an opportunity to play them against each other. I left Germany comfortable with having a solid team in place capable of handling the local situation well. A bit too optimistic as we will come to find out!

The concept of PC manufacturers opening company-run stores was unique. I questioned the profit model the two companies were using to justify what I considered a pretty bold adventure. Even after eliminating traditional distributors and retailers, they still had storefront and staffing costs. The generous margins in the early years supported this. But none of these stores was designed as destination stores or had unique merchandise like the ones Apple has built recently. In the end, hits to the German market from direct marketers like Dell and Gateway made profitability and, finally, survival tough. Vobis and Escom stayed alive longer than I had expected. The reason can be found in the Germanic reluctance to buy PCs by phone! Laugh all you like. Few Germans possessed credit cards and could not pay with check over the phone. Preferring to pay cash at retail, they enjoyed the instant gratification when taking their merchandise home then and there. The German soul ticked differently then and now.

Visiting a smaller company called Groupe Bull in France was blessedly without controversy. Her executives had long been Windows fans and supporters, but like other licensees, they were waiting for a better-performing version. Bull's main sales went through large warehouse-type retailers like Carrefour. Like any other PC manufacturer in France, Bull's team considered Apple the main competitor. French PC buyers with their entrenched avant-garde intellectualism were indeed superloyal Apple customers. Their way of life had blossomed with the appearance of the Mac. The French regarded themselves as artistic, just plain different, or more sophisticated than other Europeans. Snob appeal or not! Apple had successfully stroked their intellectual complexity, flattering their egos. Jean-Louis Gassée, Apple's first French country manager, gets kudos for recognizing and rewarding the French souls' artful and esoteric aspirations. IBM PC popularity therefore took longer in France than anywhere else to take hold and dislodge Apple from the top spot. Bull needed an improved version of MS Windows ASAP, to compete and win! As usual, we were working on one.

Visiting the Far East for the first time was enlightening, with a little culture shock thrown in. Ron Hosogi, who managed my Far East (FE) group, had planned our trip thoughtfully. A Japanese American who spoke the language fluently, Ron knew his crew well and was well respected, having been MS's first Japanese subsidiary manager. Ron was analytical, understood technology well, and had a good memory for the different political vagaries of each subsidiary. I quickly learned that office politics were a way of life in Asia.

The first stop was Tokyo, where we showed up at Nippon Electric Company (NEC), Toshiba, Hitachi, and Sharp. MS's Japanese subsidiary, ran by the smart and eccentric Susumu "Sam" Furukawa, had been MS's first foreign one. He managed to establish excellent relationships with NEC's management and her nearest competitor, Toshiba. Our market share in Japan was close to 100 percent, with only a few supersmall OEMs, comparable to US screwdriver shops licensing from DRI.

For a German, the gatherings in Japan were truly amazing. With the exception of our friend, Mr. Kaoro Tosaka from NEC, most executives preferred speaking Japanese instead of English. I came to rely mostly on translations provided by Ron. At once, my patience was stretched thin. Ping-ponging to and fro via intermediaries stretched the meetings into infinite exchanges. A congenial atmosphere was predominant. Any emotions customers experienced were hard to judge as their facial expression hardly ever changed. I was instructed by the local team to contribute only when I was handed a piece of paper—under the table—containing the proper answers the local team leader had predetermined. I ruefully obeyed. Nonetheless, this style of evasive communicating drove me nuts! Being the new guy on the block, I played along humbly and respectfully, but only on this first visit.

I later discovered most real negotiations were done in drinking clubs late in the afternoon, often lasting long into the night. You were considered a kind of kin if invited, and you were allowed and expected to cry. Yes, cry! Most Japanese managers lived in densely complex political environments following strict etiquette. Confrontation on company premises potentially endangered their jobs. The off-premise private-club meetings provided a kind of relaxation and release. They could let go, and no mention would be made of it later, ever. An unwritten but religiously followed rule. Confidentiality was held in highest regard. I came from a sharply contrasting cultural background and enjoyed a less prescriptive operating style. Admittedly, I struggled with their exotic social mechanics and delicate rules of communication. Before leaving Japan, I informed NEC of our intentions to enter the printer software market, and her management showed sound interest in reviewing our plans.

Next stop: South Korea and a totally different business culture. Many of the executives I met, such as the ones from Samsung and LG,[15] spoke English reasonably well. Antithetical to the Japanese, the customers were refreshingly direct. We had a reasonable market share, but various medium-size OEMs were licensing from DRI. The customers I met knew precisely what their competitors were up to and perpetually mentioned DRI by means of prying price concessions from me personally. Luckily, they exported most of their systems; otherwise, DRI would have taken more business away from us. The various execs I met had remarkably close bonds. Most of them knew each other from attending business schools together. The word *confidential* was unknown to them; it was replaced by foremost honoring personal ties. A handful of them displayed a touch of hostility. Around 95 percent of the discussions centered on pricing and contract terms. If you made any concessions, everybody would know within the hour and demand equal treatment. An interesting climate to negotiate in!

There was a common theme broached by our Korean customer. They envied per-processor and per-system agreements. Why then, I wondered, had they entered into them? There was a per-copy option providing maximum flexibility. Welcome to Asia. In our debriefing sessions, Ron and I probed the local team about these complaints. We concluded that per-copy licensing counted as a personal failure for the ambitious sales reps. June Park, the local manager, confirmed this in private. The behavior was typical for Far East culture; avoiding a potential loss of face in the office trumped over allowing customers to choose freely. I praised the team for their achievements but explicitly told the assembled warriors to offer all options in the future. They promised. Culturally, they had a hard time following through. There was no serious interest for Windows. The Korean manufacturers' motto was selling cheap, bare-bones consumer machines, leveraging up their deliciously low labor costs.

The third stop, Hong Kong. Our largest customer, VTech. Her main business was toys and phones, which she was producing in the People's Republic of China (PRC). VTech had ventured into the PC business, selling her own brand locally and manufacturing PCs for European companies, one of them being Vobis. A first sign of things to come! The discussions with VTech went well overall; her executives appreciated the first visit of an MS VP. Bosco Ho, her chairman and CEO, brought up one interesting point I did not want to compromise on. As I mentioned earlier, VTech was manufacturing PCs for other companies. Bosco Ho wanted us to grant him a universal DOS license for all PCs his company manufactured, to gain a higher-volume discount regardless of who sold these machines down the road. Agreeing to his demand would have led to an overall royalty reduction, effectively

---

15 Lucky Goldstar

granting her distributor status. I was adamantly opposed to entertain this. We had plenty of offices around the world to license and support OEMs locally. Our business approach helped us gain insight into regional markets and improve market share, a good reason for continuance. Bosco Ho understood my position, and we found a way to compensate VTech with marketing funds mandating her to combat local piracy. Grasping this would help both of us; we parted as friends.

Next stop, Taiwan—Acer and smaller OEMs. Under the leadership of her founder, Stan Shih, Acer had developed into a PC powerhouse. The main discussion centered on renewing her DOS license and exploring opportunities for Windows. The latter did not materialize. Agreeing on a new DOS license, we gave Acer pretty favorable terms. Again her management used the DRI threat effectively to achieve dreaded discounts. DRI had indeed been superactive in Taiwan. Several local companies were licensing her products in place of ours; I had to take Acer's arguments seriously and dared not to lose her business. SVP Simon Lin was the main negotiator. He was a reasonable, smart, and friendly man, so the two of us reached the desired compromise in a short time. We promised to stay in touch, which developed into a long-lasting friendship. Not having a local subsidiary, we conducted our Taiwanese OEM business out of HK. As soon as I returned to Seattle, I lobbied for opening one to support local OEMs better and fight DRI more effectively.

My most important stop: Beijing. The undeveloped PRC market represented an incredible opportunity. Despite consistent effort, not a single local PC manufacturer had ever bought any software from us. Yet MS-DOS could be found on 99.9 percent of PRC's PCs. How had it gotten there? Strictly through piracy! The locals had never paid a cent. What a mess! Our goal, therefore, was to explore if the government would support an effort to reduce piracy and use her clout to instill honesty into local manufacturers.

Arriving in Beijing, we met at once with Mr. Zee, CEO of the Great Wall Technology Company Ltd., who hosted our visit. After he detailed what to expect from our next-day government visit, we left for the old and disheveled US embassy building. We met the ambassador in the same office, occupied fifteen years earlier by then president Bush (41) as ambassador to China. Here we received a thorough briefing on the legal and political situation, which prepared us for our lobbying effort.

My local team insisted on renting two stretched limos to add status to our visit. I reluctantly agreed. Such pompous manner was normally not MS's style. I was accompanied by a government-cleared translator, one of our attorneys, the sales team, and Ben Hsu, our support engineer. Ben spoke both Chinese dialects fluently, and his role was to function as my trusted second ear. When we arrived at the Ministry of Technology, people were leaning out their office windows to catch a glimpse of the limos, which were two out of three in all of Beijing. Would causing such a commotion work for us?

After a formal greeting ceremony, we were seated in a large conference room with our backs to the inside wall. My translator told me this was a centuries-old custom, allowing a visitor not to fear an attack from behind. At least our host was considerate and looking out for us. A large delegation from the ministry joined us, though the founder of Great Wall and the highest government official present were the only members who actually spoke. The rest of the group functioned as note-taking scribes.

After politely talking around the issues for a while, per custom, the highest government official, a vice minister, told us flat out that he would not allow us to license any OEM directly but would be willing to sponsor a group license. He hinted that if we could strike such an agreement, the government would guarantee us a minimum annual amount paid in US dollars through the Bank of China. He further promised Great Wall would submit quarterly MS-DOS reports on behalf of all PRC PC manufacturers.

During recess I thought the initial proposal was a good starting point. My reservation: I had just refused giving VTech a comparable role. I questioned if Great Wall would be able to obtain accurate numbers from her competitors. Could the vice minister, even if the PRC government had a stake in all local PC manufacturers, really enforce his will on them? Being the eternal hopeful, I was not convinced his first proposal would be his last word on the matter. Back in the conference room, I politely detailed our hesitations and determinedly aired our laundry list of concerns. I found out the hard way: it was the communist's way or the highway. Welcome to a totalitarian regime. With no other option and wanting to at least get the big toe in the door, I eventually caved in.

In November of '89, Beijing was still recovering from the bloody upheaval of the Tiananmen Square protests. There were more important issues to address than dealing with software piracy. Soon thereafter, the verbal offer was taken back and later repeated in modified form; the details were rehashed again and again. We never gave up; they kept coming back. The wheels of the Chinese government turned slowly, inching forward, testing to see if a newly proposed deal was a politically correct one. Eventually, consensus began building between the policy makers, and two years later, a final deal was struck. Richard Fade, who by then was managing my Asian OEM business, deserves kudos for overcoming the obstacles Chinese bureaucrats threw into his path—tenaciously holding on and never letting go. To perform my signature duty, I happily returned to Beijing in '91; Mr. Zee from Great Wall signed on behalf of the government. I still have a copy of the historic document residing in my office.

Following the signature ceremony, we were required by custom to invite half the ministry to a formal reception, serving Western food. Afterward, a small core of the agency employees instrumental in concluding the deal enjoyed a more elaborate celebration. For the fancy dinner we had planned, all arrived in dark-blue Mao uniforms and throughout kept their caps—replete with a centered red star—on their heads. Our crew, in business suits, provided quite a contrast. Deviously they asked us to order rare high-proof Chinese liquor,[16] which arrived in tiny bottles. Every single one of our seven guests proposed a toast to me and me alone. Seven times I stood, seven times I tossed one back; I still don't know how I survived the evening.

When the jamboree ended, we got into our car and asked the driver to take us to the Great Wall of China. November was frigid in Beijing. Liquored up, we climbed the historic monument after waking a protesting guard and tipping him to let us pass. The stars twinkled on the firmament in a subzero night. The moon was nearly full and lit the impressive scenery. Here, as we tottered about on the Great Wall of China, the icy winds from the Mongolian steppes greeted us and cleared our heads sufficiently to propose a final victory toast with good old Bourbon. Finally!

The agreement in place, we patiently sat back and waited for the royalties and quarterly reports to arrive. The money arrived on time for two years, but no MS-DOS unit was ever reported. We never challenged the arrangement and bit our tongues, never complaining about missing information. Two years and two million US dollars later, our agreement expired, and the Chinese government allowed licensing local OEMs directly. Patience and restraint had paid off!

---

16   Called *baijiu*, approximately 120 of deadly proof!

Getting out of bondage

"Bring it on!"

In late '89, Jon Shirley announced his retirement while staying on the board for another ten years. As captain, he had steered our ship successfully through the rough seas of the last five years. Our finances were now rock solid despite the constant growing pains, market turmoil, and various nasty, competitive headwinds we'd experienced. The greatest disappointments during his reign were persistent and ongoing product delays. Unfortunately, they were mostly outside his influence. Bill had his most loyal knight, Steve Ballmer, running the OS division. My group suffered more than any other sales organization from its notorious failure to deliver products on time. The man in charge seemed to have earned the right to fail.

In '88 my group was proud of winning a big, breakthrough contract for our soon-to-be-launched LAN Manager product: developed with the goal of turning high-end PCs into small business servers. It fell into what analysts call the network operating system (NOS) category. Novell, a Utah-based company, led that category with a nicely evolved, mature, and stable product called NetWare, with Compaq being her largest OEM customer. I wanted a slice of Compaq's NOS business, and so did our product manager, Rob Glaser, the eventual CEO of RealNetworks, then reporting to Steve. My best group manager, Richard Fade, seized the opportunity and teamed up with Rob, landing the business.

Unknown to me, Steve did not have the OEM channel in his sight of aim to sell LAN-Manager. For him the enterprise sales reps were better suited to peddle our newest baby. When he heard about the Compaq deal, he came storming into my office to read me the riot act. The deal was done and the ink dry. I was shocked by Steve's outrage and responded, matching his lack of restraint. He quickly discovered his yelling tirade did not intimidate me. I remained convinced the freshly struck deal with server-market leader Compaq had tremendous potential complementing his plan perfectly.

Yet Steve adamantly insisted on undoing the arrangement. Where was his logic? I happened to have the bulky and heavy Compaq file folder containing the contract on my desk, and when the screaming did not stop, I hurled it across the room at his chest, imploring him, "Undo it yourself!" It sickened me to see that Steve's blind intervention was sacrificing a several-million-dollar deal. Angrily, he stormed out of my office.

All later arguments fell on deaf ears. In the end, the agreement got undone. I was furious and took the unusual step of complaining to Jon, recommending to him first, and a couple of days later to Bill, to fire him. Jon stared at me in awe, replying only, "Are you serious? You can tell Bill. I won't." Bill just laughed at me. I soon learned the episode would come full circle back to haunt me later.

The incident pours a direct light on two different management styles. First off, Steve's intent was misunderstood by Rob. Secondly, scolding others and not apologizing for miscommunicating hurts morale. Finally, not empowering subordinates to take significant market share away from a competitor runs counter to Auftragstaktik principles. Undoing the deal displayed zero respect for Compaq, our largest customer, you must, at times, swallow your pride. I didn't work for Steve directly then, but I knew Bill dealt differently with me. When he gave me directions, they were never meant to be prescriptions. The incident smelled of a rigid and hierarchic command doctrine combined with micromanagement, an ossified structure the Prussians had abandoned centuries ago. Would he learn from this incident?

As long as the OS development group was managed by him, the delays continued and got worse. A frustrating experience considering that we extended Steve's promises to customers verbatim without any real certainty how the company could fulfill them. Second guessing the unrealistic plans did not help much either. While management below Steve got reshuffled several times, he seemed protected. I looked on in disbelief and thought, *In order to get ahead and be bulletproof in MS, you need two stamps on your forehead—the first one saying you're smart and the second saying you are personal friend of Bill or Steve. Ideally of both!* Over the years, I observed how a lot of personal favors were dished out to this class of people. I knew I had at least one stamp on my forehead, and I never tried to get the second one from either of the guys. When running a business, only performance and respect need to count, not personal friendship!

In hindsight, the undoing of Compaq's deal was a hidden blessing for my group and our overall business relationship. LAN-Manager came out with so much delay that Compaq would have had no chance to bundle it with the designated servers. This first version was buggy like hell, and Compaq would have been utterly unable to serve her customers reliably. It wasn't until version 3.0 that we produced a mature and competitive product with NT technology. I felt lucky not having to suffer the inevitable avalanche of Compaq's complaints. Undoing the license nevertheless left a mark on the Compaq / Steve Ballmer relationship, which never fully recovered. In a way, I got my revenge, helping Richard Fade to convince 3Com to come to the rescue and assist MS in developing a vastly improved version. A fragile partnership was the result. 3Com marketed the product independently and paid OEM royalties. At least some revenue was headed my way.

On the OS/2 front, the IBM/MS relationship was unraveling fast. As we geared up for Windows 3.0, our performance in regard to project OS/2 had, in IBM's opinion, deteriorated to outright disappointing. Without being directly involved, I had to trust the grapevine's signaling and confirming that we had our best development resources primarily working on Windows. My group in turn was fully occupied selling its soon-to-be-released new version, hardly mentioning OS/2 anymore. Only IBM wore a lonely champion of OS/2 badge any longer.

Our planned ten-million-dollar launch event in NYC should have been the most glaring sign that we were no longer in sync! IBM did not have to wait until then. In March of '90, Paul Maritz, in command of OS/2 development, stopped sending her the newest code, effectively and to her utter dismay, immediately unraveling the partnership. IBM's newly demanded and far-reaching design changes were to blame. Our customers applauded. Working out the separation details took time, but at last, in the fall of '90, our divorce papers were signed.

I was jubilant. The divorce afforded a golden opportunity to transfer the IBM business responsibility to my group. When it happened, IBM no longer enjoyed the darling client status she had basked in for over a decade. No longer being primus inter pares, her future royalties would from now on be strictly based on expected unit sales. I made sure of that!

Steve took the split much harder. First of all, he viewed the divorce as his own personal failure, and in addition, he felt scorned. Never mincing words, he declared IBM was, from now on, MS's enemy number one. "I hate that company!" The second OS war was about to begin before the one with DRI had been concluded.

Bill worried if we could really win that new war. Lots of execs, including me, told him confidently we would prevail. IBM had zero support from other OEMs, and the software community trusted and followed us for now. Could

IBM, with all her vast resources, buy all the favors needed to win? If market participants would behave like paid-off Lotus, maybe!

Not sharing my optimism, Bill looked at the bigger picture. What bugged him most was that Big Blue was turning out more patents than all other computer hardware and software companies combined. Patents are easy to inadvertently infringe upon. He knew they constitute the ultimate monopoly. IBM's merciless and ruthless enforcement policies made her patent portfolio a nasty entity to fight against. We were experiencing the reality of such contest thanks to a lawsuit Apple had recently filed against us in regard to Windows. With IBM, we had entered into a five-year cross-patent agreement, obligating us to pay an annual $1 million for patent peace. Nobody could predict what would happen when it ran out.

Another of Bill's considerations was IBM's huge market power considering her dominant position with enterprise customers and the resulting mainframe choke point it had created. Would she use this, unhinge us, and make winning the enterprise battle for Windows unfeasible? I'll admit that upsetting a supergiant like IBM was scary. We would need perfect execution to win and Bill knew this was historically not our strength. Despite the breakup, he explicitly told me that he wanted peace, not war, with IBM, and he refused to give up on his wishes. Steve, on the other hand, would have loved to slowly strangle Big Blue—not that he could have succeeded single-handedly. His message to IBM's management: "Bring it on!" His message to me: "Win and obliterate OS/2!"

It takes two to tango and to keep the peace. As we were trapped in customer commitments, our newly acquired foe was a considerable force to reckon with. IBM's then president Jack Kuehler personally spearheaded the attempt to gain an upper hand. His motivation was twofold. His aspiration: not letting IBM's customers down and wrestling the lucrative OS business away from us. The result was a brutal and bloody war staged with full force for six years—in the marketplace and in the courts of the land.

Turbulence

Étude in *F-moll*

In March of '90, I visited CeBIT, the largest computer-technology trade show in the world. Adding to the allure was that the event was held in Hannover, Germany, my hometown. I had lived there for thirty years and, as a student, often worked at the fair, doing a variety of odd jobs.

I was first to meet with Theo Lieven, Vobis CEO. The upcoming release of Windows version 3.0 had made him finally come around. Unfailingly opportunistic, he viewed it as an enticing chance to create store traffic, uprooting the local market yet again. For us, the deal was an impressive breakthrough that wiped out nearly all of Vobis's business with DRI. Kudos to the local team!

I arrived early at MS booth, expecting a glorious signature event followed by a celebration. The champagne and the canapés were waiting! Our German subsidiary manage, Christian Wedell, was present while the account rep had fallen sick.

Inside a scant five minutes, our meeting turned sour. After Christian handed Lieven the documents we had prepared for signature, he sat them down in front of him and began ranting about MS, her trickstering, perceived unfair business practices, etc. His first complaint was the amount of royalties we were asking for. Second, he called our terms and conditions obnoxious. Not mincing words, he raised his voice in operatic crescendo, slashing away with other stinging accusations. I was getting very uncomfortable. The Vobis team had agreed to all terms and conditions before we arranged the CeBIT visit. What on earth was going on here? When I at last got a word in, I politely made my point, expressing my astonishment, and asked for a brief recess.

Christian and I immediately called our rep, who was equally as puzzled. I knew him well, having hired and worked with him when running Germany. A smart engineer at heart and a dedicated and honest man, always calm and well behaved. The perfect match for Mr. Lieven, legendary for tantrums and other public and often humorous behavioral aberrations. Theo was behaving like a spoiled child in need of a stern rebuke to get relocated into reality. It was every bit that outlandishly puerile. Was his behavior a last-ditch effort to arrive at more favorable contract terms, or a "go to hell with you and your company" adios?

We didn't have a clue. I was jet-lagged, a bit exhausted, and not as content as usual. I'd been watching his cofounder, who had arrived with him. He showed signs of despairing over his partner's antics. Back in the meeting room, I told our visitors what was on the table was the best deal they could expect from us. Interpreting my remark as a harsh ultimatum, Theo responded with a freshly fomented hurricane of rage. Infuriated, I told him I had no interest in wasting precious time enduring his simian tirades. For me the situation was black-and-white: Stop moaning and groaning and sign the deal. Or leave in peace. I added that the price of MS Windows could easily go up later in the year. I knew, as I said this, I should have refrained from further pressure. As the reserves of my patience fully depleted and my annoyance meter pegged, I did it anyway. The situation had gotten the better of me.

Mr. Lieven interpreted my last message as a threat to his livelihood and immediately fired back, calling me names I won't translate into English. I understood at once I'd committed a grievous error and made an effort to calm him down. With me not succeeding, both visitors bolted for the exit without ceremony, stomping off into the heraldic Hannover morning—the German way. Contract unsigned.

My second mistake was reporting the incident in my monthly status report—archived by someone else and later to be discovered, and played up in the ongoing Federal Trade Commission investigation. The Feds later would not take my explanation verbatim, insisting I had threatened a customer with not licensing Windows at a standard price and/or tied its sale together with MS-DOS. None of this bore any resemblance to my recollection.

The final outcome actually rewarded both sides. Mr. Lieven personally signed the abandoned agreement without changes except the date less than a month later. Swallowing his pride, Theo proceeded to take full advantage of the deal. Unexpectedly, his company made a commendable effort and succeeded in being the first to market with preinstalled German Windows PCs shortly after version 3.0 got released. The local OEM support personnel helped Vobis accomplish that feat. All other German OEMs followed Vobis nearly a month later. Being early to market helped Theo to create tremendous traffic and additional business for his stores and amplified our local Windows public relations campaign.

When we met at CeBIT, Theo, as DRI's best local customer, had to be aware of a planned new DOS release by DRI. Certainly he could have licensed it much cheaper than our version. Perhaps unknown to him, he would have gotten Windows 3.0 for the same price, saving serious money. He never explained why he did not act upon this opportunity. Did he fear DRI's new version would be incompatible? Or had he angered DRI's management after reading press articles mentioning that he was trying to sign on with us? Playing the press can be as complex and as layered as one of his favorite Chopin piano concertos.

I spent time walking the fair ground, meeting other customer executives and reviewing local progress with my German team. By chance I visited DRI's booth. Our DOS competitor had an enormous representation, and if I could have looked behind the curtain, I would have seen her newest baby, DR-DOS 5.0, being covertly demonstrated. Tornado-bearing clouds, unknown to us, were building up on the horizon.

Much later I learned from our sales rep that Theo Lieven had never seen the finished contract beforehand. His managers should have known better. Theo was the essence of a control freak, and allowing his people the freedom to cut the best deal possible was not necessarily his style. He always believed he could do better than his teammates. On the contrary, I trusted my local team and gave them elbow room to strive for the best possible outcome without micromanaging them. I had no interest or need for having my ego stroked—which was, for Theo, always on his carte du jour. He never buried his hatchets, and I disregarded his hostilities.

With most OEMs on board, Windows 3.0 started an incredible business cycle. In the first year alone, we sold over 10 million copies, mostly through the OEM channel. More copies than we had sold from 1985–1990 altogether! Selling Windows at such an accelerated rate resulted in doubling my OEM revenue every second year—and would continue for the next eight years to come. (A compounded 42 percent annual growth rate!) With the well-being of MS's stock price now hinging increasingly on my group's performance, financial analysts now wanted to talk to me on a regular basis. I complied. Compared to journalists, they were open, knowledgeable, and honest with me and showed understanding and respect for the limitations of my disclosures.

A SERIOUS WAKE-UP CALL

Having finally corrected our faulty MS-DOS release by the fall of '89, we assumed we had gotten away with a minor black eye for the sins of IBM's spotty work. Instead, DRI, under her new CEO Dick Williams, ambushed us with an advanced DOS release of her own just after I had returned from CeBIT. Deviating from tradition, DRI dialed up the version number, signaling she intended to leapfrog out in front. According to the computer press, OEMs, and early users, her latest version was, for the first time, superior to ours. Talk about an outraged Steve; again it had happened under his watch, flat-footed and pants down! DRI had included a text-based user shell in her release, disk-caching software, a superior memory-management system, and an advanced way of loading drivers into memory without getting in the way of application software. In short, her groundbreaking engineering freed up conventional memory and enabled applications to run faster or allow for additional or larger applications to run simultaneously. The new version ran best on the now mainstream Intel 80386 and high-end 80486 powered PCs.

Being outmaneuvered and losing our lead meant losing customers. Under the gun, I opted to offer price concessions to fight market share erosions, but even that did not always work. Each time a contract ran out, we battled. When already less-committed customers in Asia, Europe, and Latin America deemed our discounts insufficient, they did in fact abandon us. Most customers who moved away were shipping lower-end PCs. The cheaper and meanwhile better DR-DOS 5.0 was just what they needed to meet their aggressive price points.

The Mexican company Grupo Printaform serves as a typical example. This company was a well-established enterprise owned by the Espinosa family and ran by her president and CEO Jorge Espinosa Mirales. Her origins were in selling paper and paper forms. In the mid '80s, Mirales had ventured into the low-end IBM PC clone business. The demand for these came from Mexico's consumers, who were unable to afford expensive imports. Cutting cost to the bone, Printaform initially licensed from DRI exclusively. Later she partially switched over to us for a line of higher-end business PCs. Visiting the company on my first trip to Latin America I found a determined executive in Jorge Mirales. With DRI's version 5.0 now available, he saw no urgency to give us any extra business. The one caveat was we had a Spanish version whereby our competitor's was available in English only—a least for now. We could offer Printaform a clear-cut advantage, selling to non-English-speaking customers. But Jorge, shrewd as he was, nevertheless demanded a significant discount, and I listened.

Printaform's success in her core paper business was throwing off an enormous amount of cash flow and allowed her to extend generous credits to her trading partners. A key weapon in the fight for merchant loyalty in the cash-strapped Mexican commerce, it assisted Printaform in obtaining the lead in the local PC market, beating out IBM. Negotiating hard and agreeing to huge price concessions while under the DRI gun, I managed to keep sizable amounts of his DOS business and enticed Mr. Espinosa to license Windows for a few high-end PCs.

When I visited with him again two years later, he had just arrived home after having been kidnapped. With both of his legs in casts, he was an emotionally subdued and totally different man. The kidnappers—angry cops fired during an anticorruption campaign—had broken his legs one after the other to put muscle behind their ransom demands. The family had paid, reluctantly, and after much delay. Jorge still had his fighting spirit in him and was in the process of regaining control of his company. I openly admired his toughness. In his five-month absence, one of his sons, liking our brand best, had run the business, and now the majority of OS traffic had been switched to us. The old man was displeased about the prices his son had agreed to pay us. Our meeting went relatively well with me, making easy concessions Jorge sincerely appreciated. I will never forget that visit and the sad look in the eyes of the old warrior. Shortly thereafter, he resigned from his post and gave the reins officially to his son.

Once a year, Bill cleared his schedule and made time for what he called his "think week." The brief hiatus allowed him time to catch up with his readings, accrue product ideas and thoughts for future initiatives, and analyze mistakes. At the end of the week, he summarized his conclusions in an exec-staff e-mail. This year he admitted that MS had violated one of her core principles by not improving DOS significantly enough for more than two years. He reminded everyone that best-of-breed products were the only proven avenue for maintaining leadership. Competitors like DRI could sneak up on us at any time. Sales and marketing could help with public perceptions, but users would vote with their feet against us if we allowed ourselves to become complacent. I had heard his sermon before and hoped that reiterating his perception would inspire enduring change.

Counterpunching DRI, we announced making an end run for a superior version with a six-month delivery horizon. The marketing folks did a good job comparing our planned advanced features favorably to DRI's. A handful of journalists and politicians accused us of stalling DRI's sales momentum with a containment strategy of announcing early. Comparing the timing of our announcement with those of other software companies, we were just following common practice. I further believed the development team—after Bill's scathing e-mail—now truly understood the urgency to keep a promise.

That same summer, we got a new president in Michael Hallman. I decided to test the new man. As long as I had been running OEM, even when reporting to Jeremy Butler, Jon Shirley had continued signing all contracts and amendments. To find out if the new guy wanted to follow in his footsteps, I asked my admin to save all documents needing presidential signatures for one month. I then sent her upstairs with a fully loaded office cart. As predicted, within minutes, Mike was on the phone. I listened patiently as he barked away, ruminating over my provocation. What the hell? I explained what Jon had practiced then, exploiting the example at hand, proposed he delegate signature authority to me. He agreed, without hesitation, empowering me to sign all OEM contracts. A smart move! Compared to Jon, he had limited knowledge of my business and its intimate details. Jon had added value in the process. Michael opted to free up his time. Compressing the approval chain had another positive side-effect: reducing turnaround time with customers.

In general, he was easy to work with, but I received hardly any directions from him. With Jeremy constantly on the road, I met with Bill more frequently. A lot of employees were surprised when replacing Jon with an outsider. They rightfully assumed Scott Oki, my old boss, or Jeremy Butler would have been better candidates. Scott had quit the company earlier, reading the tea leaves correctly. By '91 Jeremy was surprisingly on his way out as well. Constant traveling had stressed him out and severely impacted his health. The obvious choice for the job would have been Steve. One of our board members told me later that the board did not think he was ready for the job yet.

Just after Mike joined us, I was invited to address the board for a second time. In addition to analyzing the tough DRI situation and the continuing OS/2 onslaught, the board wanted a ten-year OEM business outlook. I spent two weekends combing through our historical numbers and any forward-looking analysis I could find. In place of presenting just a simple growth chart for our future revenues, I developed a worst- and best-case scenario before deciding on a most-probable trend line. The board was impressed by the methodologies applied to back up my predictions. I still don't believe anybody truly trusted my final conclusion as I predicted nearly ten billion dollars of OEM revenue for fiscal year 2001. Luckily, nobody threw me out or assumed I would still be in the same job and could be held responsible. In the end, though, the unexpected happened.

Long-term planning like this was creeping into the company. The plan being demanded and repeated annually from then on, I considered it a total waste of time, considering the inherent competitive and dynamic business environment we lived in. Relevance and focus was asked for and not an endless contingency-planning exercise, which more than once led to politicized gamesmanship.

Rather, I signed up more Windows customers. To gain additional penetration, I started a contest for the sales reps. With version 3.0 boosting performance significantly and allowing for multitasking, existing customers increased their commitments. All of the directly selling OEMs signed on. Our main holdouts were Amstrad and Compaq. Despite having signed on, IBM refused any public support. Not increasing prices helped our crusade as much as winning the hearts and minds of reluctant ISVs. This was where Steve excelled as he whipped our so-called evangelist support group into shape. Only Lotus, WordStar, and WordPerfect remained the lone holdouts exclusively seated in the OS/2 tent.

CRISIS IN THE FAR EAST

With Windows finally on an uptrend, I visited my Far East customers again. We had succeeded in obtaining commitments from them for our printer OS. In the fall of '90, the project ran into serious trouble. Not only had we missed all promised deadlines, but also worse, the code itself was not stabilizing. Product management was seriously considering abandoning the well over one million lines already written. I decided to personally give customers the bad news. Not an easy task, considering the investments they had made and the loss of face our decision would cause for the Japanese execs who had signed on. My meetings with Canon and NEC had the singular goal of minimizing damages. Delegating the unpleasant task would have set a bad example. I knew these customers were unhappy even as we offered to give them the unfinished source code at no charge, allowing them to use it with no strings attached.

During my stay in Tokyo, I met a second time with Sony's management. Our meetings were always cordial. Their executives spoke English well, having spent mandatory time in Sony's US subsidiary. However, the electronic consumer giant was no intimate friend of MS. Sony's president, Mr. Idei, later expressed this belief as "MS wants to control us." I still do not understand how he drew such an inaccurate conclusion. We wanted cooperation and openness; he wanted proprietary secrets and unmerited advantages. The overly polite pins-and-needles atmosphere in meetings I attended did not signal trust.

Bill would have loved to have Sony on our side. He visited with her execs numerous times without achieving a relationship breakthrough. Sony was no doubt an innovative and powerful consumer electronics and entertainment company. If her top management would have opened the kimono just a little bit, the industry would have moved forward faster. Instead, maintaining distance, she included as many proprietary features in her PCs as possible at the cost of endangering the holy grail of compatibility. MS was eager to engage in media technologies with her where she had an undisputable lead. What I found was a company who reluctantly bought our OS products and was barely willing to work with us. Sony obviously had her own agenda and ambitions. MS found herself stuck in the role of being just another of her nonstrategic vendors.

Before returning to Seattle, I stopped in Hong Kong. The local OEM group had organized a small Asian customer summit. I delivered a keynote speech and met with several key Asian manufacturers. What I heard in regard to losing business came as quite a shock. Therefore, I extended my stay and gave the Asian sales team a pep talk, repeating key OEM strategies and reemphasizing our business objectives. We were under severe attack by DRI, particularly in Taiwan. Several medium-size Taiwanese OEMs had abandoned us and signed up for DRI's new DOS version. I made time to be interviewed by a local journalist who had, a couple months earlier, done the same with DRI's executive responsible for Asia. He had named twenty-nine different companies in Southeast Asia, which he claimed were licensing from DRI exclusively. The journalist gave me a hard time, chiding me over how we had lost the edge and wanting to know what I planned to do about it. I told him in no uncertain terms would we make a serious effort to bring these customers back into the MS igloo where they, as I saw it, belonged. After the interview, I asked an aide to dig out the article he had mentioned. Fortunately written in English, it contained a treasure trove of information the DRI guy had foolishly disclosed. We immediately proceeded with mapping out a counterstrategy. We took the target list of prospects straight from the newspaper article. Whoever volunteered this should surely have been fired! Bragging rights over common sense!

My last stop was Taiwan, where I attended the opening of our new facility. As part of our counterstrategy, we decided to bring in experienced support personnel from the United States to give it a running start. They made all the difference in the world. The smaller OEMs DRI's executive had supposedly won over and boasted about hugely appreciated the sudden attention. DRI had no permanent feet on the ground, sporadically dispatching personnel from the United States. Being able to nurture the manufacturers locally daily was the key to welcoming them back into our realm. The result: six months after my HK visit, DRI had lost 90 percent of her accounts on the island.

At the end of the visit, I was joined by Bill to formally open our new sub. His visit to Taiwan was publicized when it was announced that he would be the keynote speaker at a local ISV conference. The idea to direct attention to the conference and increase attendance backfired; we received a bomb threat. Traveling with Bill and visiting additional customers meant I was included in all the security measures. Sweeping hotel rooms for bombs, changing cars frequently, and never venturing alone in public were but a few. Fortunately, it was a hoax. I admired Bill for not cancelling the trip. After experiencing all the hoopla and intrigue, achieving celebrity status like Bill's was definitely not what I longed for!

Pirates of the World

On my trips, I made a point to personally investigate how small system builders were conducting their businesses and where their OS software derived from. I personally observed the incredible amount of piracy happening in Asia, South America, and to my surprise, Germany. Software theft seemed to have no meaningful moral or legal consequences. Any child could do it by just copying, and hardly anybody watched.

Copies these crooks made did not look like the original software we advertised and sold. Honest buyers recognized the differences and gave the cheaters a hard time. To overcome authenticity objections, a network of illegal replicators sprung up and produced look-alike packages. The illegals often shipped them in small FedEx boxes avoiding discovery. We, in turn, engaged customs authorities to stop them. Their motivation to stop the nuisance: when a counterfeit package changed hands, the government missed out on import duties and taxes, just as we did in royalties!

Our own product management was partially responsible for these thefts by not embedding a copy protection scheme that was difficult to break. In the early days, Bill had made his opinion about stealing software well-known. His comments caused controversy in the industry and with the trade press. In PRC, where piracy was most endemic, IP[17] theft can be traced as far back as Confucius. One thousand five hundred years earlier, together with other Chinese philosophers, he had extolled the virtues of public ownership of IP. Modern day OSs and software apps free to anyone for the taking? We had quite a task ahead of us, reeducating a country with a citizenry of then nearly a billion people misguided for centuries.

I put together a small task force to gain an in-depth understanding of our losses and challenged its team members to turn heedless thieves into honest customers. The team discovered straight off the bat that software

---

17   Intellectual property

replicators our OEMs used were part of the illicit supply chain. Many of them conspired by producing extra copies and selling them out the back door. A number of our own OEM licensees did the same. I estimated we only got paid for two out of three copies in use. This meant my business could have been 50 percent larger—a great motivation to personally lead the crusade against the IP terrorists.

How did this get so out of hand? Our OEMs inserted backup copies of our software and a user manual into every PC box. They were supplied to them by duplicators of their choosing without any control from us. Here laid the crux. So my task force recommended the implementation of a proven concept used by the entertainment industry requiring duplicators to obtain a license from us. Our immediate concern was solving the mechanics of such a supply chain. Thousands of companies throughout the world reproduced our products. How would we select and license the right ones?

Grant Duers, the team leader and an experienced logistic expert, pointed out that reducing the numbers of replicators would lead to larger production runs and lower repro costs. He had my ear. Investigating how costly these inserts actually were, he discovered every so often that they exceeded royalties paid to us. Our customers were being creamed! (In a footnote to our fact-finding mission, we enlisted the product groups to reduce documentation page counts.) This convinced me that if our moral intent would not find customers' approval, cost savings for sure would! Intending to close loopholes and increase our ability to control the flow of goods, we asked our attorneys to include auditing rights and report requirements in our envisioned replicator licenses. We then made it mandatory to put an MS designed label, containing an encrypted serial number, on all packages.

Implementing the new procedures enabled us to compare replicators' unit reports with the ones OEMs submitted. It took no time before large discrepancies were discovered. Lag time between the reports often caused this and was easily straightened out. Deviations had other reasons: every so often, OEMs served the black market without paying us or reported fewer machines than they were actually selling! Sufficiently large deviations warranted pronto follow-up, and an audit via an independent accounting organization was typically the result. It required my approval. Ironically, most caught did not deliberately cheat. They had grown so fast that their data processing and internal-reporting capabilities lacked accuracy. My controller therefore made it her task to educate customers on proper data capturing and reporting methods. To her delight, collecting the proper amount of royalties due was, from then on, much more easily achieved.

Other MS groups watched anxiously if the new measures made a difference, and they followed as soon as our lead showed progress. Aligning all sales groups and improving the safety system occupied our logistics group for decades. Next we established a network for issuing security numbers electronically, directly to the printing press. When installing our software, we demanded that end users type in the sixteen digits found on our security label. Still not adequate! The pirates soon cracked our algorithm. In turn, we made the math harder for them. The next logical step was to use a WW computer network, which could validate every single number from a central place denying multiple uses to cheaters. With the appearance of the Internet, that got solved as well. Not all holes were ever plugged sufficiently, leaving room for substantial leaks.

Internally, my colleagues labeled me the antipiracy czar. In the press I was described as MS's enforcer. None of these monikers offended me one bit. Successfully discouraging the thieving backdoor artists was all that counted. As long as I ran OEM, I continued advancing these means with passion. Eventually, our products would be delivered on CDs or DVDs. We worked with one supplier, printing unduplicable (or so we thought) blanks with holographic images and distributed them to other replicators for final production. They gave our products a distinct and artistic look and again increased our security incrementally. The pirates countered with a design mimicking ours. In a fascinating game of global cat and mouse, we indeed won over time after several incarnations of hologram technology. The last task was to create piracy-proof sticky labels. They were not only hard to replicate but also nearly impossible to remove without damaging the carrier. Applying banknote printing techniques and special glue, we upped the ante.

Protecting our products was without question a never-ending effort. Each of our actions provoked a reaction from the notorious scam artists. Even in this field of endeavor, we learned again that complacency was enemy number one.

Falling behind

MS Windows 3.0, in combination with MS-DOS 5.0, benefitted five companies totally committed to design and sell these types of PCs: AST[18] Research, Dell, Packard Bell, National Cash Register Company (NCR), and Gateway (GW). The most interesting case study of these emerging movers is GW's. Her founder, Ted Waitt, brought the company to life on the prairies of South Dakota with a strong belief in selling PCs directly to customers, bypassing the retail channel. Displaying his loyalty to the community he'd grown up in, he designed the Gateway shipping boxes with a flourish of Holstein cow pattern. Producing high-end PCs for gamers and power users, his company was soon widely recognized as bearing a strong brand. I got to know Ted well, and one day he approached me about licensing MS Excel and MS Word for his higher-end product lines. Supportive, Bill gave me the green light for a trial period. I was convinced we had found an ideal partner. Typical GW customers bought highly customized PC systems by phone. Why not try to sell them our Windows applications at the same time? GW could easily preinstall them and save an interested buyer time and installation hassle.

Teaming up, we expected to increase our Windows apps market share, reduce piracy, and give our products extra market octane thanks to Gateway's advertising. For GW our willingness constituted a first in the US market, and I was hopeful that pioneering application bundles would boost her sales and visibility. I agreed to a three-month exclusive. Afterward I would unleash the opportunity on other parties. The reason for GW's short contract term can be found in my desire to move slowly with so many inherent unknowns. After all, our Windows apps were by now part of our crown jewels—at par with our OSs. I offered GW a reasonable but not too aggressive price. The exclusive made this palatable.

Market reaction to our experiment was measured skepticism. Considering

---

18 The name contains the initials of the first names of her founders.

the additional cost burden, other OEMs told me the initiative would never succeed. I disagreed based on my German experience and what we were doing in Japan. Retailers, our own retail sales, and our application product group were initially not persuaded either. I was stepping on a lot of toes, exploring untested territory. Jealousy over the additional OEMs' revenues developed, and a kind of turf war ensued—politicking at best. Steve, in particular, wasn't terribly enthused about licensing apps to OEMs in the US. Again, he was fully committed to "his" retail channel and had the merchant's full attention and best interests in the fore. It was soon too late to be sentimental. Against all odds, the experiment developed into a stunning success. Within six weeks, we had GW's competitors knocking at the door. At the end of the trial, GW begged me, to no avail, to extend her exclusivity. In its place, I convinced Ted to increase his commitment and sweetened the pot. Other OEMs trying to follow GW timidly barked at our prices but nevertheless experimented with a limited number of their own bundles.

To my astonishment, our applications' software competitors took several years to follow us. They lost revenue, market share, and product awareness. Over time, the product group seized her opposition and, in the end, like Steve, embraced our deals. Ted told me personally that the bundling had an exponential effect on GW's growth and visibility. MS obviously benefitted by enlarging her market share on systems where our apps might otherwise not have landed on—at least not legally.

Dell, GW's fierce competitor, founded by Michael Dell in '84, was now located in Austin, Texas. Like Ted, Michael had established his company from the committed belief of a direct-sales model being superior to retail. Dell was regarded as a low-end PC manufacturer until he refocused his company on producing PCs for enterprises. With a larger ego than Ted, Michael insisted on talking to Bill instead of his underlings. I met him several times and always admired his business sense but found him otherwise distant, impersonal, and not committed to MS's range of products.

With Dell gaining share with enterprise customers and Gateway and Packard Bell being successful with consumers, Compaq's PC business in the US felt the crunch. While her server business was going gung-ho, her desktop PC sales slowed substantially. Only available at retail, she had missed the direct-sales wave. The struggle with competitors and her falling stock price caught the eye of chairman Ben Rosen, who was determined to turn the situation around. I was surprised when I received a phone call from Mike Clark, a Compaq VP, asking me if I would come to Houston, Texas, and offer the exec team my market perspective. As much as I wanted to help our "partner from birth"—as Ben Rosen characterized our relationship—the invitation made me uneasy. I was not sure if I could—or wanted to—compete with the in-depth perusals of her dedicated marketing pros and their presumably high-handedness. At the same time, I felt honored to be considered.

To accomplish the task, I had to base my projections on our database of actual OEM reports. It served as the basis for our annual planning exercises. For manufacturers who licensed from competition, we obtained volume estimates: occasionally from them, their competitors, or independent marketing research. The piracy segment was more complicated to guesstimate. Whatever the source, our statistical data supplied us with accurate-enough info to plan our business quite well. Outside analysts, on the other hand, depended on less-accurate data often obtained directly from PC manufacturers. The desire to impress investors, brag to analysts, and leave competitors in the dark often diluted realities—up or down. Understanding this, I was cautiously optimistic of being able to address Compaq's management diligently.

The confidentiality of our database was the last obstacle. Customers providing data in good faith wanted it protected from falling into competitors' hands. The solution was to combine several customers and present only calculated totals. So I set out to estimate WW PC shipments, including the screwdriver segment and the number of pirated systems. I then prepared a chart comparing the growth of directly selling OEMs with the ones going exclusively through retail channels. My analysis made Compaq's WW market share appear considerably smaller than her own propagated estimates. Expecting a push back, I hoped the attendees would come open-minded!

Admittedly, I was nervous and hoped any perceived controversy would lead to a lively debate. I was not to be disappointed. Entering the meeting room, I found five of Compaq's top echelons: Garry Stimac, Mike Swavely, Mike Clark, Rod Canion, and surprise, surprise, Compaq's European manager Eckhard Pfeiffer. I knew him from my time in Munich. Commencing with my presentation, I got interrupted multiple times. Several people in the room, most specifically Rod Canion, then Compaq's CEO, attacked as expected my total-market-size assumptions. He further was skeptical about my growth analysis and conclusions regarding the fast-growing direct-sales channel. Mike Clark had told me the night before that I could, under no circumstances, turn the meeting into a "pitching Windows" event. My conflict: the sales data of the companies I had combined in my chart were all bundling Windows on most of their PCs. Sneaky maybe. The conclusion of my presentation did spark a Windows discussion, the one I had hoped for. Maybe I had been a bit too deliberate and hopeful in my selection, shooting for the so-far unobtainable. What else to expect from a sales guy? As I left the room, Mike thanked me for a job well done; Eckhard took the opportunity of doing the same in German. I was left in the dark on what I had accomplished and how much our discussion would influence our business.

For one week, nothing happened. It was later confirmed that Ben Rosen, Compaq's chairman, who did not attend the meeting, had gotten the information I had provided. I was not surprised when he shook up top management but astonished when he promoted Eckhard Pfeiffer to Compaq's CEO post. I felt good having Eckhard at the helm. I had a personal relationship with him and hoped to influence our business relationship personally from then on. But Compaq's management did not venture into the direct-sales channel and, most disappointing to me, did not license MS Windows as a result of my visit. Instead, the newly appointed CEO focused his immediate attention to improving supply-line management. While I may have been a factor in shaking up the old team with my global presentation, my hidden objective was unaccomplished. In confidence, Mike Clark told me a couple of days later, "Don't worry, Windows will happen, just a little bit patience, please!" Me and patience. His reassurances sounded Chinese to me.

Not going straight back to Seattle, I flew into Dayton, Ohio, to meet Bill for a memorable visit to the National Cash Register Company (NCR). We were hosted in the Wright brothers' mansion, which had been converted into a guesthouse—on historic grounds as I thought. Meeting NCR's flamboyant and outspoken CEO and chairman Charles Exley Jr. for the first time was an experience in itself. The over one-hundred-year-old company had roots in the cash-register business. By now her minicomputer-based banking and accounting systems as well as her ATMs and communication boxes made up the core of her business, nicely complemented by service revenue. In '83 she had ventured into PCs. Her strength was not just in hardware but also in value-added accounting and banking software. She meanwhile was a solid Windows customer, and Chuck Exley had expressed interest in meeting Bill to receive an update on our future plans and to explore how to cooperate on NCR's home turf.

After a cooperative and in-depth information exchange, he invited us for dinner with his whole executive team in the marvelous dining room of this historic place. Right after, desert Cuban cigars were passed around, with Bill, to my astonishment, taking one of the large Churchills and Exley helping him to light up. Only once before had I seen him smoking cigars: during a dinner we hosted in '86 for Scott Oki near Cannes, France, where we bid him farewell from his international job. Like everybody else in the room, he did not dare rebuff Exley's personal invitation to enjoy a bad habit I still share.

As the blue smoke and the aroma of the Cubans filled the partially candlelit dining room, the conversation, helped by a few well-aged cognacs, got increasingly animated. Exley suddenly stood up and, as he opened one of the curtains, made his voice heard by saying, "All my guys are making way too much money." There was an immediate silence in the room. With a big grin on his face, pointing to the illuminated parking lot outside, he added, "Otherwise, they could not afford all these fancy Nazi cars." We looked at each other in astonishment. Porsches, BMWs, and Mercedeses galore—only his visitors had arrived in American-made automobiles. He was quite a character. Unfortunately, one year later, NCR was bought by AT&T, with him leaving for good. NCR subsequently gave up building PCs.

AIMing

June of '91 delivered a bombshell. *Apple*, *I*BM, and *M*otorola announced their *AIM* alliance. The intentions of the triumvirate, firmly AIMing at MS's head, sent shockwaves through the PC industry, rattling its foundation. Bill's earlier phobia, expressed after the OS/2 divorce, had been entirely warranted! The alliance was the brain child of Jack Kuehler, our old foe. It proposed a new computing platform incompatible with Intel's and based on IBM's brand-new RISC CPU–christened PowerPC. Its advantage was supposedly its price, and Motorola announced that it would act as a second-source supplier. The PC, as the world knew it, seemed doomed. Fantasy or reality, the press was already convinced.

The menacing aspect for MS was the coalition's intent to create a brand-new OS—a cutting-edge object-based masterpiece. Obsoleting MS-DOS, Windows, OS/2, and NT all together! The maverick mouthpiece announcing—in his usual aggressive style—Apple's side of the deal was Mike Spindler, my old boss and now Apple's CEO. An exceptionally brilliant guy who did not always have his temper or his mouth under control! We had nicknamed him Dynamo. Not mincing words, his speech extolled the wholesome benefits of good old-fashioned, vigorous all-American competition principles at their best—AIMed, unfortunately, squarely at our core! How many times has history described this? Among the great, the greater the challenge, the loftier the response! Go ahead, Mike, wake a sleeping titan.

The only real product the alliance had to show was IBM's prototype, the not-ready-for-prime-time CPU. On the software front, there was no mock-up cardboard cutout, or even faint smoke and mirrors. The announced OS was the purest form of vaporware. A dream and a bet! MS announced MS-DOS version 5.0 with a six-month time horizon. The alliance announced hers twelve to eighteen months in advance. MS did not run whimpering to the Feds about a wildly premature, empty, and deceptive revelation, nor did any of IBM's competitors or DRI. They applauded, and the press found nothing faulty to report.

Soon thereafter, I joined, Bill, Paul Maritz, and Steve to assess the situation. We concurred at once about the announcement's deadly intent, though we were of differing opinions on just how much true and dry gunpowder the newly formed alliance actually possessed. Paul expressed skepticism about its longevity. He did not believe in Apple and IBM being able to work with each other. We all agreed that the new RISC chip would probably see the light of day, and Apple would probably purchase the CPU for building a new generation of Macs.

The ambitious OS pronouncement sparked a heated discussion. Based on his working experience with IBM, Paul did not believe that she and her enlisted software partner, Taligent, could deliver a viable OS in the projected time frame. As usual, Steve was diving for the panic button with Bill not far behind. These two still loved to create paranoia, yet it did not work in today's closer quarters. Neither Paul nor I bought their arguments. The alliance had just too many divergent goals. What benefit would IBM receive from seriously altering the PC landscape? Wasn't having Apple as future partner a cutthroat detriment? An Apple entry with an identical architecture would definitely hamper her already-thin margins further. Would Apple really buy an OS developed by IBM and give up her proprietary advantages? Dream on! Would this alliance be able to attract enough independent software vendors and lure them away from Windows? Possible—considering the resources of the three giants! I remained as skeptical as Paul, but crafting a tactical response was not, at this juncture, in my job description.

Shortly thereafter, Bill covered our bets, announcing a special Windows NT version for IBM's PowerPC platform. IBM strangely promised OS/2 and a UNIX version. Sun chimed in that her UNIX, *nomen* Solaris, would be available as well. While many people had begun accusing MS of being an unassailable monopoly, this event laid bare our unending vulnerability. The AIM alliance had the clear and immediate potential to stand the current PC world on its ears. With her meanwhile impressive one billion dollars in annual revenue, MS was a dwarf compared to the united powers of these three entrepreneurial monsters. Competition was vigorously alive and vitally well, absolutely threatening MS's livelihood. The Federal Trade Commission's investigation of MS simply continued. It quietly, powerfully, and malevolently swept through the MS campus, unbidden and beyond the reach of logic or reason, like a ghostly ground fog, unleashed by the futility of our competitors and our very government itself. Taxpayer funded. As if AIM were nonexistent.

I took the public relations circus around the alliance with a grain of salt, and so did the always-pragmatic Paul. We had experienced plenty of other threatening announcements before; they did not necessarily translate into timely execution or genuinely competitive products! The paper they were written on was cheap and expendable.

One year later, the alliance derailed when ambitious OS plans did not materialize and Apple couldn't agree with IBM on hardware specifications. The Power RISC chip did make it into the market in proprietary IBM and Apple systems. MS withdrew NT. IBM never finished an OS/2 version and focused solely on UNIX. Apple bought her CPUs only from Motorola. Mike Spindler lost his CEO post. No one ever heard from him again. Amen!

The other company who announced an OS for the new architecture was Be Inc. She was run by Jean-Louis Gassée, Apple's former French country manager, who had helped me contract Multiplan for Apple's IIe European launch in early '83. After leaving Apple, where he succeeded Steve Jobs as leader of the Mac division, he entered the OS fray. Eager to compete with and dethrone MS, he developed BeOS for IBM-compatible PCs. Unable to find sufficient customers for his baby, he blamed what else? My business practices. Jean-Louis, a smart man, should have recognized his core strength—great French showmanship. Not enough to pull his company through. Bill and I applauded him for his admirable efforts to gain a foothold. He earned a second chance, but IBM's proprietary UNIX versions eventually ate his lunch.

Still worried about the importance of IBM's new architecture, Bill initiated, right after AIM blew up, several informal efforts exploring a renewed partnership with IBM. He was rejected. Tenacious Jack Kuehler made sure of that and responded, courting our competitors and forming hostile alliances with Lotus, Borland, and Novell. I felt bad for Bill, but IBM was by no means ready to give up on OS/2, and it showed. The wounds of her current management team were too deep and needed time to heal and, really, a new leader to arrive. The only good news: few of our OEMs reported serious demand for OS/2. Observing this, I noticed IBM's close allies were betting on the wrong horse—ultimately destroying themselves.

Assignments and Threats

After my boss Jeremy Butler left the company in the fall of '91, I again reported to the president. In addition to managing all non-US business, he had also been in charge of our product-localization group. The main task of this group was devoted to publishing rules and educating developers on how to write easily localizable code.

To improve our localization efficiency and reduce delays, management wanted to reengineer the current process by integrating localization specialists back into the core-development groups. Jeremy did not finish the integration project before he left. Lucky me! I was the gifted soul the burden now fell upon. I shouldered the task as a thought-provoking challenge. Keeping my eyes on OEM, I delegated most of the gritty work to the group's current management team and accomplished the integration in just six months, well ahead of the twelve-month goal.

Soon thereafter, another strange assignment fell into my lap. Oddly, Steve asked me to visit with DRI's executives, exploring a technology exchange. The idea was to obtain the rights to her allegedly superior DOS technology for use in future MS versions. Steve had reportedly met Dick Williams, DRI's CEO, at a conference in the UK. He seemed open to pursue such an exchange. Promising to follow up on their conservation, he made me the point man. I considered my assignment an odd and unwanted mission. My attempt could be misconstrued as apparent collusion between competitors. In using DRI's technology in future MS-DOS versions, antitrust lawyers might accuse us of eliminating a competitor. I did not understand why we wanted this technology in the first place, or why I was picked to consummate such a deal. With a lurking suspicion, I departed for Utah to visit the hyena's den.

As I arrived at DRI's facilities, my doubts did not disappear. Dick Williams was present together with one of his attorneys. I was alone, and the right thing for me would have been to turn around and return with my own legal support. Instead I stayed on and made a vague proposal along the lines I discussed above. I found out that the technology in question was available for $30–$40 million. Steve had instructed me that he would not pay DRI in cash. In its place, he was willing to extend DRI a reseller license for MS-DOS as payment in kind. The total trade-in value he envisioned was $5–10 million. Hinting that their asking price was roughly five times higher than I had expected brought laughter; the deal was off.

I loathed the idea of having DRI as a trusted second source. Against all popular sentiment, I wanted her to be around and compete. DRI's independent effort was the best insurance for my customers to receive constantly improving MS-DOS versions. A compelling reason to keep her alive and our development guys on their toes!

Did Steve send the wrong guy? You bet he did. He should have gone himself. His intentions were only loosely communicated and had too many inherent shackles. In addition, I was biased about DRI ever becoming a reliable partner. My compromised initiative can be summarized as denying DRI a deal bordering disobedience. Allowed under the doctrine I was operating under, considering the ultimate edict voiced by Bill: beat DRI fair and square in the market—much easier to follow and fun to execute!

I cannot exit '91 without mentioning two other key events taking place. The expansion of American Online (AOL) is one; Novell buying DRI the other. AOL's key contribution was to expand its online services everywhere, from serving the gaming community to enabling PC home users to exchange e-mails and access rich online information. Networks of this kind already existed, but most of them either were not as easily accessible or charged relatively high fees. Initially, we applauded AOL—just one additional reason to buy a PC. As reality set in, top management viewed AOL's endeavor as a lost opportunity for cash-rich MS.

Novell's buyout of DRI that year was a serious and potent threat. Raymond Noorda, Novell's CEO since '83, allegedly coined the word *co-competition* in describing the PC software industry. It suggested that even rivals enjoy a natural concordance to attain ultimate successes. His company originally sold servers. Novell's breakthrough came when she developed a network interface card for PCs and corresponding software. The software, later called NetWare, enabled PC servers to easily connect to each other. During the '80s, Novell transformed herself from a mostly hardware company into a pure software company, an impressive accomplishment. Until then, other server and network card providers regarded Novell as a competitor. Dropping the server hardware from her sales portfolio, Novell succeeded in making NetWare the leading PC network operating system. By '91,

NetWare's lead appeared unassailable with MS-LAN-Manager and several UNIX versions together having less than 10 percent market share.

Thanks to my group limiting DRI's success as an OS supplier and after the failed buyout, her CEO, short of cash, looked for a suitor. As a perfect merger candidate, Novell had the money and her CEO the zeal to make MS's life harder. When I heard about Novell's purchase of DRI, Bill and I were profoundly alert. Ray's famous vindictiveness toward MS accompanied by Novell's cash hoard and her much larger sales force strongly suggested a prolonged and more vicious firefight, for DOS market share was about to commence. In selling NetWare to our OEMs, Novell already had existing relationships with a lot of my customers. I feared her sales force could easily leverage these and prove plenty successful selling DRI's product line in unison. Bill agreed.

As the merger unfolded, I was urging the MS-DOS group to finally deliver a version that could be embedded and function in read-only memory (ROM). DRI had successfully sold one for a long time. I considered this product segment her last bastion to fall. Not needing a hard disk to function, it had found its ways into the industrial controller market and emerging handheld devices. Not a huge segment then, but with the accelerating miniaturization of electronic components, its expansion was a foregone conclusion. The reason why I eagerly wanted to create a foothold, establish our brand, and take business away from DRI. I convinced Bill that this was an urgent, exciting, and obtainable opportunity, and with his help, the product group eventually delivered.

Luckily, my nightmares did not become reality right away. I never fully understood why Novell did not hammer the holy hell out of us. The machinations of executing the merger were undoubtedly one reason for her failure. Novell's core business had mostly retail focus. A smaller amount was done through enterprise-focused OEMs. DRI had hardly any retail presence and sold directly to smaller non-brand-name OEMs. The salespeople in Novell and DRI had different skills and served different distribution channels. The grand tactical error: her sales management missed the opportunity to create synergies between them.

As we attacked Novell/DRI in her last stronghold, we found scant resistance. One and a half years later, despite—or because of—the merger, we had neatly swept a significant portion of the embedded business away from under her. No serious effort was made to stop us. No discounts, no marketing, nothing. It was too easy, I thought. Frankly, I was disappointed, having expected a bloody battle.

No attack on Windows either. No taking out a license, as IBM had done, from Apple to get over the patent hurdle. No cloning attempt of Windows either. Novell certainly had the resources and the skills to attempt such a feat. She could have succeeded in launching an alternative and nicely polished Windows-like product. Did Ray Noorda have other motivation for acquiring DRI?

As we were entering 1992, a stand-alone DOS for PCs without Windows-like features was no longer in popular demand. The weight had undoubtedly shifted. With Novell/DRI seemingly out of the race, any serious competition for Windows would have to come either from OS/2, UNIX, or Apple's Mac platform. And last but not least, the software pirates were still threatening the high seas of success for all of us.

Disruptions

Ups and downs

Finally, in mid '90, the Federal Trade Commission (FTC) notified us of the secret probe her commissioners had started months earlier—by now solely focused on OEM business practices. A fishing expedition as I thought. The commission muzzled us, swearing us to secrecy. Over the next year, requests for information piled up on our attorneys desks. With the FTC refusing to fully disclose what and why it was investigating and restraining MS from sharing details, the nation's highest law enforcers of the Department of Justice (DOJ) filled the void. Apparently, there were different rules for them. The leaks they provided fueled speculation, impacted our stock price, and were just a forbearer of what type of damage ill-informed press people and slimy informants are able to cause.

Steve and Bill concluded that the Bush 41 administration had gotten us into the mess. The upcoming '91 election presented a golden opportunity to get us out of it. So our top honchos set out and openly promoted Democratic contender Bill Clinton. With the country sinking into recession, they considered a personal engagement pretty low risk. One of Clinton's advisers neatly defined the moment in history with "It's the economy, stupid!" when describing what primarily motivates an electorate to lean toward a presidential candidate.[19]

It was not just a pocketbook choice. The two candidates portrayed rather distinct differences. The leadership style of 41 was perceived old-fashioned despite liberating Kuwait and witnessing the fall of the Berlin Wall and the Soviet Union under his watch. Bill Clinton, in contrast, gave voters hope for a more financially sound future, exuding a deft decisiveness accompanied by well-received folksy humor, optimism, and plenty of

---

19   Let's see how this holds up in 2012.

promises. A generational shift was about to occur in American politics, and an older-style incumbent helplessly watched as history was made.

Steve went on an overt crusade enlisting CEOs from other IT Companies. Bill joined him—MS employees watching with astonishment two avowed capitalists showing flag for a Democratic contender, short of a probusiness agenda. After Clinton's election, the hope of the top guys evaporated. The government continued pursuing the case against MS unabated. The assumption of a Democratic administration dropping antitrust allegations, odd and not too promising from the beginning, had not panned out.

The resignation of Mike Hallman in '92 after a brief two years' stint as president came as no surprise. With his prescriptive management style, he had failed to see eye to eye with tech-trajectory and direction-centric Bill. Not replacing him with a single person, the board created an Office of the President. The following people were named to help Bill run the company: Frank Gaudette, Mike Maples, and Steve Ballmer, signaling a major shift in leadership. Mike was assigned management of all development groups, Steve was appointed head of worldwide sales and marketing, and Frank, the current chief financial officer, continued his responsibility for finance and administration. Steve moving to the top position in sales and marketing impacted me directly.

Effective immediately, I reported to the man I'd tried to get fired a couple of years ago. With awkward feelings, I looked forward to our next meeting. In view of his normal style, I assumed he would give me a hard time; Steve did not disappoint. Preemptively sharp as always, he did not summon me to his office; he showed up one day in mine unannounced and rather early. After accepting my congratulation, I asked him point-blank how he viewed the overall performance of the OEM group, and mine in particular. Smart or not, I genuinely wanted to know where I stood. In typical Ballmer style, he didn't miss a beat, roaring out a swift, straightforward answer. "I consider your performance barely a 3.0!" In MS's culture, his remark amounted to nothing short of an insult, period. Let me explain why.

Twice each year, MS employees convened with their superiors to discuss job performance. Normally, the employee was asked to write a review in advance listing achievements and failures. In response, management graded job performance on a scale of 1–5. The ultimate 5.0 grade meant you walked on water; a 3.0, on the other hand, was the lowest acceptable rating—anything below meant you were about to be shown the exit. Having constantly achieved ratings at or above 4.0, Steve's off-the-cuff assessment was nothing short of "Take notice, you are not bulletproof." I could have asked why such tepid ranking; in its place, I laughed it off with a cocky reply of "If you think so!"

He grasped immediately I was not taking his from-the-hip comment to heart. In the ensuing discussion, he brought up my former recommendation to fire him. No reason to look for any other motive. At fourteen years his senior, I intuitively knew my former actions had wounded him. With loyalty a most coveted priority when he evaluated employee performance, I had stepped over the line, and in a nutshell, he was still pissed.

Ignoring smoldering emotions as much as possible, I launched into an in-depth discussion of the OEM business. Ironically, we were in total agreement on how it should be run and what goals and objectives needed to be accomplished. I had expected nothing else. Otherwise, knowing his influence with Bill, I would have been fired long ago. Arrogant though it might sound, deep inside I suspected he had no intent to unseat me. His mediocre performance comment was a random volley, intended to commandeer at last my attention and respect. It undoubtedly did—without me offering any apology for past behavior.

My intuitive conclusion: any type of cowering at such pivotal moment would have backfired. The man was testing my core and unwavering attitude. In an instant, we had renewed our basis in how to work with each other for the next nine years. He actually never formally graded my performance again. In future performance reviews, we analyzed my business in close detail and established top priorities and objectives. Without exception, our powwows were respectful exchanges from which we each came away having learned a great deal. Had I tamed the tiger in him early on? In hindsight, I conclude we came to value each other's judgments, though not without—as in a healthy executive relationship—agreeing to disagree a few times. Respectful disagreements, which strengthened our relationship and advanced how we proceeded with the business at hand.

From then on I never experienced the rants and tirades Steve sometimes leveled on people. He was, in general, supportive of my management style, and any criticism was posited constructively and in a professional manner. His cooperation and praise, often expressed publically, made it much easier for me to manage my group and left second-guessing of my decisions to the lesser informed. To this day, I own a mostly positive memory of our working relationship. Steve, if you happen to disagree, please drop me an e-mail!

Steve, always profoundly aware of distribution deficiencies, asked me to take another look at our lingering system-builder business. As the market was shifting toward Windows, our old agreement with Phoenix had become obsolescent. Commencing, I tasked my marketing group to investigate if hardware-component distributors would be interested in broadening their product line by selling OSs to this community. They responded positively. Encouraged, I gave the green light to hire a new breed of sales reps. We made them responsible for helping our newly selected distribution partners to promote and sell OSs through establishing relationships and creating demand with system builders.

After logging experience in the United States, we expanded our business horizon internationally. We were at once off to an imposing start. Adding system-builder conferences, direct mailings, and later, e-mail campaigns to our marketing arsenal accelerated and manifested our feat. But a fresh set of unanticipated challenges kept popping up. Several subsidiaries outside the United States complained that we were diverting portions of their current retail business into the OEM bucket. They failed to note that our packages were not sold to end users but to low-volume hardware manufacturers. The OEM group was responsible for this market segment. Instead of closing license agreements via signature, we had expanded our business into selling licenses in boxes.

Considering the bigger picture, the subsidiary managers should have simply shut up. In talking to them further, I learned their real desire was to control the pricing of these OEM products locally. I considered abandoning central price control a foolish move. Through a single WW price list, equal treatment was guaranteed for all OEMs around the globe. Local price control for OEM products made no sense to me. It would inevitably lead to price shopping and uncontrollable discounting and enable local favoritism. I ignored the arguments.

As time went on, a kind of subsidiary coup arose—a chorus of voices pleading to take the low-volume OEM business away from my group. Simple greed for my taste! By nurturing the low-end market segment, it had become the fastest growing portion of my business. Our centralized management and unified marketing was an essential component in cultivating it. With no new directions coming forth—from either Bill or Steve—there was no need to cave in to emotional pleas from people with little or no understanding of the intrinsic nuances of this business. By offering no concessions, my popularity with subsidiary management did not exactly flourish. As the lobbying went on, Steve got annoyed and eventually tried to convince me to give in: "Be a nice guy." My answer to him: "Let's continue with what makes the most sense to ensure success. I am not running this business to win a popularity contest!" I carried on.

In '92, the day for my first court depositions had arrived. I never thought this would ever become part of the deal. Our lawyers and I spent endless hours to prepare for this methodically, page by careful page, combing through each and every document. Ad infinitum, valuable workdays disappearing forever amid the mindless dry riffling of paper! At first I resented the long, tedious hours. Later I came to appreciate them as useful and clever exercises helping me remember events from the past, put them in context, and sharpen the logic behind my later arguments. Amazingly, most issues we covered did in fact come into play during my depositions. These attorneys were real pros. After my first testimony, they grew in stature for having engaged me in mock battles to learn the cat and mouse game opposing lawyers would eventually engross me in. I grew a mantle of radar, sensitized to the blindsidings and the character-ambush attempts.

During the FTC deposition, the government lawyer chased me through our written policies and procedures, the standard OEM contracts, and our price guideline. His main game was to introduce e-mails or documents and asked me to interpret the intent of the originator. My approach for handling the situation was to not engage in second-guessing or trying to get into others people's heads years after the fact. In time I grew a little overconfident. I began noticing my counterpart was badly prepared. He had not logged enough time understanding the OEM business or PC technology. I soon became bored and began playing mischievous games with him, trying to take him off his envisioned path. In the process, I tossed him a prize, a real windfall, inadvertently volunteering information he would never have gathered on his own. At the first recess, one of my attorneys took me aside and insisted I change my behavior, to answer the most unsophisticated questions with patient robotic restraint; playing mind games was not what he approved of. I was thankful for the reprimand, and the deposition continued with me being more disciplined.

I quickly learned: whatever somebody wrote can be located even if the originator purposely didn't keep it, and secondly, all of it could be used against you. Well-intended, complex, and hard-won policies could be misinterpreted by an opposing lawyer, turned against you to make you look like a miserable fool. E-mails you had completely forgotten might be archived by other people for no apparent reason. Years later you remember hardly any of the contexts they were derived from. Count on a federal investigator to dig them up. Be aware he will put words in your mouth to prove fictional evil intent, and you will be required to defend yourself. You will be made to appear flummoxed and dumbfounded, devious and inept. Sullied and ill-sorted and scandalous! The list of things to watch for in my interrogation was a long and a dirty one, a snake-infested labyrinth of trapdoors and snares. After two long nine-hour days, the ordeal ended with me exhausted and delighted to return to my familiar business. One down!

I had no clear idea why I was being deposed in the still-simmering Apple patent lawsuit. I was not privy to the peculiarities of the patents MS allegedly was infringing upon. What I remember from the otherwise boring and utterly meaningless deposition was how it began. When I entered the conference room where I was scheduled to appear, I first noticed a high-end video camera Apple's legal crew had positioned opposite my chair. The camera rolled as soon as the deposition started. The system was equipped with blindingly bright lamps. As soon as the light hit my eyes, I reached for my sunglasses—yes, even in Seattle. The opposing attorney at once decried my reaction. During the ensuing recess, I assured my lawyers that staring in these lights would soon cause severe headaches for me. The other side needed to put them out. It took nearly thirty minutes to resolve the situation. In the end, the other side relented. The camera was used—*sans lumière*. Curtains were opened, allowing sunlight to illuminate the scene. It mattered little—the interview ended abruptly. Both sides considered my short deposition a waste of time.

What I learned from both experiences was to reduce mail stored on my PC. From now on, my e-mails were archived after four weeks max only. I made certain my filing cabinets contained no information older than six months (OEM contracts were never stored in my office, in case the reader wonders). I asked my managers to follow my example. Unfortunately, not all of them did. Reducing what we kept was smart. The company would have been miles ahead if management had made my personal habit a shop-wide policy.

FRONT LINE PARTNERSHIP

In his new job, Steve at once began beefing up MS's enterprise sales force. Over the next few years, he increased the size of this group at a staggering rate and made certain our consulting organization worked in unison with it. As that group was tasked to create increasing demand for our server and office productivity products, it continued to be decentralized and regionally driven. Marketing, on the other hand, got streamlined, and strategic initiatives were now designed centrally for impactful worldwide executions. Was my boss learning from how I had organized OEM?

The key product introduction in '92: Windows version 3.1. A much-improved and nearly bug-free version with exciting multimedia features was soon followed by 3.11, adding crisply performing communication protocols. Shortly thereafter, a protocol stack was added that enabled users to access the fledging Internet through corporate networks. Still missing were long file names and a reliable shield against ill-written applications able to collapse the whole system.

The impressive improvements were due to SVP Brad Silverberg's leadership. He had come to us from Borland a year earlier. Brad, a bit of an introvert with dry, caustic humor, was admired for his business acumen and technical skills. An avid biker, he was often referred to as "lean, mean, and hard-core in all matters," a characterization I readily endorse. He had an excellent understanding of the software industry and an incredible talent managing what I decried as MS developer prima donnas. He was nicely complemented by Brad Chase, an outgoing marketing executive with highly creative ideas and a can-do attitude. Brad C. was a super guy to work with and a truly gifted listener. The "two Brads," as they were often called, did not arrive a second too early. Their keen energy and astute market knowledge was desperately needed to win the battle against OS/2. I loved working with them, strategizing our competitive raids. Both true pros!

Eckhard Pfeifer, Compaq's CEO, had meanwhile succeeded in streamlining product lines and manufacturing systems. Back in the black and growing, Compaq had maintained her position as the leading PC server company and was approaching Eckhard's goal of overtaking IBM as the largest PC shipper. In March '93, our marketing teams ironed out a synergetic alliance for mutual customer benefits. Without our revamped enterprise sales force, the Front Line Partnership (FLP), what we entered into would probably never have come about. It delighted Steve. The joint announcement of this far-reaching pact by Bill and Eckhard at the Federal Trade Show in Washington, DC, was an enervating accomplishment and a major milestone for my group. The partnership was not just another product deal; it was a pledge to promote each other's technologies "primarily" but not exclusively—explicatively leaving room for other close relationships. From now on, the two companies considered each other partners in arms with similar marching orders, marketing missions and harmonized product developments. The implicit vow was to mutually develop products that were "the easiest to use and the simplest to install with the best performance and value in the industry." Most importantly for me, Compaq at last signed a five-year Windows license for all her desktop PCs. All in all: awesome! Big Blue, you better watch out.

Customers from both companies greeted the news with enthusiasm, anticipating enhanced IT performances. Several of my OEM customers, and IBM in particular, though, were shaking in their boots. They recognized we had picked the best company for such an encompassing technology and marketing bond. Right off, they suspected the pact rewarded Compaq with better terms and conditions, centering the ensuing parleys with them on price, price, and again, emphatically, price. They got it right. Going forward, we had agreed to a most favored pricing clause with Compaq. She in turn had committed to purchasing the highest volume of Windows licenses ever. As long as Compaq occupied that position, she undoubtedly earned this concession. I never imagined how much resentful fervor the announcement of our alliance would create! Particularly inside Dell, Hewlett-Packard, Digital Equipment Corporation, Acer, and the large Japanese manufacturers! IBM's response turned out to be the most demanding and eventually the toughest to accommodate. This surprised me the most, considering the signature under the OS/2 divorce papers had hardly dried.

Over the next few years, the temptations to consummate comparable agreements with other OEMs mounted. Doing a second or a third comparable pact would not have been impossible but certainly delicate, considering our commitment and desire to keep the relationship with Compaq vitally alive and well. After careful considerations, MS never seriously contemplated another FLP except with IBM. In its place, we got creative and developed a novel strategy to mollify all other customers. Nevertheless, I was pleased to see the FLP survived the manifold changes resonating through the industry. Hewlett-Packard, who bought Compaq in '99, decided to renew the alliance in 2010. We must have done it right eighteen years ago!

For the next couple of weeks, I committed considerable time to our larger and historically loyal customers, putting the newly formed alliance in the right light. They all felt excluded and left out in the cold, accusing us of complicity, extending Compaq an unfair advantage. Fortunately, our product offerings were strong, and despite a cacophony of objections, no major OEM abandoned us. A few threatened to make a revengeful run to the competition. Responding to the challenge at hand, I dared my marketing folks to help alleviate the situation. They started by putting the marketing section of the FLP under a microscope. Under the guidance of Carl Gulledge—my ace marketing guru—they indeed came up with an interesting plan, providing me with a rich variety of options to address the lingering and thorny concerns.

Compaq, in the meantime, used the FLP to the hilt. Her aggressive public display created extra pressure for us to offer some form of marketing cooperation to other OEMs. When consolidating our ideas, we borrowed a page from Intel's playbook. To promote her CPUs, Intel had convinced OEMs to insert her logos and promotional slogans into print and TV advertising. In return, Intel paid for selected or all advertising cost. The savings earned were enthusiastically welcomed, allowing beneficiaries to extend budgets through placing additional ads or simply saving some money. Intel further convinced OEMs to put "Intel inside" stickers on their PCs, an idea that really hit home with me. AMD copied her as usual, so why couldn't we initiate something equivalent to promote the MS Windows brand better or get our antipiracy messages across? Education over legal actions—what could possibly work better with end users? OEMs selling PCs with a genuine OS should have a vested interest in differentiating themselves, especially from the side-winding, cobbled-together screwdriver outfits.

My marketing folks knew our OS gurus were unfailingly displeased with overall PC quality and performance. As an example, booting a PC took far too long—one of Bill's favorite topics! The Apple Mac booted in half the time. Could we inspire OEMs to design swifter-booting PCs? The pact with Compaq intended to promote each other's advancements. Why not encourage and incent other customers to work on comparable product improvements? Far more beneficial for end users and pungent than merely adding sticky slogan labels on PCs. Intel was satisfied; we went beyond. To retain our Windows logo would require a minimum system configuration and performance testing, thus upping quality for purchasers.

Rewarding OEMs for designing superior features—sure-footed and better-performing PCs—and promoting Windows and enlisting their help in reducing software piracy sounded intriguing to me. Joining us, PC manufacturers should increase their competitiveness versus Apple Macs and IBM PowerPCs—an easily acceptable and common goal as I thought. We discussed our ideas with the product groups and got their nod. Involving our attorneys, we asked for a first draft of what would later become a market development agreement (MDA). Its final version specified royalty discounts for OEMs as rewards for joint marketing efforts and more user-friendly PC designs. OEMs could choose how much they wanted to participate. The more engaged, the higher the rewards. The newly brewed concept should help level the playing field with Compaq without the need to appoint additional frontline partners. OEMs now had a simple method of positioning themselves as our partners and ultimately enriching the lives of PC users. Weren't we all in the same boat, competing with Apple, IBM PowerPCs, Sun's workstations, and other proprietary computing solutions? According to surveys, MS had just become the second most-recognizable brand worldwide. Why not share our newly acquired image with our customers?

Engaging the product groups and MS top executives in a lively debate preceded the selection of the top priority items we eventually included in our first MDA. Bill, much engaged, commended our approach. After finalizing the item list, we determined the rewards of each—based on priorities, hotly contested internally. To respond to technology changes and emerging marketing trends, we made it a habit to reevaluate the listed items annually. It allowed us to drop activities, add new ones, and readjust individual monetary rewards.

Introducing the MDA concept to our customers was nevertheless extremely challenging. For the most part, the smaller OEMs reacted positively. The larger ones were harder to please. As much as they liked gaining reductions, a few resisted joint product promotions or adopting requested hardware alterations. In turn, we showed flexibility. I couldn't understand where the resistance originated. Pursuing a proprietary design agenda without our well-intended guidance contradicted with the strong desire to be on a level playing field with Compaq. OEMs felt we were babysitting them and ultimately wanted hand-outs without performance considerations. IBM led the pack. Convinced we were doing end users a favor, we stuck with the concept.

Rescuing little critters

The computer mouse was first mass-introduced by Apple for the Lisa workstation and the Mac. In '83, it made its way to the PC—thanks to a point-and-click interface we designed into MS Word and MS Multiplan. The appearance of Windows in '85 made it, together with the keyboard, the number one device operating a PC. When mice usage took off, so did its design. The device lightened up (no more steel rollers!), advanced ergonomically, and came with better resolution to make pointing on the screen highly accurate. MS had taken an early lead in this hardware category ahead of Logitech, a Swiss-based company, helped by superior retail marketing and advanced driver software—the program instructions making the little critters function to perfection. Logitech's retail presence was modest, yet she was the leading manufacturer selling inexpensive mice to my OEM customers, ahead of a multitude of Taiwanese competitors.

Back in '88, I saw an opening for my group to add OEM mice sales to our growing basket of successful products and compensate for shrinking Windows revenue. With our mice rated best at retail, I boldly predicted we had a chance to unseat Logitech in the OEM channel. The VP of our hardware group, Rick Thomson, listened and was at once enthused about venturing into the realm of sticking it to what he considered his main competitor! Selling the critters nevertheless was not an easy task. As we did not own a factory, our production cost and our prices were therefore up to 100 percent higher than Logitech's. What helped was a combination of quality perception, our top-rated brand image, and OEM's desire to differentiate and improve the quality aspect of their system offerings. Soon we were on our way to selling 20 million of these little critters annually, expectedly annoying the reigning Swiss-based company.

I received a flurry of calls and letters from her CEO Pierluigi Zappacosta, who couldn't comprehend why his salespeople were losing business. Considering his attractive prices and alluring design, he firmly believed in some form of foul play. He accused my sales force of giving favorable

Windows terms to OEMs in exchange for buying mice from us. I repeated again and again that only our high-brand recognition and better quality accounted for our winning hand. Several times I personally investigated his claims, and each time I failed to find any wrongdoing. To be dead certain, I had my legal team investigate the matter. Eventually, after accepting my explanations and having no further proof of malfeasance, he decided to accelerate his design efforts along with his cost-reduction programs and wholeheartedly revamped his marketing. His renewed, vigorous competition eventually began impacting our sales efforts. Though as long as Logitech's perceived image did not visibly improve, we grew at a reasonable pace. Compared to executives of other competitors, Pierluigi was a true gentleman. He did not run to the government for regulatory help but made us struggle by competing fairly and vigorously. I commend him!

By '93 our mouse business had suddenly stalled despite vast increases in Windows units. My group had a hard time locating new customers or keeping existing ones. Customers were no longer willing to pay our premiums. The golden era in which we had enjoyed a fountain of unremitting rewards was drying up. I had met with the mouse product group several times, outlining my concerns and demanding forceful actions. Its response had remained tepid; cost reductions or exceptional new products were not forthcoming.

I deployed e-mails to my managers requesting a detailed competitive update. Most surprisingly, they all no longer considered Logitech our largest competitive threat and identified the danger coming from the smaller mostly Taiwanese companies. Computer mice had gone from novelty to commodity. Brand recognition no longer vital, price had become the key factor winning a bid. With an unresponsive product group, I faced a discouraging outlook.

On my next trip to Europe, I visited a midsize customer called Viglen in the UK. She had been a longtime Windows and mouse customer of ours. The visit turned out to be superfortuitous. Her CEO Vig Boyd and his right-hand man Diran Kazandjian personally showed me a mouse designed and manufactured by a nondescript company in Taiwan. When they mentioned the price the little critters could be purchased for, I was flabbergasted. Graciously, they allowed me to take three prototypes back to Redmond. As I left, Vig casually mentioned that the mouse manufacturer was hunting for additional business. Back, I pried one open then did the same with one of ours. Patiently and deliberately counting the number of parts, I found the tally immensely revealing: our mouse contained twice as many pieces as the prototype obviously achieved by using higher-integrated components. In addition, I admired the lighter plastics and the slick ergonomic design the manufacturer had exploited. In comparison, our mouse looked like a dinosaur. Our product people had been asleep at the wheel by not reducing cost through integration and improving eye appeal through design and advanced materials. Without being an engineer, I knew intuitively that our design was costly, antiquated, and no longer competitive. When I confronted our product manager, he told me he had nothing close in the pipeline.

He took two of the mice along with him, promising to check them out. I sent him a couple e-mails, eager to learn what he'd concluded. After two weeks, the guy was still looking into the matter! Not responding well to silence, I went over his head and engaged Rick Thomson, VP in charge of MS's hardware division, one more time. Rick was an engaging guy with great business sense and an excellent understanding of doable hardware technology. Performing the same screwdriver exercise in my office, I got the response I had wanted in the first place. Rick had heard about my newly found mouse though had never seen its innards firsthand. He promised to get engaged at once by lighting a fire under somebody's behind, understanding immediately that his retail business could be next to fall.

Two weeks later, he personally visited with the Taiwanese manufacturer and closed a sweetheart deal. Purchasing the design, related patents, and manufacturing rights, he placed a one million mice order with the company. Our run rate would allow us to sell them in less than two weeks. The new design had cut our cost in half and resulted in an up-to-date, most reliable, and sexier-looking product. When Logitech and the rest of the bunch woke up to the news, they found themselves outgunned. We had regained enough of a competitive edge to win additional critter business. During the hardware division's midyear review, Bill commented that I had saved our total mouse business. He never told me personally, but I made sure the sales rep who had invited me to visit her UK customer received extra stock options. Vig, who had gifted me the prototypes, was granted free internal use of MS Office for his company. Unfortunately, my rescue did not guarantee success for long. When new technology emerged, I had to repeat the bailout a second time.

Monopoly accusations

"MS's responsibility is to keep her customers happy."

And that might have an opposite effect on MS competitors, as our general council once remarked. I fully agreed with him, and let me add—so be it.

In February '93, MS got good news: the five FTC commissioners had deadlocked; no antitrust action would be forthcoming. One of the commissioners owned stock in the company, and the other four were divided. No majority, no action. We nearly popped the corks but were cautioned by our general counsel, Bill Neukom, who knew how to interpret political vagaries. He had come to MS from the law firm Bill Gates's father had helped found. Dartmouth- and Stanford-educated Bill N., in his trademark bow tie, stood well over six feet and was an accomplished marathon runner and fly fisherman, indicating to me that he could be a patient man. I considered him an astute and extremely smart attorney, fully capable of managing the over three hundred lawyers we employed along with the myriad of outside firms MS retained. Bill N. was warm and good-natured but tough as nails when need arose. He was aided by Dave Heiner, who headed up the antitrust section. He made sure the OEM division was getting sufficient training on antitrust issues and our contracts would survive government scrutiny. I liked working with Dave. He was an established expert in his field, a bit on the dry side, but always with a good joke in hand.

The Federal Trade Commission being deadlocked didn't mean the case against us was in the dustbin. Far from it! Janet Reno, a bumbling misfit of a US attorney general (AG), and her crack teammate Anne Bingaman were still purring over the case. The two ladies toured Capitol Hill long enough to convince equally brilliant and presciently motivated politicians they possessed a solid case against us. Funds were made available, and in August '93, the Department of Justice (DOJ) announced that Anne Bingaman, the head of its antitrust division, had eagerly taken over the stalled investigation ready to lead a tough pursuit.

More depositions were conducted, and more subpoenas were issued. Our attorneys raided my office and my e-mail storage sometimes several times a week, though nothing of much interest was ever found. People working for me, other execs, and product group personnel experienced similar invasions. The government was treating us as criminals. The raids executed on behalf of the DOJ created a distorted prism of anger, violation, and helpless despair. I counseled employees to not feel belittled and guilt ridden or distracted when suffering random searches or reading vilifying press articles. Innocent until proven guilty? Certain politicians, the press, and the DOJ turned that noble principle on its head. Our competitors happily chimed in and joined the witch hunt. Public opinion labeled us a monopolist, and competitors and their lawyers branded us the evil empire, a slogan that stuck.

Resenting the uproar most was Bill. Once upon a time, he had been the darling of the press and Wall Street. He had been honored by President Bush with the Technology Achievement Medal, no small feat but counting no longer. Being characterized as a greedy monopolist was tough on his stomach. How can tireless hard work—the deepest of all possible commitments and sacrifice—and enviable success or overachievement become punishable by law? Bill once told me in private that he feared for his legacy if the world would ever remember him as just another Rockefeller. It scared the shit out of him! Bill had, of course, basic legal education from his stint at Harvard and whatever had rubbed off from dad. A conviction under antitrust laws, he knew, could cause considerable damage to companies and individuals. Deeply worried about the unfolding story, his sleep came less easy at night. Even so, he assumed that the government did not have a solid case. The potential stigma was what bothered him foremost. Blindfolded Lady Justice, that allegorical personification of truth and fairness, bearing her delicately balanced scales of impartiality into the brave new future of PC software tech, shook her head in incredulity.

With the danger of being sued by the Feds quite imminent, the two Bills made it their top priority to settle the case by shuttle diplomacy, visiting the Feds in their lair in Washington, DC. In public, wrongdoing was adamantly denied. Behind the scenes, MS's team was adroitly probing how a fair and favorable compromise could be reached.

Meanwhile, the government was painting MS as a company "in a position of immense power in the software industry." People working there disagreed. From the day I had stepped into the company, I personally observed our absolute and nearly daily vulnerability. Stop innovating? You would lose. Was this no longer true? Had our leaders created paranoia simply for fun? Using emotional outbursts, feeding fear, and projecting doom too often had won few accolades with me, but I never doubted we indeed lived in a hypercompetitive and exposed world. We had to earn our success the hard way, producing and marketing innovative software people wanted to buy and use. As the saying goes, good products win. Mostly! Management knew no other principle cause for success than striving to deliver them.

Admittedly, version 1.0 product realizations were mostly visionary and imperfect, which left us vulnerable, offering competitors dangerous openings if they attacked early enough. After version 3.0 appeared, a counterstrike was harder. After we gained the lead, competitors either envied or hated us! An opportunity missed. Carpe diem, dear opponents! Build a stronger team, work a few extra hours, and make essential personal sacrifices. Finding themselves unable to beat us fair and square, they exploited the legal route. Stirred by legal arguments of overambitious and tea-leaf-reading lawyers. Self-identified losers never quite acknowledging their inability to compete or how badly they timed countermeasures. Shame on them, and welcome to a US system of jurisprudence allowing these political and legal escapades! Garry Reback, starry-eyed high priest of antitrust litigation, among others of his lowbrow ilk, built empire careers teaming up with marketplace losers trying to publicly flog victors like us in the US court system.

At the very end of the nineteenth century, Congress had passed so-called antitrust laws extending powers to the DOJ for reining in superstars like us. The way I read these laws: a successful company had to produce winning products and compete rigorously for buyers. Any company surviving such competitive warfare and emerging in a leadership position was, as result of these laws, free game for ambitious DOJ attorneys. Shady politicians regularly joined con gusto, cheered on by the press stirring up public animosity, using sensationalistic hysteria aroused by unproven allegations. Gotta sell print and attract viewers! Nothing had changed since medieval times or Salem; religious dogma hunts had just been replaced with antitrust pursuits.

I look at antitrust allegations and dealings from a business perspective—disgustedly. Ayn Rand, the phenomenal philosopher, said it better in the headline of one of her early essays[20] about the topic: "Antitrust. The Rule of Unreason." Government exploits in the past, regulating meteoric superstar companies, had mostly damaged consumers. If changes in competitive behavior of punished leaders were ever achieved, they never produced regulators' desired results. The ultimate punishment under antitrust laws was to break up a company more often than not, only leading to needless pain and suffering for her employees and clientele. In undertaking antitrust actions, the government punished her people and did not protect them

---

20 In February of 1962

as the constitution intends. Politicians exploiting the spirit and intent of the law should all be ashamed. Why not allow the vibrant wellspring of bottom-up, wholesome, and creative human endeavor—what could easily be argued as the very divinity of commercial evolution herself—to freely and naturally flow forth? In its place, spending energies on top-down judicial smothering of beautiful and powerful human enterprises had become not only acceptable but also celebrated. Government for the people and by the people through excessive regulations?

Fair or not, interpreted properly or not, the antitrust laws were on the books. Rockefeller's Standard Oil experienced antitrust regulations; so did AT&T, IBM, the NFL, and the MLB organizations. These laws, according to regulators, existed to protect consumers first and foremost, not necessarily competitors. The government was supposedly doing her citizenry a public service when applying them. In reality, these laws provided cover for reward-losing competitors. Any notably successful enterprise with high market share was suspect of being a monopolist and, therefore, might be accused of hurting consumers.

The legal community was split on just how genuinely effective government antitrust actions really were. If an antitrust case wound up in court, a sitting judge was no guarantee for justice being served. Judges were charged with having to be not only absolutely unbiased but also smart and wise enough to see right through the phony arguments and theories the plaintiff's lawyers and economists presented. They were often not well equipped to handle antitrust conundrums, lacking sound understanding of underlying economic theories and, certainly in today's warp-speed world, up-to-the-second technological insight.

As a mathematician, I consider economics inexact, leaving room for speculation, guesswork, unproven assumptions, and political maneuvering. Any court having to deal with antitrust issues was therefore in a tough situation. Trying to pass a fair judgment would therefore be, to some degree, subjective and certainly prone to appeal. Flaky career lawyers at the DOJ and biased politicians like Janet Reno and Anne Bingaman did not share my views of antitrust bias. In their high-handed view, they embraced the opportunity to rein in MS on what amounted to a grossly misguided mission to serve. Glorifying their political prowess by taking over the case from the FTC proved too tempting for them. MS's aim was to confront the damage head-on, dissect the dynamic duo's questionable motives, and limit any harm unleashed by them.

Let's quickly examine what constitutes a monopoly. Certainly there is such a thing. According to USLegal.com, a monopoly is a control or advantage obtained by one entity over the commercial market in a specific area. The two elements of monopolization are the power to fix prices and exclude competitors within the relevant market and the willful acquisition or maintenance of that power as distinguished from growth or development as a consequence of a superior product, business acumen, or historical accident. Firms with monopoly power have, in general, little or no competition—indicated but not proven by high market share—and are able to completely dictate the terms and conditions of how her products get acquired. Ironically, an enterprise can obtain monopoly power legally. The US postal service or state lotteries come to mind. Consequently, a firm labeled of being a monopoly remains, against common belief, an honorable company encouraged to compete vigorously.

Economists then argue that monopoly power is typically derived by creating or using barriers to entry, effectively barring competitors from the marketplace. Barriers of entry can be formed in various ways. To judge this properly, the relevant market, meaning the environment a firm is operating in, has to be firmly determined—one of the most contested subjects in antitrust trials. According to most economists, market power of a monopolist is often demonstrated in setting prices independent of any competition. Consequently, in a monopoly situation, as the theory goes, the usual supply-and-demand model fails to function.

Take a look at your local electric power company. Assuming there is only one around, she could, in theory, extract any price from customers not possessing their own generator or a naturally fueled energy source. Certainly a valid reason to watch her carefully.

Excess profits and profit maximization are other signs of a monopoly-governed marketplace. The laws in the United States require any enterprise to not conduct her business in an abusive way, defined as limiting supply, engaging in predatory pricing or price discrimination, refusing to deal, entering into exclusive dealings, or engaging in unreasonable tying/bundling of products. Some examples: consequentially, MS could not have made the sales of MS-DOS conditional to buying Windows or refused to sell MS-DOS to a single OEM while supplying all others.

According to economists, a typical monopoly charges unjustifiably high prices. Let's examine if this could have been a reason for labeling us a monopolist. OEMs typically paid MS 1–2 percent of total system value for DOS systems and twice that much for the ones powered by Windows. Prices for OS/2 and the Mac OS resided in the same neighborhood, with UNIX being at least three times as expensive, while DRI charged approximately half. Nothing indefensible here, as we had always kept prices in check!

My discussions battling MS top management certainly bore that out. I was consistently told to compete by lowering prices and be fearful of the pirates who operated at zero.

We had earned our position with considerable sweat and successful products, converting first-time customers into happy loyalists. Improving our product line constantly, we kept them hooked. Fair-minded economists normally would admire the way we played the game, remaining in the lead thanks to fabulous work ethics, great products, superior marketing, and contented clientele. Sounds legal to me!

Not so fast! Evaluating whether or not a company had obtained or maintained monopolistic power, legally, leaves a tremendous amount of room for judicial interpretation. The laws and the applied standards and tests are by no means ironclad, leaving for plenty of subjectivity. At stake are egos, reputations, careers, and future earnings. Most companies accused of becoming or maintaining an illegal monopoly will be convicted in the public mind and by the press long before judgment is passed, branding the accused with an unjustified and unearned reputation and leaving a sitting judge with at least a hypothetical pretrial bias. Good reasons to avoid any such trial altogether and make a dedicated effort to settle all allegations out of court—exactly what the two Bills were now engaged in.

Weighing the above carefully, government attorneys knew they had a weak case against us but could not drop it for political reasons. Three issues caught their eyes: our minimum commitment clause, allowing processor-based license consolidations, and the duration of our contracts.

I had made the CPU-based licensing option available to make customer reporting easier without removing any other existing choices. I was in full agreement with the government: variety makes happy customers. Why, then, was introducing this option even an issue? Customers were never coerced to commit to this variety, and actually only 40 percent selected it for Windows and around 60 percent for MS-DOS. The DOJ had this on record. The minimum commitment clause making OEMs pay for systems they sometimes did not ship while enjoying larger discounts cannot be considered anticompetitive either. A more realistic commitment schedule translated into paying one to two dollars extra per unit and left no money in limbo. Customers had that choice! Offering longer term contracts benefitted them through locking in prices. Customers demanded them; I hated them. They allowed me less flexibility in adjusting terms, conditions, and prices. Did the Feds really ever tried to understand my customers' motives?

If these contractual terms had to go to placate the Feds, luring them into a settlement, only the minimum commitment option needed analytical attention. It was the basis for the volume discounts we extended for firm purchases. One option was to do away with it; higher-volume customers would feel cheated. Not a valid path to pursue. I needed to find a clever and government-acceptable replacement system. After sleeping on it, though, the mathematician in me took up the challenge. Any new discount system I came up with had to be simple, pragmatic, and based on mutually-agreed-upon forecasts. It had to allow for real-time adjustments and treat performing OEMs better than laggards. Using these principles, I developed a model using past revenue data. I soon discovered that my envisioned scheme would hypothetically provide increased revenues per unit without jeopardizing my overall business. Let the Feds have their cake!

After probing my assumptions carefully with my team and accountant, I informed Bill, who ferociously challenged my ostensibly arrogant-sounding assertions and my conviction of not losing customers or any business to competition. Hearing me project more dollars per unit as result of the adjudicated changes made him come around, and he conceded, agreeing that the DOJ's demands would do us no harm whatsoever. As long as the company continued to produce winning products, used her profits to blaze the trail deeper into the dizzying universe of tech advancements, and created higher standards, she would emerge out of the current mess not only unscathed, but with increased profit.

The two Bills meanwhile had succeeded in easing the DOJ's concerns about exclusive dealings, tying products together, and any form of customer coercion we were suspected of. As they inched closer to settling the matter, the Feds insisted on including an exhaustive list of forbidden activities into the final settlement agreement. With zero proof in their hands, they wrote us the equivalent of an unwarranted speeding ticket, making us, for political reasons, appear as villainous as possible.

I was in Bill's office when the last remaining objections were overcome, Anne Bingaman on the other end of the phone line, all-ears Janet quietly listening. By agreeing to cease and desist demanding minimum commitments in the future, the deal was sealed—both sides exhausted and relieved. I was all smiles after reading a clause the two Bills had been allowed to sneak into the covenant. It allowed the future integration of unlimited consumer-benefitting functionality into Windows as we saw fit. Or at least we believed so!

After the two parties finalized their agreement five years after the ordeal had started in July of '94, a judge needed to sign off on what was now labeled a consent decree. This was a step in the legal process designed to ensure no collusion existed between the Feds as plaintiffs and us, the defendant. Laugh as much as you want! When the DOJ announced that the two parties had settled the matter, the headline read "Microsoft Agrees to End Unfair Monopolistic Practices." A catchy phrase meant to impress the public but grotesquely distorting and patently untrue! The DOJ had never proven—in any way, shape, or form—that we had a monopoly, yet we stood condemned. The agreement itself stated that the parties "having consented to the entry of this final judgment without trial or adjudication of any issues of fact or law; and without this final judgment constituting any evidence or admission by any party with respect to any issue of fact or law."

Anne Bingaman proudly sucked the following out of her thumb: "MS has used its monopoly power in effect to levy a 'tax' on PC manufacturers who would otherwise like to offer an alternative system." So spoke a public servant whose education and career was allegedly derived from a command of the facts, details, language, and integrity. Microsoft the "IRS" of the PC world OEMs could not escape from—how ridiculous! With me as the chief tax collector, oh my, oh my, where had I landed?

The fastidious and insightful Janet Reno patted herself on the back with the following stupid comment: "Today's settlement levels the playing field and opens the door for competition." Her leveling bulldozer at work! Protesting the ill-worded DOJ press release did little to revise the projected image. The government had the larger megaphone. Politicians applauded, and the press, having condemned us long ago, fell in line. The ladies from the DOJ celebrated on Capitol Hill. Politicians believed they had served the public well. In reality, they had extracted meaningless concessions from us.

Embarrassingly for the DOJ, Judge Stanley Sporkin—appointed by the DC district court to sign off on the decree—shared my opinion, arriving at it from a much different angle. Garry Reback, an outside attorney, reinforced his belief by submitting an amicus curiae[21] brief. He had been hired by several competitors with the goal to stir the pot against us; to me his activities in themselves appeared conspiracy driven, with him as the mouthpiece.

In February of '95, after an unusually lengthy hearing, the judge ruled that the decree was not in the public interest. First round lost! Disagreeing, MS and the DOJ immediately filed a joint appeal. To our delight, the appellate court overruled the judge three months later and removed him from the case. He had read a book called *Hard Drive*, adopting the opinion of its authors, James Wallace and Jim Erickson—avowed MS opponent—as his own. Justice by bedside reading?

---

21 Friend of the court

The DC district court then assigned Judge Thomas Penfield Jackson to the case. Late in '95, he finally rubber-stamped the decree. Signing it had consequences for him. As long as he stayed on the bench, he was deemed its legal guardian, enforcer, and babysitter if you will. If any person, company, or government agency believed that the consent decree had been violated, a petition could be filed, and he would determine what action to take.

Voluntarily, we implemented and followed the agreed-upon rules two years before his court formally approved them. Returning to business as usual, we soon discovered that the new rules were causing no harm whatsoever. The struggle with DRI continued unabated. The battle with IBM's OS/2 intensified. Apple persisted in luring away PC buyers. We continued to live, as now and ever shall be, in a competitive world. Justice sighted, or justice blind. MS trusted the government fulfilling her part of the deal, expecting the matter was settled once and for all. Dream on trusting one's government!

Patent horrors

Against MS tradition, the good news first: In '93 we won the Apple patent infringement lawsuit. For Bill, this was a wake-up call and a reason to educate executive staff on the deceptive threats patents constituted. Particularly wary how OEMs could make use of them, Bill personally set out on a crusade to find ways to reduce the risk for MS and the industry as a whole. Violating just one inadvertently could result in a lawsuit and an injunction, freezing OS shipments and leaving the whole industry out in the cold.

Releasing the next DOS version in '93 was again a race against DRI. Her mother ship, Novell, had finally succeeded in giving us the long-anticipated trouble. To remain competitive, we were required to include a hard drive (disk) compression program. Today's affordable hard drives storing terabytes of information have made these programs basically obsolete. Back then, most PCs came with hard drives having a hundred thousand times less capacity. No wonder compression programs enabling users to store up to twice the data on a hard drive were popular. A math algorithm was used to condense the data and later reversed to recover it, a process technically somewhat comparable to encrypting/decrypting information.

Novell had licensed and used a compression algorithm derived by Stac Electronics, a small and rapidly growing software company located in California. Since 1990, this had been her livelihood. Attempting to license the same software, we soon ran into difficulties dealing with Stac's execs. Unable to negotiate a reasonable deal, we eventually bought a comparable algorithm from another company. During the initial negotiations, two of our programmers inspected Stac's algorithm for due diligence reasons. It left them impure and impugned. To remain on the safe side, they should never have been part of the team that would eventually develop our own version. Not following these sterility safeguards left the door open for infringement accusations.

After we released prototype copies of our new DOS version, Stac immediately asserted that we had stolen portions of her meanwhile-patented algorithm. After careful study of Stac's claim, we initially ignored the allegations and then made several attempts to settle the dispute. When that failed, Stac's CEO, Garry Clow, took us to court. With both parties stubbornly entrenched, the case ended up in jury trial while our product was already installed on millions of PCs. The technically challenged jury with an inadequate grasp of the salient facts found us guilty and convicted us to pay $120 million in punitive damages. To make matters worse, Stac succeeded in obtaining an injunction from a disgruntled judge, preventing us from shipping the malingering products until we paid up. The industry trembled under a doomsday scenario.

The product group scrambled, removing the compression code from the product. Until a new version was ready, OEMs were stuck. Bill's darkest prediction in regard to potential patent impairments had come true. When an internal investigation discovered MS Word also included the infringing code, our Office application suite could no longer be sold either. Wall Street got the message, and our stock began abruptly trekking southward.

We stubbornly maintained our innocence. I honestly didn't know what to believe. My problem was of a different kind. OEMs had stopped shipping PCs. Millions of systems were sitting in warehouses and at retailers around the world, unable to be legally sold until thoroughly reworked. The injunction would inevitably cause a tsunami of damage to the PC industry in general and to MS earnings in particular. Inside the company, denial was rampant. A bury-your-head-in-the-sand attitude dominated. I was convinced OEMs and retailers would hold us responsible for huge monetary damages. We were in deep shit!

Calling our chief financial officer, Mike Brown, I alerted him to the grave potential the injunction imposed. We agreed that speed was of outmost importance in settling the matter. A short time later, we walked into Bill Neukom's office and proposed to resolve the matter amicably and urgently. He encouraged us, and my boss concurred at once. I didn't inform the product group; entangled individuals were still licking their wounds and might sabotage our mission. Stac's CEO, hoping we were serious, agreed to a meeting and offered to pick us up the next day at San Diego Airport. Upon arrival, we lunched with Stac's execs and found them amenable to avoid further escalations. Against considerable odds, we soon agreed upon a balanced and reasonable compromise. I had to leave to attend my annual OEM meeting that Sunday. By then, Mike Brown's second in command had joined us with one of our attorneys, and by Tuesday I received a call in the wilderness informing me we were out of the woods.

Our settlement saved the company $40 million and spared my customers incalculable heartaches. The industry sighed with relief.

The incident rang the alarm bells, prompting us to take a harder look at how to avoid similar situations. During the summer, Bill and our patent attorneys convened and decided on a plan to approach large computer manufacturers with the goal of mutual patent exchanges. I reluctantly became involved in these negotiations. Our patent attorneys were far better suited for these engagements. Over time, they succeeded in buying selected patent portfolios as well as closing patent exchange deals. As additional defense, MS aggressively stepped up her filings for patents. Over time, I got several patents co-awarded for work I had done previously protecting our software from being pirated.

During 1994, under the leadership of Dave Heiner and Bill, we explored new avenues to further protect our core products. A simple concept but not at all easy to implement! We always defended and indemnified OEMs against patent challenges brought against products licensed from us. Could we, in return, ask OEMs licensing our products to refrain from asserting patent infringement claims against us or anybody else distributing or using them? The desired result: patent peace for Windows distributors, commercial licensees, and end users. I immediately embraced adding such a truce-like condition to our standard agreements to protect the Windows ecosystem.

It was later called a Non-Assertion of Patent (NAP) clause and turned out to be much harder to implement than I ever imagined. Japanese companies, led by Sony, were the most outspoken opponents. Fair values for patents are hard to determine. Google, Samsung, and Apple are experiencing this today. OEMs, possessing remotely related ones, argued that we treated them unfairly by disallowing them from extracting value not only from us but also from their competitors. Their old-school patent attorneys had a marked lack of interest in patent peace. It threatened their livelihood. Sticking to our principles was extraordinarily time-consuming and made licensing each successive product harder while earning us few new friends. The idea of granting competitors immunity when licensing Windows was against the industry's belief of fairness. The contentious NAP clause was later reviewed by several government agencies though never challenged. Its derivation was defensive in nature and protected everybody for over a decade.

By 2004, a new regime of MS patent attorneys determined that the company had acquired enough patents to drop the NAP clause from OEM contracts. In their book *Burning the Ships*, Mike Phelps and David Kline explained how they arrived at the decision. The reasons were political as well as economical. MS was still fighting the Feds in court, and her larger customers had gotten bolder, finding mischievous means thanks to industry group patent pooling arrangements to assert patents claims. One of these attacks was successfully fended off on an appeal involving the US Supreme Court. Another larger one was looming. The old idea of using patent cross-license agreements as a defense against the newest legal tricks was therefore revisited and adopted. It favors patent-rich OEMs and leaves the patent have-nots in jeopardy.

Patents will always be landmines for any technology company. Start-ups, in particular, are most vulnerable to the misfortunes of stepping on them. Until patents are abolished, no company is safe from enforcement attacks by ambitious and watchful patent attorneys. Obliterating patent protection, on the other hand, is unfair to inventors and would endanger future innovations and progress. Therefore, companies will continue chewing on patent challenges and need to strike a balance. During my time, MS successfully avoided a Windows patent war by employing the defensive NAP clause in license agreements, a harmonious approach compared the aggressive methods our competitor IBM deployed. To this day, I am proud of having done the industry a tremendous service against a lot of odds and resistance.

THE SAGA OF BIG BLUE

A SWEETHEART DEAL

Fighting IBM for domination of the OS market influenced my professional life profoundly. Now responsible for the IBM account, I soon discovered the mess I had inherited. To disallow the divorce settlement from spinning out of control, Steve's team had made troubling concessions, and in my mind, he had given the shop away! The current version of MS-DOS had been licensed under a flat-free agreement, and IBM was therefore paid up. Compared to similar volume shippers, Windows cost her one-third of the normal royalty. For OS/2, she paid just $2 per unit and had the right to give a huge amount of copies away for free. Having performed the bulk of its development work, I found that price borderline appropriate. My particular grief was with the DOS and Windows deals. Not only were they unfair to other OEMs, but they also made IBM a nearly unnoticeable customer from a revenue aspect. I felt justified and empowered to clean up and, over time, bring her onto equal footings with my other clients. My hair was standing up! Blame it on your boss.

I still do not comprehend why IBM was fighting us so relentlessly. Bruised egos at work must have been the key reason, because PCs and software sales were only a minimal portion of Big Blue's overall business. She was predominantly a mainframe and IT services company that experienced growing pains by '91. After painful attacks from Japanese competitors, her mainframe market share dropped rapidly, which slowly turned a trickle into a river of red ink. Losing market share in the PC market to the likes of Compaq, Dell, and Hewlett-Packard, among others, appeared of lesser importance in the overall picture.

John Akers, a thirty-year IBM veteran who was running the company, had taken the helm when IBM's stock price was at $32. Wall Street had originally looked favorably upon his reign. Now it seriously questioned IBM's business model, her ability to respond to mainframe competition, and Aker's overall leadership style. The stock price was now at $22 and supposedly warranted calls for resignation. Unthinkable for the CEO of mighty Big Blue! The proud, astute, and buttoned-down IBM culture was thrown into an absolute state of shock, having to relearn the meaning of humility.

Analysts traced IBM's misfortunes back to former chairman and CEO John Opel, who in the early '80s had dreamed of IBM becoming a $100 billion company. I was working for Digital Equipment Corporation (DEC) when these projections leaked out. In an interview, DEC's CEO Ken Olson called them simply "nutsy." IBM's CEO, guilty of not foreseeing major IT paradigm shifts during his reign, had his work cut out. Build-up in payroll and manufacturing plants eventually created a nonsustainable cost structure, steadily eating away IBM's substantial reserves. Struggling to turn the Big Blue tanker around in the recession of 1991/93—by then an Exxon Valdez of bloated delusions—appeared impossible even for Hercules!

Declining sales—caused by comparatively poor mainframe performance, sky-high maintenance cost, and mediocre services—and ever-increasing restructuring charges added to the worries. Until then, IBM had given the impression of being an unimpeachable empire and reliable supplier. Now, displeased IBM shareholders and Wall Street promptly dumped her shares. Watching a merry-go-round of changing plans, a carnival of confusion from within, a sharply disorientated sales force, and deteriorating development efforts, IBM customers swiftly reconsidered. At the height of what analysts called the second OS war, IBM was suffering from self-inflicted wounds and had become a structurally weakened giant. Good for us, as I thought.

But by spring of '92, OS/2's PC-access technology, called Workplace Shell, was at par with Windows. Additionally, running Windows apps under OS/2 had gotten easier. Against all odds and to my disgust, the hurting Goliath had managed to technically keep up with us and seemed on track to increase market share. The desire for corporations to limit the various OSs they deployed due to support costs made it an uphill battle. After Lotus and WordPerfect products became available for Windows, there was even less reason to change. Another obstacle was the growing customer uneasiness to commit one's IT future to a shaken-up company like IBM! Adding to the misery, OS/2 was still too resource demanding to run well on low-end consumer PCs. With MS a seemingly safer bet for enterprises accompanied by more sex appeal for consumers, the resounding trajectory for Windows just continued!

To overcome these obstacles, IBM's top management formed a personal software division under John Soyring, a Michigan Tech grad. It was fully responsible for marketing OS/2 to OEMs and enterprise customers, and its activities soon made our life harder! Giving the division an enormous amount of money, IBM's president Jack Kuehler hoped bruised egos would work harder and guarantee a return on capital. To turn up the heat to the detriment of MS and promote PC sales, he went so far as to sanction the bundling of free Lotus SmartSuite—our Office software competitor—with OS/2-powered IBM PCs.

A declaration of war, as the furious Bill-and-Steve duo interpreted it. I sensed a serious competitive move. Lotus had never before sold her office productivity suite through OEMs. Watching diligently, and with none of my other customers entering into a similar deal, I concluded that IBM had obtained exclusivity. Her management had given us no chance to bid, obviously not wanting to feed a foe. Foolishly, all of my later inquiries for a competitive bid were turned down. IBM was not interested in a better price or a better product. Hurt was on her agenda. Our relationship had hit bottom and turned icy.

Soon IBM breakup rumors reverberated. With her stock price careening southbound toward $12, about 25 percent of what it had been five years earlier, Akers resigned. He was replaced by Louis Gerstner, former CEO of RJR Nabisco, stunning the IT industry and Wall Street! Being an outsider who was short on the type of tech background presumably required to run the world's number one IT behemoth brought further damage to her stock price. The voices in the press demanding a breakup, on behalf of shareholders, grew louder. Aspiration over reality, but the brand of logic Wall Street adores. Gerstner took his own sweet time coming to terms with the destabilized components of IBM's dilemma and stubbornly resisted.

For the financial press, the fight with MS was mostly on the back burner. Journalists covering the IT segment found the transition made IBM step up her anti-MS efforts, positioning OS/2 as "a better DOS than DOS and a better Windows than Windows." Marketed in short as "OS/2 for Windows"! It read like semicohabitation, but no strategy of peaceful coexistence was ever intended. Instead, it depicted Windows running as slave under the auspices of superior OS/2.

Gimmicks as I saw it. Customers did not buy the slogans either. Windows sales increased by 50 percent yearly to 15 million copies sold—no suffering there. As IBM followed through with her strategy to install both OSs on most of her enterprise PCs, I savored the additional revenue but never believed that there was enough room for both of us.

IN THE SHADOW OF CHICAGO

To impress analysts, IBM began disclosing quarterly OS/2 shipments. A marketing ploy asserting OS/2 was alive and well, with dialed-up units fueled by shelf-ware fantasies. Let's talk about bragging rights. Microsoft, on the other hand, made certain Windows sales numbers were correctly publicized. Creating the impression that IBM was making huge progress in the fight for OS market share. The tech journalists, always supporting the underdog, relished it. Hand-fed by IBM's cleverly scheming marketing department, even reputable commentators got caught in the propaganda net, signaling IBM's unmitigated success.

It fell on me to calm our by-now-famous internal paranoia. With access to IBM's royalty reports, I knew the numbers IBM's marketing augurs were concocting and bragging about were wildly exaggerated. I later found out that they included a healthy dose of not-completely-functional demo copies. Eventually, I won the internal arguments. There were no other possibilities. Numbers, as they say and I believed, speak for themselves.

In May of '93, IBM's new CEO hosted Bill in NYC. Consistent with his never-ending desire to improve our relations, Bill eagerly agreed. According to Bill, a cordial meeting ensued, though, to his great regret, became one with no concrete results. Much to my surprise, Bill shortly thereafter exhibited a glaring lack of restraint by disclosing details of their discussion. An enormous faux pas! The reaction from my contacts at IBM was not far behind. Lou was flabbergasted by the liberty of Bill's exposé of the dialogue and questioned if he could ever trust him again. By chance, Bill met him again later in the fall on a golf course and apologized for the mishap. He was to meet Lou a third time in a colorful, clandestine meeting.[22] Meanwhile, Gerstner continued to resist all calls for a breakup and, unknown to outsiders, had already begun implementing plans to transform IBM into a premier IT services company.

---

22 I will reveal this for the first time later.

In his book *Who Says Elephants Can't Dance?*, Gerstner reported that immediately after he joined, he questioned if IBM should continue the PC business—software and hardware. Calling it nonstrategic, he was the first IBM CEO to ever seriously question the plausibility of maintaining it. Yet Lee Reiswig, GM of her personal software group, contradicted his writings during an interview with *Research Board* in '94 by indicating that Lou was fully behind the efforts to beat us with OS/2. He went on saying that IBM's meanwhile $11 billion software business was profitable enough to support investing in OS/2. Considering Mr. Gerstner approved the budgets for IBM's software group—including its nasty marketing efforts—and for sure cherished the stinging OS/2 ads, I have a hard time disbelieving Mr. Reiswig. This convinced me that his big boss changed his mind later in the game when it was obvious that the operating system war could be lost.

Proving this beyond any doubt is the flood of OS/2 versions that followed, using as revamped name WARP, borrowed from the popular Star Trek series. This was supposed to create the perception that WARP-driven OS/2 power would lead to an intergalactic PC universe ruled by IBM! Anybody that naive? Hearing the new name, I assumed we got leapfrogged à la DRI. Not so, but its use heightened paranoia levels inside MS, assuming IBM had finally delivered a damaging—maybe fatal—blow. Our next countercapable version of Windows was at least a year away. We were fighting for our lives.

MS countered, announcing her next version of Windows, code-named Chicago, without setting a firm delivery date. Yet all of '94 IBM's public relations and marketing hoopla made us look like a sad second-place finisher. We sold hollow promises while IBM shipped a hip-named OS and annoyingly reported its ever-growing units. My group felt considerable internal pressure in our mission to keep our customers firmly seated aboard the Windows train. Fortunately, they lent scant support to a competitor whom they felt should duke it out on its own.

Digging in our heels, we continued selling our current Windows version while promoting the hell out of the mysterious Chicago, later named Windows 95. A brilliant containment strategy emerged. Releasing early copies to ISVs and helping them to design powerful Windows applications bought time and diverted them from supporting IBM's WARP drive. Engaging audio and graphic-card manufacturers achieved the same. With every advancing Chicago demonstration, we gained support not just from the industry, but also from formerly skeptical journalists like Walt Mossberg, who agreed that 95 was a great leap forward. Successfully positioning Chicago as the panacea for the PC world created an immense vaporware shadow, public relations headaches, and a lot of darkness for IBM's WARP culture.

Like OS/2, Chicago integrated MS-DOS and Windows functionality making installation and operations simpler. Its new interface was sexy and appealed to Windows, MAC, UNIX, and OS/2 users alike. Even Apple was getting worried: had we finally caught up with her? IBM was in a pickle, and in the fall of '94, she knew to annihilate us meant doing it within the next twelve months or her warp drives would be rendered inept.

Alliance Round Two

Shortly after we announced our Front Line Partnership (FLP) with Compaq, I received a "me-too" signal from Big Blue. Like others, IBM had set up a tech lab outside MS's campus. In the run-up to Windows 3.0/3.1 and Chicago, it offered key customers real-time access to development resources and sped-up system designs. To establish a back channel with our adversary, I met once a quarter with IBM's lab manager, Roy Clauson, discussing outstanding issues and taking the pulse of both companies. Speaking our minds frankly with no eye toward polite diplomacy, these exchanges, often over a nice bottle of red, developed into an extremely useful dialogue. Instrumental in keeping a minimum of cooperation alive—while his division management and mine still suffered through the pains of divorce flashbacks.

Out of the blue, Roy enthusiastically suggested mighty IBM to join the FLP club. Understanding her management feared Compaq's rise and success strengthened by our recently announced alliance, his thinking was not surprising. Not wanting to add another company to that equation, my response nevertheless was guarded skepticism. But I reluctantly agreed to meet with his division management.

This is how I met Bruce Claflin, GM of product and brand management for the PC company, and Tony Santelli, his counterpart for IBM's PowerPC business. I found them both cordial. They expressed their darkest disdain about the Compaq situation, adamantly requesting equal or better treatment for IBM. In danger of losing the number one position to Compaq, both executives hoped a renewed partnership could reverse that trend. Acknowledging they had the best prices and terms and conditions in the industry, they demanded adding a close marketing partnership at a time when IBM's software division was relentlessly attacking us. A puzzle these two had to solve internally. The timing for their yearning seemed odd.

Back, I engaged Bill and Steve and persuaded them to at least give the two gentlemen the benefit of the doubt. Bill agreed immediately. Steve was hesitant and sharply skeptical. Deep down he believed IBM could never be trusted again. Not vetoing a cooperative attempt, he remained pragmatic, telling us, "Surprise me!"

Concluding a pact between us hinged on the need for IBM's execs to stop fighting Windows and to offer it without prejudice to customers! There was no need to give up on OS/2. The constructive way out of the current dilemma was for both sides to accept a rational form of OS impartiality. Only that promised to restore some trust between the former partners. Would IBM's PC division have the freedom to act in its best interest, or would the current WARP-centric corporate policy dictate the outcome of our negotiations?

In the summer and early fall, we became fully engaged in hashing out such an alliance. In a number of clandestine meetings in the wooded vastness of Montana and Illinois, we met, avoiding publicity. Few people knew. Our first proposal was dead on arrival. We were told that IBM's corporate executive committee (CEC) would veto it. I grew suspicious at once from the tone of the rejection. Her negotiators felt that IBM deserved Compaq-equivalent recognition and all accompanying rewards without too much in return. Their hands seemed tied. Disregarding my first impression, my team kept at it, probing for a balanced deal. After four iterations, both sides believed we had arrived at an acceptable and internally sellable compromise. Tony and Bruce were all in and suggested that their boss, Rick Thoman, would come on board as well. Now it was early September '94 with hostile OS/2 WARP drives revving up further and making more noise than ever.

As expected, the alliance needed the approval of IBM's CEC. Three month went by, and with patience as one of my lesser virtues, I made several attempts to peek prematurely behind the curtain. Nothing was forthcoming, not even from my local contact. The next stop was November '94 at the COMDEX trade show, where IBM finally agreed to a meeting between Rick Thoman, SVP of the PC Company, Bill, and the negation teams. Our expectations were not terribly high. IBM had been stalling far too long for a positive outcome. Thoman point-blank let the cat out of the bag, informing Bill the deal was dead—killed by IBM's CEC. Steve breathed easier. The war would continue, and IBM had no intent to decommission her warp drives. Yes, we would have liked it the other way around and viewed the decision an unfortunate one, but we weren't afraid, only disappointed. Life would go on with battles brewing, blood to be spilled.

What we did not know and what neither Thoman nor any of the other IBM executives dared to tell us was that IBM's CEC, with Gerstner's approval, had already put plans in motion to escalate the war. No doubt the main reason of the long waiting period! One month later, it became apparent. IBM embarked on a fierce "IBM First" campaign, forcing the PC Company to load OS/2 on extra PCs, signaling an acceleration of Big Blue's push for dominance.

In hindsight, luring us into the negotiations appears to have been an attempt to uncover market intelligence in the context of the Compaq engagement. I never expected that much sneakiness and dishonesty from the two gentlemen. I doubt they or their superiors ever had the intent to enter into any partnership in the first place. I stopped the back-channel talks. We had reached a dead end.

Trust and Verify

We publically locked horns, vigorously blocking and tackling each other's ambitions. No side gave in. IBM continued to impress the market, touting huge WARP shipments and giving the impression the pendulum had swung her way. I was furious! The announced units were three to five times larger than the ones IBM was reporting to us. At least show me the money if true!

I got together with my controller and IBM sales rep. We compared royalty reports with the publicly announced numbers going back two years. The discrepancies were mind-blowing. In a second meeting, my controller offered further details on how IBM reported results. I was shocked to hear that all of her reports were undergoing constant corrections going back as far as three to four quarters. We confronted our friends at IBM with our findings and asked for clarifications. The unpleasant job fell to my controller, who received only evasive answers; additional follow up inquiries were blatantly ignored.

I instructed her to no longer take no for an answer. In the event of more evasiveness, I advised her to warn her opponent that we could feel compelled to use our auditing rights to get to the bottom of the mystery. My accountant, Nell Miller, was a nonconfrontational wonder, considerate to a fault, and in my mind, the best person to work out inconsistencies without raising a stink. The one at IBM, however, rained admonitions upon her about IBM being an honest and above-board company, how she'd never been audited by any suppliers. How could we dare? Culminating in a statement, any audit attempt would be construed as an insult and an act of outright war by upper management. Quack, quack, quack.

Our contract absolutely allowed an audit, and we were swiftly approaching the moment to enforce it. So far we had tried to solve the situation amicably. I was not sure why our simple request for accurate and honest numbers had been so strongly rejected. Was IBM's administrator covering up his own sloppy work, or was IBM concealing something darker from us? We gave him an ultimatum to remedy the situation within two weeks. Upon receiving no additional explanations, we sent a notice containing our intent to audit.

I did not ask permission for my action, considering it part of my job running the business. The first time I actually mentioned to Bill what we were struggling with was in preparation for the COMDEX meeting with Thoman. He was surprised and agreed after understanding my reasons. As a rule, all audits needed my approval. Aware that audits had the tendency to cause temporary conflicts at best and enduringly soured customer relationships at worst, I never ever permitted them lightly.

With IBM, we undoubtedly had probable cause as further indicated by analyzing her replicator shipment reports. Stalling our requests was in flagrant violation of our agreements. After receiving the notice, IBM's management initially showed no signs of accommodating us. A couple of weeks later, rational thinking set in, and her execs agreed to let an independent auditor look at the books. We proposed to use the esteemed accounting firm Deloitte & Touche. IBM rejected this on the grounds that she was our corporate auditor. We settled upon Ernest & Young. Meanwhile, nearly three months had passed. When the auditors visited IBM for the first time, they were, to our great surprise and dismay, staunchly refused entry. A phone call from me rescued the situation, but when the auditors asked for relevant numbers, IBM stubbornly refused.

Getting on the phone with Bruce Claflin broke the impasse. During subsequent visits, Ernest & Young experienced no hostile resistance but reported IBM felt threatened by the inquiries. All of IBM's reporting generated from mainframe-based accounting systems. She had installed these all over the world at numerous manufacturing plants. As the auditors dug into the scant data, they discovered its desolate state, unearthing the rationale behind the numerous reporting corrections.

After three months, the auditors informed me in January of '95 that not-so-Big old Blue would need another six to eight months to deliver what they needed for completion. Stunning! And this from the number one IT Company in the world. I assumed there was a correlation between the recently rejected alliance proposals, IBM's marketing attacks, and the mule-like stubbornness and resistance in regard of the audit. So I asked the auditors to estimate what they thought IBM owed us. Hesitantly, they mentioned 10–40 million dollars, exceeding, several times over, my grandest expectations. I insisted they disclose the amount to IBM's management. Playing with an open deck and waving scary numbers, I hoped to turn the situation around, injecting earnestness.

IBM's response arrived promptly in the form of an angry phone call. Point-blank, my accountant was told that if we did not drop the audit and agreed to accept, as final, IBM's interpretation of the existing license agreements along with payments made, IBM would no longer be interested in obtaining a license for Chicago. In clear text: we owe you no money, so stop bugging us or go to hell with Windows! If one stops to listen for the moral and business resonances within: the hell with you, period. Wow!

The call came seven months after we had sent IBM the delicate notice. I could not believe the caller's audacity, which eventually led to his dismissal. His statement was outright stupid. The company owed us a hefty chunk of capital, and by denying us what was due, her management was playing with fire, risking a high-profile and low-percentage lawsuit. So far we had been patient, but the last clear threat left no room to continue the game civilly. The caller had exhausted my patience at last. Time to accommodate IBM! I asked my accountant to document the content of the phone call verbatim. She then wrote a letter to her counterpart in IBM, summarizing what she had understood. In particular, she agreed with his request to discontinue our ongoing Chicago negotiations until the audit was completed and we had received all monies due.

A few days later, Bruce Claflin called. He was an energetic person, in general finely mannered, restrained, fair, and polite to a T. This time, though, he shrieked at me: "You can't do this to me. This will kill us" I held myself in check. After he burned off much of his anger—halfway through the call—he started listening. I told him that despite the desolate state of our current affairs, I considered his call a sign of goodwill. We would therefore continue to feed him Chicago code so he could get his systems launch ready. I then implored him to support my position inside IBM to finish the pending audit before launch, and in return for his complicit goodwill, I would instruct my team to continue negotiations. Not compromising beyond, I firmly insisted on having the monies owed paid before I would sign a Chicago license. He swallowed hard, agreed, and promised to accelerate the completion of the audit. With no interest in conflict escalation, I was relieved and assumed he would keep his word and follow through as promised. My IBM sales rep Mark Baber, not trusting him at all, remained sharply skeptical.

Until then I had considered IBM a reputable company. Stalling the audit and insisting on her tainted position despite clear-cut agreements did not sit well with me. A second faux pas, within months after mischievously closing the door on us and starting the in-your-face "IBM First" campaign. I was concerned that MS would never receive the royalties due without applying leverage or filing a lawsuit. No appetite for the latter on my part! Why add yet another agreement to the ones IBM wasn't living up to?

We had nearly six months to finish up before Chicago was scheduled to ship. So as promised, negotiations were back on, and we continued to provide IBM with all necessary code and technical support to ready her systems. I crossed my fingers. I wanted her at the launch, and after talking to Mr. Claflin, I came away with the impression that he genuinely wanted the same. The clock was ticking.

Start me up

"Disconnecting people from the Internet is a human rights violation."[23]

Microsoft nearly missed it

While I was entangled with IBM's stubborn hierarchy about contract compliance, MS's development group struggled with the fledging Internet. The desire of the academic research community and some government agencies to exchange data and ideas beyond the bounds of existing networks brought it to light. Its realization nevertheless needed a protocol so WW messages could be properly directed and received and a wiring and signal standard to enable physical connections. The first was invented in the '70s and the latter in the '80s. The government seized the opportunity and funded the implementation of a worldwide-spanning Net but initially restricted its use to the academic community and government agencies. It took until '91 and an act of US Congress[24] to finally open it for individual and commercial use.

Another technological breakthrough was needed to make this network popular with unsophisticated users. It arrived when Tim Berners-Lee[25] invented a software protocol and combined it with a programming language allowing software programmers to access files and data over this emerging communication vehicle. His effort led to a software program called a browser, which enabled any computer users to easily surf files and data over the Internet. A polished and user-friendly version, sponsored by the National Center for Supercomputing Applications (NSCA), saw the daylight in '93 and was surnamed Mosaic. Limited in functionality, it was nevertheless an immediate hit within the superfast-growing Internet community, signaling a communication revolution that was about to erupt.

---

23 According to the United Nations
24 The so-called Gore Bill
25 A scientist working at CERN

At about the same time, our gurus spiritedly debated how to include Internet connectivity into Chicago. Boldly predicting: an advanced version of the Mosaic browser would one day become the next killer application for PCs! OSs that supported Internet protocols best were UNIX and the Apple's Mac OS.

The gurus proposed three steps to propel Windows ahead and keep it ultracompetitive:

1. Offer basic Internet services in Chicago.

2. Integrate a sophisticated Internet browser into the product.

3. Make the integration tight so end users enjoy Internet surfing as a seamless extension of Windows-powered PCs.

They labeled this go-to-market strategy: embrace, extend, and innovate. In absence of a hard-to-write business plan,[26] the gurus pleaded with management. When our fanatics defined the once-in-a-lifetime opportunity as the most effective way imaginable to "deliver an enormous blow to our OS competitors," they finally captured management's attention.

Intellectually, Bill comprehended full well that this opportunity constituted an inflection point as expressed in a later document headlined the "Internet tidal wave." The missing links for him were the business model and how to time our entry. Insufficient and costly Internet access for consumers was another hindrance he had to weigh. Worse yet, would this cause Chicago to slip, and how much would a delay set us back in beating OS/2? The company's top development executives, including Bill, were still vacillating when an emerging competitor forced their hands. Thank you forever for that!

In April of '94, Kleiner Perkins Caufield & Byers invested in a start-up later to be called Netscape. Jim Clark—her brainchild, CEO, and cofounder—at once recruited members from the NCSA Mosaic team to develop an advanced commercial version of that browser. Within an instant, the Internet-development community, ever connected and closely knit, leaked the news. Get used to that speed!

MS moved up her timetable and immediately obtained browser code as foundation for her own development. The ticking clock for Chicago determined that a bare-bones browser version would have to do. Additional

---

26 Everything on the Internet seemed to be free of charge.

bells and whistles and tighter integration would be added later. Hedging our bets, we embarked on this alluring adventure with a from-the-hip product plan for a me-too browser. Would this be enough to stop a well-funded start-up and turn Chicago into an attractive-enough Windows to the world product? I had my doubts.

Jim Clark's company had her browser in the market for IBM PCs and Macs by December of '94, calling it Navigator. Using the software became popular overnight, and being the only game in town, the company swiftly achieved a de facto 100 percent market penetration. MS echelons, still on the road to the finish line and unable to compete, became nervous—sorry, extremely paranoid. As our own gurus had predicted, the Internet had indeed caught on, and Netscape's browser was causing a killer application like firestorm.

Despite relatively high monthly interconnect fees, users opened their wallets, connecting in waves and exploring the new frontier. In a short time, and notwithstanding the initially relatively slow connection via phone lines, millions got addicted. The rest, as we know, is history. Netscape's clever use of the Net, bypassing traditional retail channels and providing instant user gratification and real-time update services, made the start-up look extremely cool! This sizzling consumer entity left MS in the dust and made her Wall Street debut as the new tech darling. Bill was fuming.

What bugged him most was the potential computing paradigm shift he immediately spotted. Thanks to Berners-Lee's invention, programmers writing Internet applications were able to do so completely in disregard to any OS running a computer system. In his view, a competitive platform and a core threat to our livelihood had been created. In typical fearmongering fashion, he prophesied that the end was near for Windows because ISVs could now create fully cross-platform portable Internet applications and abandon Windows altogether. I did not share his belief in that much doom and gloom; he rallied the company around it.

Welcome to the world of "Information at Your Fingertips," as Bill titled his November '94 COMDEX keynote speech. I was in the audience that fated day. All of us were impressed by his futuristic-sounding Internet vision. Sadly, ironically, shockingly: it was only a foretelling. A paper tiger was talking. MS's CEO possessed no marketable product. Netscape was kicking the holy hell out of us! The audience nevertheless applauded him for a well-delivered speech. MS, since forever ago—as long as collective entrepreneur memory stretched—was in grave and immediate danger of losing her lead once and for all.

Hurdles before launch

How to finally christen Chicago spawned an intense internal discussion. Initially favored was Chicago or Windows 4.0. I introduced calling it Windows 95. That name would put a time stamp on the product and make it look obsolete one year later—the major reason my name of choice was so robustly disputed. In the end, though, it stuck. Labeling our integrated browser Internet Explorer was most explanatory. Its inclusion had meanwhile become a dire necessity.

Before my group started the Windows 95 licensing drive in November of '94, my management team and I held a strategy session. We defined objectives by region and customer. I insisted on attending several regional meetings over the next couple of months to personally instruct my people around the world. I wanted the sales reps to go all the way and report any obstacles to licensing 95 immediately. In return I promised my personal attention and a lightning response. I didn't have to consult with Steve or Bill about our objectives. Nothing short of 100 percent customer penetration would do. This was the photon torpedo I needed to get WARP off its trajectory!

With the next WW OEM team meeting scheduled for Alaska, we christened the 95 penetration campaign "Gold Rush." Soon, my conference room got converted into an information hub, referred to as a war room. Well put! The admins were made responsible for keeping its info up-to-date. Like tenacious bulldogs, the brave women circulated daily among sales reps, reminding them to get their licenses done. Never letting go, they managed progress better than their bosses. To the dismay of my personal rep and against corporate policy, I included them in the sales bonus pool. One of the best decisions I ever made—they earned and deserved all of it! Signing up customers early was in true gold rush spirit, rewarded with a plaque containing a quarter ounce of an American eagle gold coin. Most of my crew got one, and mine still decorates my desk.

Our contracts had been restructured to make customers' lives easier. At least I thought so. First we asked them to sign a so-called master agreement defining how to do business with us for the next decade. While it would make negotiations easier in the future, it made them harder for near term. All parties wanted to make sure that they could live with its content for a seemingly infinite time span. The second contract they signed was product specific and contained a one-year license for a chosen product. Separating the two agreements enabled us to comply with the consent decree and allowed us to change licensing conditions for individual products on shorter notice.

Three fresh challenges laid ahead for my group. Let me start with pricing. I had come up with a new price list for 95. The normal price for OEM customers for the MS-DOS/Windows combo had been in the $35–$55 range. My goal was to charge an additional of $10 per unit for this superior product. Bill and Steve predicted that this would jeopardize our success when they finally learned about my decision three to four months before launch. I disagreed and promised to deliver. They let me try. Convinced of our seemingly invincible momentum and supported by product management, I felt emboldened and found only minimal resistance except from IBM.

The patent peace clause was another tough pill for customers to swallow. For Dell, Compaq, DEC, Acer, and most of the Japanese manufacturers, it was nothing short of a red herring. It took months to overcome counterarguments and settle upon reasonable compromises.

We had proactively implemented all rules of the consent decree voluntarily. Therefore, we no longer had any purchase guarantees; the so-called minimum commitments were gone. This left us with the dilemma of how to fix volume discounts. To do so, our sales reps were instructed to agree with their customers on a realistic sales forecast for year number one to determine royalty rates due. For year number two, we would then adjust these based on achieved sales results. Up or down!

We applied a similar principle to the market development agreements (MDA). Remember, they enabled customers to gain extra discounts when participating in certain quality, product design, marketing, and promotional activities. During the first year, we trustingly assumed that customers would do all proposed activities and therefore applied all available discounts. Next year's rewards would then be adjusted based on actual participation. Pretty fair and pragmatic as I believed! In case customers disputed the revisions, a review board handled their complaints as leniently as possible. Hardly ever did I get personally involved.

My group showed excellent sales progress. As usual, our friend Theo Lieven in Germany had publically gone ballistic about the new pricing and MDA details, again trying to negotiate terms and conditions through the press. With no one paying much attention any longer, he signed the proposed deal late but unchanged. Approximately eight weeks before the planned launch event, the press figured out that we had two major holdouts—Compaq and IBM. With Compaq we were tangled up over the patent peace issue. She had taken several companies—one of them being Packard Bell—to court over hardware and software patent infringements. With ambitious patent attorneys working for her, stirring the pot, it became most difficult even between partners in arms to settle a dispute like this. A long phone call between Bill and Eckhard eventually ironed out the wrinkles, resulting in a partial cross-patent agreement. One down, one to go!

The hardest nut to crack was IBM. I was still hopeful we could finish this nasty marathon audit and get paid what was due well ahead of the launch. The longest any OEM audit had ever taken was six and a half months. Could IBM's be resolved in eleven, or were we overzealous, insisting on such a timetable? I had no idea what Bruce Claflin had set in motion after agreeing to my request for acceleration. I refused to believe he had only paid lip service; he just wasn't that type of a guy, and he knew me well enough. I would follow through and refuse to sign a 95 license until this was settled. For reasons unclear to me then, either the Big Blue ministry declined to help him or our friends over there believed I would budge.

Complicating the 95 negotiations with IBM further were the sweetheart deals she had been accustomed to in the past. Windows pricing was going to rise, and as for any other customer, it would only be based on projected sales volume. Compared to the old agreement, she faced a $37.50 increase per unit for year one, which could increase further in year two if she did not participate sufficiently in MDA activities. Still convinced that OS/2 would prevail, her negotiating team nevertheless swallowed its pride.

Nine weeks before the launch, Bruce called and told me IBM would not be able to complete the audit in time. He begged me to let it slide—me swallowing hard. It was inconceivable to me to not have the company who had accelerated the acceptance of PC technology at that launch event. And I would for sure land in hot water if that ever happened. I am still convinced that with some goodwill from that exalted monster data-processing company, the requested info could have easily been supplied. IBM's management certainly had the means. In its place, her team chose to engage in a childish and costly power play.

I relented again despite me growing sick and tired of the delaying tactics and of another refusal to square up with us. During a tête-à-tête with Bruce and Tony in July, I proposed the following deal: let's move forward by agreeing to settle the audit for $25 million—a middle-of-the-road estimate according to my auditors. Let's further concur to complete the audit, and regardless of its outcome, no money will change hands later. MS will assume the risk in case the audit will come in higher, and IBM will do the same in case the sum to pay will be lower. My message: send us the check, and we will sign the meanwhile completed 95 agreement. Both gentlemen tried to dicker by offering a ten-million-dollar guarantee—without advanced payment. Lacking trust in them any longer, I insisted in my proposed amount with cash up front. In the end, they both nodded.

The next morning, I had second thoughts about my proposal. I was concerned about the difficulties both might encounter in selling the settlement plan. Calling Tony, I made a strange and unexpected offer: if Lou Gerstner would sign that agreement personally, I would get Bill to do the same. If they could make this happen, I requested three originals. Yes, I wanted to keep one of them for my own library as a historic document. I offered a five-million-dollar discount for Lou's signature.[27] Was I overstepping my bounds? Perhaps. Foolishly, I banked on good old-fashioned raw courage for IBM's hierarchy to approach its chairman and tell him the bold-faced truth. Yet the cowards in IBM did not fess up. I had played a game of compromise and redemption in the spirit of reciprocity, progress, and easing pain. I lost. The final response signaled excessive pride or principle. Stupidity did not cross my mind, though in retrospect, the word neatly applies. Continuing animosity—irreconcilable differences—toward MS still appeared deeply entrenched. Until the complicit and afflicted souls left IBM altogether, we would continue to struggle with any and all cooperative efforts.

What I did not know was that Bruce Claflin's days in IBM were coming to an end. Two months later, Tony Santelli stepped in to fill his gap. As I understand it, Bruce was being blamed for having lost the lead to Compaq in '94. He went on to run 3COM as CEO. No wonder he no longer possessed sufficient clout brokering an internal deal—I had unknowingly and unfortunately made my pact with the wrong devil.

In hindsight, the man in IBM who was responsible for the inflated OS/2 numbers was probably Lee Reiswig, head of IBM's personal software division. As he talked to *Research Board* in '94, he bragged about having shipped 4 million copies of OS/2 the year before, confidently boasting to ship 6–10 million copies during the next twelve months. He further claimed that 40 percent of all enterprise customers were using OS/2. Knowing the reality better, *Research Board* people mentioned in their book *The Limits of Strategy*, "Wasn't this just 'shelfware' shipped but neither bought nor used?" It was worse—most of it was never shipped!

---

27   Maybe this is the first time he will read about my long-ago offer?

ON STAGE AND BEYOND

To achieve maximum attendance, the Windows 95 launch would normally have been hosted in NYC or San Francisco. Helped by the circus-like frenzy, we arrogantly chose a different site, the new software mecca: Redmond in the state of Washington. And everybody showed—lock, stock, and barrel. The MS campus got converted into an amazing tent city, a teeming spectacle accommodating all global sultans of tech. My key OEMs, including IBM, showcased Windows 95–powered PCs, a multitude of component manufacturers demonstrated new graphic and audio cards, and our software competitors eagerly showed off their newly created 95 applications. An imposing and never-repeated industry occurrence! The main tent where the official announcement would be made held just over five hundred people and was packed to the seams. Invitation only!

Two hours prior to the main show, I hosted my customers for breakfast. A lot of them knew each other from our annual briefings. Bill and Steve joined us, thanking them for standing behind us at this critical, competitive junction. I echoed their sentiments. License negotiations had not been easy, and together with our guests, I was happy the hard work was behind us. Only IBM was missing. The last company yet to sign on. Reminding the audience of the master agreements they had all painfully negotiated and signed, I expressed my sincere hope that one day a license would fit on a single postcard. I expected these to be signed and returned without legal review. A robust chorus of applause and laughter was the response. Inwardly, I deeply hoped we could one day deliver on what I had just joked about, and somehow, the company did.

Right afterward, at 8:35 a.m., just twenty-five minutes before Bill and Jay Leno went on stage to launch Windows 95 to the roaring fanfare of the Rolling Stones' "Start Me Up," I met with IBM's team. Bruce handed me a check signed by Mr. York, IBM's chief financial officer, along with the 95 license agreement and audit settlement. We signed them simultaneously, shook hands cordially, and wished each other luck. No smiles anywhere.

I then handed the check to my controller and went into the event tent letting Bill know we had the money and the last holdout on board. He offered me a perfunctory "Great job," not a hint of a smirk. Like me, he was concerned that this was a Pyrrhic victory,[28] which is considered the successful conclusion of battle with a formidable competitor inflicting more long-term losses than immediate gains. Would we file this away under the same category?

The signs had been on the wall since three months earlier, after IBM had bought Lotus to obtain her office productivity suite, including Lotus 1-2-3 and a program called Notes: an innovative and extremely popular server-based groupware product used for network collaboration. MS had no equivalent. Like a browser, the product included its own programming language and had the potential to upset the OS balance. Notes was the brainchild of Ray Ozzie, who shortly after the merger left IBM and much later, after I had already left the company, actually joined MS as chief technology officer.

In purchasing Lotus outright, IBM was taking serious aim at another set of MS crown jewels. While IBM charged for Notes, she expanded the bundling of Lotus's office productivity suite, called SmartSuite, to all her PCs, enticing businesses to adopt it as standard and preventing consumers from buying our version. Fortunately, she stopped short of licensing it to her rivals at rock-bottom prices, which could have hurt us profoundly.

I never understood why IBM bought Lotus for such an incredible price. Nearly four billion US dollars. Lou Gerstner, in his book, mentioned that it was all about Notes and not because of SmartSuite. The Notes concept and the surrounding hype were relatively short-lived. It was neatly eclipsed by client-server computing and collaboration over the Internet. Most Notes clients were later given away for tokens or bundled free with WARP. IBM again demonstrated that she had no well-grounded experience selling PC software. Buying Lotus—though never admitted—served only one purpose: to further nettle MS. The Feds happily approved the merger with no strings attached. Fortunately, as IBM gave her newly acquired software away, mostly for free with little appreciation from her customers, zero revenue streams were created, and improvements were defunded over time. The Lotus gnat therefore developed into only a short-lived annoyance and wound up as a nice little tax write-off for IBM.

Obviously, it soured the relationship between us further. IBM's history of bundling software with her PCs started with World Books encyclopedia

---

28 King Pyrrhus of Epirus defeating the Romans at Heraclea in 280 BC suffered irreplaceable casualties.

in '94, a competitor to our Encarta. She temporarily bundled IBM-Works, an integrated and inferior MS Works–like product, with her consumer PCs. OS/2 got enhanced with a free Web browser of her own making. The behavior, while predictable after our separation, irked Bill and Steve tremendously. Bill, still looking for opportunities for fence-mending and synergies, was the most disappointed. Only he could eventually solve this puzzle, and he would get his chance.

For now he felt outright rejected and disrespected by IBM's management, and when he had any contact with IBM representatives, he vented his feelings by offering a solid dressing-down. I never understood why he was so compelled to reveal his true emotions. His adversaries were in no mood to change their way of doing business, and with—or because of—IBM's new boss, the animosities between the companies continued to flourish. My job was to conduct normal business with her PC branch and not to get defocused because of mother Big Blue's hostilities. Bill in his heart—despite being furious about IBM's current attitude—never quit dreaming of her as a one-day potential partner. For the time being, this was only in his head, and on hold! All the same, I had to be ready for the day sentiments changed and the Rubicon separating us could be crossed.

I found an interesting comment in Lou Gerstner's book about the "high-performance culture" he was desperately trying to create: "Losing to a competitor—whether it be a big fight or a small one—is a blow that makes people angry." Touché. Bill felt the exactly same when he called Richard Thoman, IBM's group VP in charge of the PC group, right after IBM had bought Lotus. I was in Bill's office when the phone call turned ugly and Bill became angry, fearful of an abrupt loss of market share and describing to Thoman how licensing and bundling competitive software was a detriment to our overall relationship. He acted exactly like a man living in Gerstner's high-performance culture. Not appreciating what his boss was preaching or how Bill treated him, his counterpart took it personally. Unknown to us, one Mr. Norris, another IBM employee, was allegedly sitting in Thoman's office, listening to Bill's comments. While I was listening on the speaker phone—and Rick was made aware of my presence—Norris's presence was not revealed, and Thoman did not use a speaker phone. Hmmm! How could the secret listener really understand the content of this exchange as he would later claim in court? I still wonder.

The dust-mantled audit was finally completed in November of '95,[29] showing that IBM owed us approximately $10 million, including interest. I was surprised at the amount, having anticipated, together with the audit team, a far greater number. IBM at once attempted to retrieve the residual millions.

---

29 Lasting a total of fourteen months!

Bruce by then had moved on, and so had Richard Thoman, enjoying a brief and unsuccessful spell as president and chief operating officer of Xerox. Tony Santelli, replacing Bruce, could not or did not want to remember what we had agreed upon, though he should have. So we kept the money, investing only a token in a new PC model intro with IBM. My sincere hope was that the long and winding road of this episode had offered IBM the following insight: the wisdom of compliance in the event another vendor ever had the guts to audit her. Goodwill and an expedited process could have saved a lot of pain, suffering, and company funds.

I meanwhile was off to learn an extra job. VP Chris Smith, who had followed in Jeremy Butler's footstep and ran what we called the rest of the world, decided to retire just before we launched Windows 95. Steve asked me to take over his territory. Now I was responsible for the WW OEM business and all other MS business for Latin America, Australia, Southern Africa, and all of Asia except Japan. The next year I spent five and a half months on the road—the road of a jetliner—seeing customers, giving speeches, talking to analysts, and visiting subsidiaries. I could not ever possibly have performed this wide new range of responsibilities successfully without the invaluable aid of an additional business manager, Kathy Weisfield. She knew the people in my new territory intimately, and I empowered her to make many decisions on my behalf. I never once regretted it. I have to admit that the extensive travels did get to me. I was conditioned to do them for two to three months per year; doubling that made managing OEM difficult. Succinctly put, my life now raced by in alternating phases. One week on the road and the next week in my office, preparing for another trip.

For my OEM business, the Windows 95 frenzy translated into a nice uptick in PC units bundled with Windows. OEMs distributed Windows 95 complete with IE. Not up to par with Netscape's browser, we followed up with solid improvements. With us not sleeping at the wheel, Netscape delivered an improved version as well just after 95 shipped. What journalists called the browser wars was definitely on and raging. MS and Netscape went head-to-head, matching each other's browser features and functionalities with passionate and unrelenting conviction.

Most interesting for me was how OEMs started complementing their business model after discovering the Windows desktop as a lucrative advertising media. Interested vendors paid OEMs to put promotional icons there. In their mind, they constituted in-your-face and well-targeted advertising that produced better results than magazine adds. Several of my colleagues nevertheless believed it threw design aesthetics overboard and cluttered up our original desktop layout. Yet we did not object. Our standard contracts stipulated that OEMs could freely add icons to Windows's desktop, including the ones from our direct competitors. But they were not permitted to remove any of ours, making certain Windows was presented and preserved as a whole.

With the appearance of 95, purchasers found a newly designed MSN (Microsoft Network) icon on their desktop. Pointing to and clicking on it steered them to an MS-owned subscription service providing online information and e-mail services. Entering into this business meant competing directly with AOL. Right away, AOL accused us of leveraging Windows and getting a free ride to obtain subscribers. Not having to pay OEMs a bounty to be present—like she did—seemed unfair to her management. The ever-present Internet Explorer icon on the Windows desktop made Netscape equally unhappy. Her management accused us of competing unfairly by illegally tying our browser to Windows—a potential antitrust violation. The Feds were listening again.

Eager to entice and induce end users into using IE, both Steve and Bill abhorred it when OEMs placed other browser icons side by side with ours. They were undoubtedly on a crusade, subscribing to the competitive culture notion, à la Gerstner. I wanted no part of it and told my people explicitly to make placement of any non-MS icons never an issue with customers. They were just distributing product options. The browser-usage decision was ultimately made by end users, influenced by recommendations from colleagues, friends, and family. OEMs populating the Windows desktop with alternatives had no control over user habits. Yes, IE was Windows and Windows was IE, as Steve had expressed to me once. IE for sure was a convenient choice. A trap to rely on. As I saw it, we needed to provide the best browser to beat Netscape. My order to my troops: Don't be confused. Sell Windows, guys, period, and not IE!

Expanding their advertising options, OEMs cleverly added promotion sequences into the Windows boot-up process. But here they were on thin ice. To accomplish this, they had to change Windows code, which they had promised to refrain from in their license agreements. Another quarrel was brewing as we carefully addressed these violations.

At the end of '95, a number of proprietary user shells emerged, intending to make operating Windows PCs easier. The first one I ever saw was introduced by Packard Bell (PB) and soon followed by creations from Hewlett-Packard and Compaq. They rivaled our own design. I remembered that PB adamantly refused squandering capital. So I applauded her for investing in 95 and understood it as a positive sign. As long as these shells were just another icon on the Windows desktop, few people in MS took notice. We were convinced we would easily win with our internally considered superior interface. When PB and others went one step further and modified Windows to boot straight into their proprietary shells, bypassing IE completely, we took notice. They were not supposed to alter Windows code, but they selfishly ignored what they had formally agreed to. Ignorance or bliss?

The computer industry had grown up fostering and cherishing proprietary solutions to make switching between hardware vendors of the utmost difficulty. OEMs, creating proprietary shells, viewed them as an opportunity to go back to the future. Their old-school marketing folks believed that if customers would get used to a certain way of operating a PC, turning to a different vendor for their next purchase would become less likely. The underlying assumption of PCs becoming a commodity amplified that belief. I understood that OEMs were trying to create repeat customers, but in my mind, the chosen path could not accomplish that feat.

Too many other considerations when choosing the next PC came to mind. There was price, performance, reliability, support, and so on. A proprietary shell weighed against these elements paled. MS was light-years ahead of smaller software companies OEMs sourced their interface designs from. Few had the grit and endurance to fund such an adventure long-term. I was convinced that attempts to complement our embedded solution, while being a nice gimmick initially, would fade. I used these alternative shells and never found that they exceeded the functionality and convenience of our seamless design. My take: let the market decide and not waste time on legal arguments. I instructed my sales reps to remind OEMs that booting directly into shells violated their agreements and that they needed to remove that feature. Not enforcing this right away legally, OEMs interpreted my inaction as us looking the other way. I concur.

As this conflict was lingering, I followed the progress of quarterly Windows 95 shipments closely. By late 1996, MS-DOS-only PCs were less than 15 percent of my sales, and the same amount was sold with the aging combo. The rest constituted Windows 95 units. By then, OS/2's version 4.0 warp drives were gasping for antimatter—our torpedo had fully struck, causing a deadly impact. IBM was now reporting her royalties timely and with great accuracy. The second OS war had basically been won. The browser war continued with all passion and legal ramifications as we will see.

Overall, '95 was a highly successful year. In November, Bill Gates published his first book, *The Road Ahead*, showcasing his IT vision outlining the role of the Internet. This year's crescendo was Bill's December 7 keynote speech at the annual COMDEX trade show in Las Vegas. Firmly in the driver seat again, he committed the company to integrate corporate networks tightly with the blossoming Internet. The speech, considering its timing, was billed the "Pearl Harbor" speech. Highly aggressive and no longer a paper tiger, he sent a clear message to Netscape: we will beat you! Our employees were thrilled to follow their chairman into an unchartered and wide-open territory as he embarked on investing money left and right in new ventures, protecting our flanks.

Business "as usual"

Of opportunities and threats

Compared to four years earlier, the '96 reelection campaign for President Clinton was a nonevent. Stung by how Janet Reno's questionable justice department had treated the company, the big boys didn't get involved publically. They used their influence behind the scenes, and as a result, MS's Political Action Committee (PAC) contributions went at a 2:1 ratio to Republican candidates. Mostly wasted, Bill Clinton's second term was a foregone conclusion.

The year would bring several key changes to the company's organization, her products, and the competitive scenery. The hard-won lesson that one's own government was as likely to ambush you for political gain as the competition did for market share had left a bitter taste and a reason to recalibrate the radar. Life's ironies persisted, as always, as even the most mature among us must learn anew nearly daily.

At the end of '94, Novell's core product NetWare tanked, and Novell's board ran Ray Noorda, her CEO, out of town. Deeply hurt, he blamed MS. In reality, aging NetWare had been beaten fair and square by a meanwhile matured and superior NT server. Ray swore in public that he was not yet done with us and surprised us two years later by acquiring DRI. Novell was happy to get rid of these nonperforming assets, knowing that the time of pure DOS systems had passed a long time ago. There was only one reason for purchasing them now. And vengeful Ray did not disappoint. He immediately filed a civil antitrust lawsuit claiming we had ruined DRI's business over the last fourteen years with illegal means. The allegations stemmed from warmed-over accusations based on the 1989–94 DOJ investigation. As Ray's word-swiveling lawyers commenced sifting feverishly through the collected ancient debris, they constructed a case that aroused him into salivation. His long-dreamed-of revenge was coming full circle, and a huge chunk of capital might be heading his way! I believed the lawsuit had a snowball's chance in hell of succeeding, but the seventh-ring-of-Hades attorneys Ray had hired thought differently. They knew how to access the thermostat down there and believed they had the help of American jurisprudence to reset it.

Starting with BASIC back in '75, computer programming languages had initially been MS's main domain, and the company had continued nurturing that domain, building a steady lead. Twenty years later, a new rival emerged. One of Sun Microsystems's founders, James Gosling, revisited a programming concept many programmers had long dreamed about: "Write a program once and run it everywhere"—regardless of the computer system and regardless of the OS making it tick. To realize and make popular his bitter brew of code—christened Java—he needed two components: instructions that programmers were familiar with and a clever OS-like underpinning enabling their execution. For the instruction part he, used elements of the familiar C and C++ programming language; for the underpinning, he constructed what IT experts call a virtual machine. To make Java's applications portable, its programming language component remained identical for all systems it was hosted on. Its virtual machine, on the other hand, needed rewriting and adaptations for each and every host.[30] When properly implemented, Java applications executed independent of the host computer's OS. Could this put the need for Windows on PCs in danger or, worse, obliterate it totally? Again Bill had reasons to be scared.

The technique had been tried before in early BASIC and Pascal interpreters. What made Java's implementation palatable and slick were two features: replacing the relatively slow interpreters with a very fast, on-the-fly compiler[31] and restricting the new language to a bare minimum instruction set. The compiler improved the execution speed of Java applications, known as applets or scripts, while the reduced instruction set made implementing Java's virtual machine on any system a less time-consuming task. As a lean and mean tool, Java suddenly allowed access to any computer via the Internet. The way this works is quite easy to understand. The applet you want to execute is sent from your PC to the computer in question and immediately executed—bypassing the main OS—as long as the targeted system hosts a Java virtual machine.

Investigating the Java concept, MS fully recognized its potential and licensed the product from Sun with the understanding of developing an improved version. To make it more useful, our developers extended its instruction set and invested heavily to improve performance. In August

---

30 In computer system specific assembler code
31 A tool, which, like an interpreter, translates instructions into executable computer code

of '96, we embedded our superior version into Internet Explorer 3.0. A browser that finally caught up with Netscape's in features and quality! Sun immediately objected vigorously. Working on standardizing Java, our extensions crossed her plans. Sun's burgeoning pride as inventor and standard setter and Scott McNealy's, Sun's CEO, scalding rivalry with Bill Gates left no room for compliments. Programmers using our version to its full extent could no longer write once and run anywhere except on Windows 95 PCs. Sun's universality dream was shuttered until our extensions were retrofitted everywhere!

Unwilling to embrace them, Sun eventually filed a breach of contract lawsuit. In her view, MS was deliberately destroying the Java standard. In our opinion, we were merely following a now well-entrenched principle, the one our Internet fanatics had once proposed—"Embrace, extend, and *innovate*"—which Sun maliciously replaced with "Embrace, extend, and *extinguish*."

The Feds watched the dispute with great interest, firmly believing we were out to destroy an alternative computing paradigm—by embracing it? The lawsuit was later settled but not before creating another injunction anxiety for MS and my OEMs who were shipping the Sun-challenged code embedded with 95 powered PCs. Nor did the settlement between the two companies end the dispute. The Feds later exploited the incident to smear MS by labeling her well-intending development and marketing efforts "anticompetitive behavior." You be the judge!

By now, MS had gotten serious with NT. One of its design goals had been easier portability to other computer systems—just in case a non-Intel platform would challenge the incumbent one day. The AIM alliance threat had made such an undertaking an absolute necessity. To achieve the lofty goal of easier portability, its designers adopted a layering-technique approach. By separating the hardware specifics of an implementation from the application programming interface (API),[32] comparable to what Sun had done with Java, only a very small portion of NT had to be rewritten in case it ever needed to be ported. Going beyond, Dave Cutler, the intellectual force behind the design effort, ingeniously enabled modular implementations of more than one API set for NT. Including from the get-go the ones for OS/2 and Windows, and later the ones for MS-DOS and POSIX, popular on UNIX systems, he attracted a far broader spectrum of applications to run on NT systems than anywhere else —creating a clear competitive advantage!

The technology he employed helped MS to remain flexible and competitive

---

32 The one software programmers need to write executable applications for a computer platform

in the wild mood swings of the ever-changing IT industry and save time to market in response to a disruptive platform event. The same benefit applied in case an emerging API set became prevalent and had to be added quickly. NT further excelled through a secure operating mode by introducing a sophisticated kernel structure to prevent it from failing when a badly written application collapsed. Adding key networking features such as communication protocol stacks for Internet and intranet access, Dave Cutler delivered a most sophisticated and reliable OS, targeting OS/2- and UNIX-using enterprise customers alike.

First released in '93 and by now in version 4.0, NT had fully matured—most ready for prime time. Intel's new CPU, introduced in November of '95, called Pentium Pro, a powerful successor to the 80486, made it shine further. There were two flavors of NT in the market: a server version, which had evolved from the original LAN-Manager product, and one designed for workstations. These were the higher-end PCs used for graphic design and complex analytical and engineering tasks. My group focused solely on selling the workstations derivate.

NT, being more capable than Windows 95, justified in my opinion a higher price. To determine its value, I studied competitive products as I had done for 95. One was OS/2. Its pricing was hard to determine. IBM had published a retail price of approximately $250. Yet that version was hardly selling. IBM's OEM pricing was unknown to me. I estimated it to be between $60–100. We further knew that IBM gave lots of OS/2 copies away for free or sold them in bulk at extremely favorable conditions to enterprise customers. Therefore, the real price fluctuated probably between $0 and $250. The other competitive OS was the one running on the Apple Macs. It arrived installed without being priced out separately. Only retail upgrade pricing was obtainable. Last but not least, there were several UNIX systems in the market. One was Linux; it was given away for free and mainly used on servers—not what I was aiming for. UNIX versions, bundled with workstations, were not priced separately except their upgrades. SCO's version had a retail price tag of around $400–$550 depending on the included libraries. With scant information and such huge price fluctuations, I asked Bill for advice.

We started our discussion by investigating future OEM business aspects and immediately agreed that growing my business was tied to two main factors: overall PC sales and a steady increase in NT penetration, assuming it could demand a higher price than 95. The first depended on the economy, affordability, and the PC's hopefully enduring usefulness—which we could only indirectly influence and which in hindsight took over ten years to be seriously challenged. Increasing NT penetration, on the other hand, was directly correlated to how much enterprise workstation demand we would be able to create. I had a job to do.

Pricing NT higher than Windows 95 was a foregone conclusion for us. Just how much was the next issue to attack? Taking OS/2 and UNIX pricing into account and knowing NT contained most of the same goodies and was capable of running a more diverse number of applications, we eventually arrived at a $100 price; it sounded like a bargain to us. It sure is an easy one to get accustomed to if used as a from-the-hip multiplier, but could we really extract that much?

Our so-called bargain turned out to be extremely hard to sell. For every hundred units of Windows in 1995/96, only two to three units of NT were sold. The reasons: a combination of price resistance by the OEMs and so-far tepid customer demand because UNIX was favored on workstations. What made switching enterprise customers even harder was that they were still suffering through the OS/2 disaster. After IBM's late solo had failed, they weren't willing to easily trust MS's intrepid lead. Stuck, I asked my marketing guys to break the deadlock and create a program incenting OEMs to crack the enterprise workstation segment with NT in mind.

With NT-powered systems capable of supporting higher-end solutions and their potential to increase revenue and profit margins, OEMs should have jumped on that opportunity a long time ago. Instead of helping us to actively convert UNIX users into NT fans, they waited for us to initiate the move. A typical reaction considering they often saw themselves as mere distributors of our OS—proving a point I often emphasized: OEMs do not create demand for any OS! Ultimately, applications do that only when running on attractive computer platforms, and the top-notch ones were indeed running on NT workstations. This gave us a fighting chance to succeed.

My marketing group soon identified a list of OEMs who each had their own corporate sales forces and should be interested in our endeavor to increasingly push NT for workstation usage. One of them was IBM. It would have been nice to have her on board, but even with repairs on the way, the company was still vying for warp-crowned revival and immortality. Our frontline partner Compaq was another ideal affiliate, and we discovered a great deal of interest in Hewlett-Packard (HP) as well.

Encouraged by OEM feedback the marketing folks created a special NT workstation, market development agreement, including huge incentives for NT-specific promotional activities. We decided to measure its success on the sales ratios between Windows 95 and NT. We stayed away from asking for firm quarterly unit commitments to avoid any conflict with the existing consent decree. When executed to its full potential, this program could lead to a short-term 40–50 percent NT royalty reduction for participating and well performing OEMs. It took less than three quarters to reveal the desired trend. To our surprise, HP was the best performing program participant followed by Compaq. Dell hardly made any effort and achieved only small rewards, though they managed to complain the loudest about the structure of the campaign. By mid '97, NT workstation sales resulted in nearly 10 percent of all Windows sales—a nice revenue boost for the company. We were on our way to set yet another standard.

In parallel to the high-end workstation market, a new challenge appeared for my group with the appearance of Windows CE in '96/'97. CE—for Consumer Electronics—was designed to capture the unique facets of the rapidly evolving electronic device opportunities. They had undergone meteoric changes over the last four years as semiconductor components became increasingly affordable, tightly integrated, and miniaturized. A new clientele was waiting to be conquered. The producers of industrial controllers was one of them, mobile phones another, with handheld devices and car-computer manufacturers completing the picture.

CE intended to target them all. A functionally stripped-down cousin of Windows 95, it was designed to run in small memory footprints and on a variety of CPUs called ARM processors. ARM stood for Advanced RISC Machine, representing a low-cost CPU with reduced instruction sets and low power consumption. Combined with reduced memory prices, the appearance of cheap graphical displays—replacing text-based ones—and higher-integrated circuit boards, manufacturers were suddenly able to produce handheld devices with more computing power than the original IBM PC. Just fifteen years later!

Compared to the other sales opportunities for CE, the industrial controller market was the most fragmented. It had its origins in replacing manual labor—first with reliable mechanical systems, then with electronic ones. Companies like Honeywell, Bosch, and Kontron come to mind when talking about key manufacturers. Beyond these larger companies, we encountered a huge array of smaller ones focused on highly specialized applications in particular in Germany. As early as the mid '80s when digital technology had evolved enough to present fertile ground for new application opportunities, they all started to experiment. Unfortunately for us, the larger companies had meanwhile developed proprietary OS for their controllers—translating into an uphill battle for us. The smaller ones tried out Wind River software or DOS versions from DRI or us. Neither DRI's nor ours were designed to address the needs of this environment adequately, yet they beat Wind River because they were less expensive. Windows CE was a much better fit. We got its feature set right in its second release and delivered a componentized OS, which industrial OEMs could pick apart and modularly adopt for a variety of applications.

When it came to mobile phones, CE had intense competition from a company called PSION along with a parade of other OSs developed within the cell phone industry. In version 1.0, it lacked essential communication functionality, but when added later, CE eventually possessed enough compelling features to compete effectively. Handheld or pocket PCs were just emerging. The market was in an embryonic state, and Palm's OS was our most riveting rival. Car computers were available already; they were expensive and mostly used for navigational aids that left us with no reason to assume such a narrow focus would not be expanded.

In studying the opportunities for CE, I soon concluded that we needed people with different types of expertise for a successful run. Ideally, they would possess specific industry knowledge and act like design engineers. We delivered source code to our targeted customers when selling CE. In order to be successful, our new breed of sales reps needed sufficient tech knowledge about how to utilize its components for key industrial applications. Success further depended on the availability of a high-quality development kit. Without it, CE was rendered useless. Complicating matters further, its components had to be extremely stable and reliable to pass industrial strength muster.

For the next two years, we experimented and struggled with selling what was a premature offspring of Windows. Until version 2.1 arrived, the software was not well-enough componentized or stable and its development kit not as capable as required—nothing new here. The other issue plaguing CE had to do with its initial performance. Never giving up, we overcame these obstacles. But to make matters worse, we had a tough time locating the right sales specialists, and we scrambled to select the right distribution partners to serve the smaller customers.

As indicated above, our fiercest competition was often an in-house-developed OS. Our potential customers employed talented tech personnel intimately familiar with their own software. For them to change and use CE meant extensive retraining and often an internal turf war. Using our readily available system was judged as giving up hard-fought-for competitive advantages based on proprietary solutions. Trust seemed to be another factor not favoring us. With all the bad press and the antitrust saga raging on, companies had grown hesitant to buy core technology from us. They understood that we were not licensing a monolithic and unalterable OS as we did for PCs. Instead we were offered components for different devices and applications, which they could customize as they saw fit. Several prospects still remained reluctant to bet any part of their destiny on our solution—in particular the phone manufacturers. Growing the business was therefore harder than expected. Considered a breakthrough, one of the great success stories was an early deal we closed with my help with Bosch in Germany. It set a trend and made some of competitors follow her lead.

As always, in good old MS style, we kept plugging away with unflinching perseverance and never gave up despite product-group management reshuffles, design changes, and notorious delays. Five years later, the CE business grew into a $150 million annual business with less than twenty dedicated sales engineers. I referred to it as a barely profitable hobby. We suffered from insufficient resources, not developing a consulting arm supporting customers sufficiently at the engineering level, and an underdeveloped distribution system. When I left my job in 2001, we had barely laid the foundation for a sustainable business. By then I had hired an industry expert in Philippe Swan, who applied a passionate dedicated effort, hired experienced people, revamped the distribution system, and managed to increase annual CE revenue more than threefold five years thereafter. A great accomplishment in itself after a slow start but, compared to MS's core PC business, far less significant than I had originally expected and hoped for.

Run by committee

In the summer of '96, I relinquished my rest-of-the-world sales job and was full-time and happily back at the OEM wheel. I made senior VP that year. The current Office of the President was replaced by a newly created leadership group consisting of nine executives. I considered the arrangement a weird setup and never believed that a group of nine could effectively act as one. Bill, as the largest shareholder, with an amiable board on his side and Steve as collaborator, ultimately held the power and called all the shots. Committees at that level create confusion and nurture unwieldy politics. The reason for the new structure was explained to me by Steve as an opportunity to prepare the future leadership of the company for increased responsibilities. Why not send them to management training? I had a hard time investing much credibility in his explanation. He approached me after the announcement was already made. As we discussed the possibility of my participation with apparent frankness, I could not determine what Steve genuinely wanted. Was I too old or too controversial to be considered earlier? Perhaps he was presenting me the opportunity of self-determination. I made it easy for him. I just wanted to continue doing an excellent job in OEM.

Sooner or later, Steve would run the company anyway. There was little room for anyone else at the top, especially considering Steve's boisterously ambitious and turbo-bold nature. Why create false hope? Declining his invitation was one of the better decisions I ever made. Five years later, only two out of the nine selected execs were still with the company, including Steve. Preparing for which future?

Bill's announcement of a revised company direction that summer was again a result of his annual think week. This wasn't the first time he had outlined something similar internally, though making his thoughts and conclusions public was a first. The new mission was to deliver a comprehensive set of products encompassing all communication and computing platforms. He wrote about seamless integration of the Internet with legacy systems such as mainframes and wide-area networks, which was probably the key reason why he still wanted to mend fences and team up with IBM again. His vision was to make the development of client-server applications easier—his indirect answer to Notes—and to implement a new navigation paradigm between computing environments. The message, loud and clear, was engineered to alert the outside world: MS was once again laying the foundation for broad-stroke global leadership. On the face of it, the propagated direction sounded fine, though for my taste it was not crisp enough and lacked finesse, an initial and troubling indication that the company was run by committee and nobody wanted to step on anyone else's toes. An early sign that our leadership style was changing!

As a result, the system-development organization got split into four main groups under Paul Maritz's oversight. One group got focused on developing Internet Explorer. Another was working on the rest of Windows. The other two dealt with NT server and NT workstation technology. All were supposed to orchestrate together and implement the outlined strategy. For the two NT groups, things worked out well, while Windows gangs ran into trouble caused by a leadership crisis.

Splitting up the Windows effort made sense considering that Internet technology then, still in its infancy, was moving along faster than the rest. The need for an ever-better browser to decisively challenge Netscape was obvious—meaning imperative. Tying IE releases solely to new Windows versions would have been competitive suicide. The IE group solidly reflected that burgeoning competitive zeal. On the surface, dividing the two worked well. Behind the scenes, things got messy. The IE group and the core Windows group soon engaged in turf wars unable to coordinate each other's work. Most of the contention centered on the Windows desktop design. It had to be modified whenever a new version of IE was released. Who was responsible for changing the software IE needed to hook into? Unable to find common ground, the ensuing quarrels of their two leaders stretched even Paul Maritz' rock-solid management capabilities. Both he—and Bill—had to intervene several times and mediate ego-driven disputes, though the rivalry raged on unabated. I refused to take sides and hoped the two prima donnas would bury their hatchets and turn their energies to timely deliverance of their creations.

For an end-user, upgrading to a new version of IE felt that only the IE code itself was being changed. Therefore, many people believed IE and Windows could easily be separated. In reality—and invisible for the upgrader—our clever update technology always refreshed a host of related core Windows code in the process—the leading cause of the aforementioned conflict. Windows and IE were indeed joined at the hip as one.

MS had therefore always shipped Windows and IE together, except for its first Windows 95 release, targeting the retail channel. For less software-knowledgeable people, this further strengthened the argument of IE and Windows being two independent and barely cobbled-together products. A total misconception. For that retail-specific variety, the development group had simply missed a production cutoff date—a minor hiccup. The Feds later drew the conclusion that Windows and IE were in fact separable, and rashly and wrongly explored the incident mentioned to prove the tying claim Netscape had alleged all along—an erroneous judgment, with severe legal consequences to follow.

Compaq going astray

As explained earlier, OEMs, including Compaq, created extra revenue by charging for third-party-vendor icons placed on Windows's desktop. In late May of '96, Compaq decided to go one step further. The VP in charge of her consumer product line, Celeste Dunn, boldly eliminated the MSN and the IE icons from her consumer line of PCs, replacing them exclusively with the ones from AOL and Netscape. Both companies paid Compaq amply for their newly won exclusivity. We were shocked and immediately voiced three strong objections. As frontline partner, Compaq had promised to promote our solutions primarily and by removing our icons she reneged. Second, altering our Windows code violated her license agreement. Third, it violated our copyright. Compaq, as distributor for Windows, needed our consent for any and all modifications of the original!

Celeste Dunn had simply presented us with a fait accompli. Upon confronting her, our account team leader was told she would look into it. Six months later, no effort had been made to remedy the situation, and what I considered bad news had not been escalated fast enough either. I heard about her blatant act by the end of that year and immediately contacted Compaq's relationship manager. A Compaq old timer and a great guy, but in this case, he demonstrated zero influence. AOL's and Netscape's money continued to talk.

Between December of '96 and February of '97, the account team and I received several remedy promises. Nothing happened. Our relationship, our contract, and our copyright were all being neatly eclipsed by third-party bounties. With no change of heart in sight, I was left with no option but to send Compaq a *cease and desist* notice. The notice would give Celeste thirty days to comply and indicated we would revoke Compaq's Windows license in the event she chose not to abide by our unambiguous contractual terms and reinstall our icons. We did not ask, ever, to remove the others! As a last resort, I warned the Compaq team a legal missile would be incoming. My shot across the bow was ignored.

In March, I followed through, documenting the violation officially. After trying to solve the situation amicably for nine months, I believed we had exhibited plenty of patience. Now mine had finally run out. My letter had the desired and immediate impact. The situation was promptly resolved, and the person responsible was removed from her post. If I would have let Compaq continue to screw around, other OEMs would have followed suit. AOL and Netscape would have made sure of it. I had no intention of hurting Compaq, but like any other customer, she needed to respect our copyright, and from thereon out, she did. I made sure we kissed and made up, and again we forged ahead in support of one another, as demonstrated shortly thereafter by extending our Front Line Partnership for another five years.

A STRONGHEADED RESPONSE

One of the toughest internal battles I had to fight two months later concerned the alternative Windows shells OEMs still had not given up on and the liberties they had taken in regard to Windows boot sequences. Mistakenly, I had let both issues slide when I first encountered them. They nonetheless agitated Bill far more than I could have ever imagined. He therefore ordered the product group to buy a variety of PCs from different manufacturers and called a meeting. Confronting its attendants, he demonstrated how alternative shells bypassed the Windows desktop. He was adamant about stopping this nuisance. Claiming our programmers were proud of their creations, he insisted in their work not being trashed or rendered invisible. He lectured his boardroom audience, venting, "Like artists developers are proud of their creations. Nobody should be allowed to alter their handiwork!" It was absolute for him that OEMs had a legal obligation to change neither the Windows code nor its visual appearance.

Considering that you could find the engraved signatures of each member of the original development team inside the early Macs back in '84, Bill's argument was not that far-fetched. If our software engineers would have done something similar, his argument might have been even more compelling. Right away, the attending attorneys and product-group managers unanimously agreed with him. I was the lone wolf in the pack, trying to put a positive spin on what my customers were doing. In response, I was lectured that OEMs should be focusing on hardware and not rivaling our well-designed and copyright-protected software. I agreed with the assembled group on one point: bypassing our Windows desktop and hiding it from end users was undesirable behavior and should not be tolerated. Ever the deal maker, I cautioned the group against overreacting.

During the demonstrations, we discovered how various OEMs encouraged Internet service provider sign-ups through advertisements during the Windows boot sequence. We further observed that outsized icons had been placed on the Windows desktop—three to five times larger than ours.

With most poorly constructed, they lacked graphical-display resolution and aesthetic value. I didn't appreciate their appearance, but I didn't find them overly offensive either. Bill, on the other hand, was getting worked up. For him, they degraded the appearance of our nicely presented, visually astute Windows desktop images. He adamantly repeated his insistence on precisely preserving its look and feel for end users. A red-faced raging bull he was.

Properly proportioned icons being added to the desktop were not of his concern. He understood that they provided OEMs with additional revenue and competitors with visibility—not an issue. He insisted that added icons should not overpower ours, demanding self-same size for all. He had a point, yet I was not too keen about having that argument with my customers.

Not interrupting the Windows boot process by any advertising was his next contention. A fairly heated discussion developed about this alleged infractions. To my great irritation, Bill took all of this way too personally.

My firm belief: the shells, a novelty now, would disappear over time. OEMs would lose interest and finally stop funding them. Our superior standard desktop would gain the popularity it deserved and win out. The rest was not too terribly alarming. Let the OEMs have certain freedom and make money through advertising and placements. Why argue about artistry and taste?

My urgings to exercise patience and restraint fell on deaf ears. The pride of the product design team, together with the legal arguments regarding our copyright and clear-cut agreements, gained traction. No turning back, we were inching toward escalation!

For the boot-up process, we found a reasonable compromise. Booting a PC consisted of two stages. During its first, BIOS software, not written by us, was loaded into memory. Then, in the second round, our Windows code was inserted. As long as OEMs used the BIOS boot sequence for advertising, we had no valid reasons to object. However, at Bill's insistence, the group decided on disallowing any interruption of our Windows boot sequence: explainable by pointing out contract and copyright violations and potentially rendering Windows less functional if engineered poorly. I took it with pain.

The most sensitive part of the discussion centered on allowing OEM shells to bypass the Windows desktop altogether. I proposed an educational approach for changing behavior and design. My proposal made no impression. Bill, growing increasingly annoyed about my arguments, gave me an emphatic lecture about copyright laws and the binding content of our agreements. I was painted too stupid to understand and too timid to enforce these. Though not enjoying his outbursts, the attorneys nodded with complicity. The discussion escalated to a fiery personal level. Bill in attack mode and me losing all my respect as I pushed back hard!

After we both calmed down, I reluctantly agreed to take a look at our current agreements with an eye toward clarifying and specifying the rights to modifications. The attorneys promised to come up with revised legal language, which I would take to Bill for approval. Again, at Bill's sole insistence, it would include explicit legal instructions disallowing the removal of icons and demand any added ones being same sized as ours.

I tried one last time to caution Bill. The Feds might scrutinize our new requirements to death and find another angle to invoke yet another investigation. Haranguing him about potential antitrust consequences broke the camel's back. He exploded with fury. Raising his voice, he shot back, citing our ironclad legal position again and going out of his way to make me look foolish and reckless. How could I, with zero legal education, even mention the name *antitrust*? He had not completed his either, but as usual, the wannabes are the most vicious and dangerous. In my gut, I knew I was on the right track with my comments, but no one else in the room had the audacity to come to my aid. Our self-anointed master ruled supreme.

This was the first time I experienced such a significant change in the valence and tempo of Bill's leadership style. In working with him all these years, he had mostly provided directional guidance, taxing me to solve the details, the often tsunami of minutiae within the tectonic-shifting macro, to accomplish well-defined goals. Suddenly he had shifted into a prescriptive command style. He obviously did not trust me to step up to this challenge and left no room to adjust terms and conditions. From now on and for the first time in my MS career, I had to operate with my hands tied behind my back. I sure despised it!

We could have easily achieved the self-same results by verbally educating our customers. Should that have failed, nothing prevented us from revisiting the matter with an eye to a sterner approach. In its place, the lawyers went to work to clarify our restrictive requirements. As I was leaving the room, Bill Neukom took me aside and confided that he was proud I had stood up to Bill. Where was he when I needed him? He found our compromise reasonable and vowed not to go overboard with the ordered adjustments. Knowing the verbose proclivities of lawyers and the CEO's determination, however, I took his comment with not just a grain of salt.

Our memorable meeting happened on late Friday afternoon. Looking Bill N.

straight in the face, I replied, "Maybe I was courageous, but come Monday, I might no longer be with the company." My behavior toward Bill had for sure not been gentlemanlike, and I was uncertain of how, in retrospect, he would frame my counterpunches. I did not get fired, however. In fact, over the weekend, Bill sent me an apologetic e-mail. His personal attacks were unwarranted. I apologized in return. The heat of the moment had gotten the better of us. Reluctantly, I carried out the task assigned to me. I reviewed the legal text with our attorney for antitrust Dave Heiner and met with Bill, who approved our handiwork. My meeting with Bill took less than fifteen minutes. Returning to Dave that fast, he was curious how I had received Bill's buy-in so fast. He joked by saying he would call on me next time he had to get a legal text approved by him. I just smiled, deed done. My smile, though, did not last long.

As suspected, I was soon to learn that my customers, to put it mildly, resented the adjustments Bill had uncompromisingly pushed for. They desired extra freedom to modify Windows so they could extract additional bounties and increase advertising revenues. Some of their requests were so far-reaching that I considered them truly unreasonable. I still believe the true intent of our clarifications was to preserve Windows, and as product creator, we should have had the unrestrained right to do just that.

Our OEM customers were in many ways our partners in arms when it came to expanding the PC market and win against other platforms like Apple's Macs, IBM's PPCs, and game consoles to name a few. They served a useful distributor function, and I believed we were on strong footing, demanding they remain loyal to their voluntary agreements. Their attorneys disagreed. Asserting they had no viable alternative to Windows, they therefore did not consider licensing Windows a voluntary act. Massive trouble with Feds was brewing if their assertions prevailed.

A few OEMs complained to the press after receiving the new legal text, and the Garry Reback gang made sure the Feds stayed well briefed about our contractual perseverance. The fight over the right to Windows desktop alterations soon developed into a crazy situation for all of us. As long as Bill was unwilling to relax the rules, OEMs would simply have to follow them. I did not, however, relish the amount of time wasted on arguing our position with smart-ass lawyers employed by our customers, who eagerly flourished harebrained abstractions of already overworked controversial opinions. Once their executives embraced these with passion, it tended to lead to bruised egos on both sides of the fence. This was, what I believe, happened to our relatively smooth relationship with Gateway. As soon as her calls for additional freedom were denied, Gateway joined the chorus of others and sang "woe is me" to the Feds. I felt trapped and forced to defend and enforce our unpopular regulations.

Internet Explorer to Win

At the beginning of '97, most market observers agreed that OS/2 was on its last leg and Windows had won the battle for customers' hearts and minds. It was easy to use, now had a reasonable Internet browser, and carried a relatively low sticker price. Windows application programs were being steadily improved, and new ones were published almost daily. The latter was not true for OS/2 but remained definitive for the Apple MAC. With Steve Jobs back at Apple's helm, our two companies had finally settled their differences. MS took an equity position in Apple and agreed to continue writing Office productivity applications for her platform. Our new alliance was announced in August of '97 and was there to last to this day.

With IE version 4.0 on deck for shipping in the fall of '97, I went on a tour accompanied by Jusef Medhi, one of our brilliant product-marketing managers. Our goal was to visit ten US customers in just four and a half days and convince them to invest in our latest IE technology. A blitz to finally win the browser war!

Our persuasion campaign focused on a newly incorporated design element called Active Desktop. The important feature for OEMs to recognize was the so-called channel bar. It displayed the names of some of our partners. If end users chose to click on one of them, they were instantly connected to that partner's website without having to know its exact name. Making navigating that convenient was not a bad idea, considering users at that juncture were still inexperienced with exploring the Internet. We were already in beta test when my group's lobbying effort convinced the product guys to allocate one entry exclusively to OEMs distributing Windows. The ensuing direct communication link promised OEMs to reduce support costs, advertise add-ons or new products, and increase site advertising revenues. Our pitch: if you take advantage of this feature, we are prepared to offer development assistance and funding.

I chartered a Learjet so we could visit all customers in one week. Our presentation was just one hour long, followed by a one-hour timeslot for discussion. The trip went off without a hitch, and nine of ten customers joined and began the required development effort two weeks later. To my surprise, IBM was one of them. When Steve, however, heard I had chartered a jet for the blitz, he called me out on the carpet, reprimanding me that I should have asked for his permission beforehand. Not MS style. Aye, aye, sir! Next time. How had he found it? A person in the controller's office who had shared a leg of the trip had told her boss, who had nothing better to do than tell Steve. There was no rule ever expressed or written against chartering a jet, and considering the tight schedule we operated under, there was no other way to accomplish our objective in the allotted time. Micromanagement at its best. Nevertheless, we got the ball energetically rolling and were well on our way for a terrific launch. With nothing comparable to offer, Netscape found out too late and was left in the dust.

IE 4.0 provided another unexpected headache for me. AOL had meanwhile conquered about 30 percent of the online-services mountain. She was still running a proprietary network but felt the urge to transition to the Internet to stay competitive. To move faster, she needed an AOL-specific browser. Not wanting to start from scratch, she simultaneously entered into secret negotiations with MS and Netscape to acquire browser code. Both companies made attractive offers, with both solutions offering unique advantages. Winning the deal was fiercely contested, understanding the tremendous customer base AOL presented and the usage share the winner would gain. At last, Netscape was announced the winner, only to be recanted twenty-four hours later. There was a sound technical reason for AOL to go with us. In contrast to Netscape's, our browser design was modular and therefore easier to pick apart and adapt. After unexpectedly losing the bid in the second round, Netscape cried foul. Posturing and ranting in public about MS pulling imaginary evil strings yet never admitting we possessed indeed an unequivocal technical edge.

To close the complex deal, we consented to a last-minute sweetener by allowing AOL to promote her services in all Windows packages, including the ones OEMs inserted in their PC boxes. I understood our desire to win but objected right away to how far we had gone. With OEMs desiring to extract money for such favors, I knew my customers would be extremely unhappy with our arrangement. As I waded into the uproar, I discovered the deal was ironclad, and we had to honor it. As can be expected, our ill-guided promise added to the already rapidly growing pandemonium. Who had the right to desktop or boot-sequence modifications had now morphed into what OEMs needed to insert into their Windows packages! Hands tied again, I endured new complaints. Having made an exclusive deal with a different Internet service provider, Dell was the first to raise a stink, followed by Gateway, who had started her own online services. I fully shared their objections, unable to provide relief.

PRELUDE IN THE SENATE

Earlier that year, politicians got back in the ring after competitors like Netscape, Novell, Sun, IBM, and Oracle raised public and behind-the-scenes complaints about MS business practices to an unprecedented crescendo. As the head of the Senate Judiciary Committee, Utah's Senator Orrin Hatch felt obliged to schedule a hearing, inviting Sun's CEO, Netscape's CEOs, RealNetworks's CEO, Dell's CEO, Bill Gates, and other computer-industry executives to wash some dirty laundry in congress.

Representing the state of Utah—where Caldera, Novell, and WordPerfect were headquartered—he should have recused himself or refrained from scheduling the hearing at all. All three companies had supposedly contributed generously to his election campaigns, and he now felt obliged to dress down a competitor of theirs—potentially greasing the skids for future donations. I still believe he abused his position, and I considered his behavior bold-faced sleaze in full and open public display, never noted by the free press or criticized by any of his colleagues.

His views on antitrust enforcement were well documented in a speech published in "Competition, Innovation and the Microsoft Monopoly: Antitrust in the Digital Marketplace." According to him, antitrust enforcement needed to focus on paradigm shifts in the IT industry and in particular regulate successful market leaders who supposedly exploited their market power in preventing start-ups to introduce disruptive technologies. Start in Utah, Senator! He theorized that such behavior would inhibit innovation and harm consumers. What a contradiction! Truly disruptive technology does not need protection from regulators regardless of how aggressively an already-entrenched market leader defends its turf. In particular, as a Republican senator, he was just plain wrong in recommending to artificially protect start-ups and crush the taxpaying establishment. There were no laws on the books to intelligently corroborate any of his awkwardly antagonistic policies.

No wonder the hearing quickly turned into a MS witch hunt, with Hatch and MS competitors calling us a monopoly. "Ei incumbit probation qui dicit, non qui negat." Innocent until proven guilty, the senator, himself an attorney, failed miserably.

Watching the proceedings of the showlike trial on TV, I admired Bill for attending the devilishly choreographed charade of pseudojustice marching toward a resolutely staged outcome. Dell's CEO saying something nice about MS was roundly ridiculed by politicians for not offering any non-MS OS for his PCs. The proof was presented based on a single phone call one of the staffers had made, the content of which didn't reflect Dell's policy at all. As usual, the predisposed press was all over the blatantly manipulated event, repeating for weeks thereafter the mounting alarmist accusations leveled by competitors and Hatch-like politicians. Some called the grilling of Bill a public service in the infamous public interest. I shook my head in disbelief, thinking of all the time and taxpayers' money they wasted!

Another Alliance Attempt

With the demise of OS/2 WARP, clearer heads in IBM's PC division finally prevailed. During my reinstituted quarterly dinner meetings with the local IBM lab manager, I indicated that Bill would be interested in reconsidering a broad partnership. My message got passed directly on to the right people. Bob Stevenson, Thoman's successor, wanted to seriously explore it. Negotiations were back on. My team and I met several times with different players from within IBM, exploring the opportunity further. Following up with Bill, we outlined a plan. Bill's interest in working with IBM now extended far beyond the PC platform, covering database technology, mainframe issues, intranet- and Internet-integration topics, PowerPC ambitions, etc. Competing and winning against OS/2 or Lotus products had faded away.

In turn, I detailed his assessment and interest at large with IBM's division management and asked to explore Gerstner's willingness to engage. His people came back with "He does not like to waste his time." I interpreted their message as "I don't think I can really trust Bill Gates." We refused to give up, cultivated influencers, and in the end succeeded in arranging a private assignation between the top management teams of both companies.

The meeting took place in the early afternoon of a late and humid summer day in Chicago and began with sitting down for lunch together. The IBM guys were dressed in their rigid pinstripes while we had just changed, last minute, from Cabela's sportswear into casual business attire. Straight off, I could see that Bill was in guarded mood. Steve had expressed skepticism on the way over. Paul Maritz and I kept an open mind without being overly optimistic about the outcome of the potshot gathering. Both parties had flown in on private jets and used black stretch limousines—curtains closed—to reach the designated meeting location in a North Chicago suburb.

In marching up the bricks of the alleyway, we used a greasy back entrance and filed into a Tattiano's-like Italian restaurant to get into a compressed meeting room. Surrounded by IBM handlers, Paul and I couldn't help but compare the moment and the mood to how Mafia bosses may have cautiously approached their clandestine rendezvous.

After introductions, we stood patiently around a long table as Lou made a great display of symbolically removing and then folding his cape and sword. The chairman of the largest IT company in the world was in an impressive form; Bill matched him—just kidding.

Sworn to secrecy, none of its attendees ever disclosed that the meeting ever happened. MS was represented by Bill, Steve, Paul Maritz, and me. Lou Gerstner had brought his software chief, SVP Jon M. Thomson; Ned Lautenbach, a sales SVP; and Bob Stevenson, who was SVP in charge of the PC division. I had met him twice before and had not been terribly impressed. Unfortunately, the meeting took place prior to Sam Palmisano's appointment as head of the PC division, which came about two to three months later. Its outcome may have been much, much different.

The two big boys got right into the topics outlined in the relatively loose agenda. I remained mostly a silent observer. There was a lot more gray hair on the opposite side of the table. The meeting was cordial, sometimes a bit tense. I concluded early on, and so did Paul, as we had a quick exchange during a break that any real action would not be easily forthcoming. Lou and Bill were living in different worlds. Gerstner had been a consultant for a long time in his business life. He understood how serving customers diligently could make a company successful. An insight most instrumental for him as he engineered IBM's turnaround!

Bill's career had been a brilliant stroke of luck, betting on the superiority of the PC concept. He had truly lent a hand in creating, growing, and nurturing the industry. Possessing no sympathies for IBM's mainframe approach, he had nevertheless concluded that bridging both worlds promised rewards for both sides. Could he charm—or better, lure—Lou into an alliance for the good of all?

Steve understood that customers paid our bills and that only satisfied ones guaranteed our future. In this regard, he was closer to Lou than Bill and understood Gerstner's reason for insisting on a humble customer-driven service culture. I had briefed him and the other MS participants on this subject in-depth and in detail. But none of them fully perceived the radical transformation IBM was painfully suffering through right before our eyes. To save and transform IBM and make his new service concept stick, Lou Gerstner had surrounded himself with a number of minions he knew and trusted from former work experiences. Most of his new team had no axe to grind with MS; the old crew who had lived through the divorce and the later rocky relationship was nearly all gone by now.

So there we were, expectations brimming. Would the titans clash or understand each other and bridge the gap? I noticed no personal hostility between the participants; everybody behaved. But real bridges of understanding weren't built either. Lou came across as an empathetic GM type contrasting Bill, the futuristic visionary. Bill engaged him several times in technology discussions, but neither Lou nor his entourage could or wanted to follow along. Surprisingly, IBM's white-haired software chief, Mr. Thomson, did not engage. For Paul and me, he was a true disappointment. He appeared to know budgets better than technology. Lou was talking in broad terms outlining business principles. His statements were kept ambiguous and not concrete enough for the MS team to seriously engage. The smallest passion for an alliance waned as the ticking clock started running out.

Steve behaved well and did not reveal his still-undiminished animosity; he tried hard to understand how these broadly worded and nonspecific suggestions could lead to a new form of cooperation. The meeting ended after approximately three and a half hours, and Paul and I got assigned the task of following up. We did so over several months with different groups inside IBM, but in the end, nothing concrete ever came out of our enduring dialogues.

No wonder we failed to find common ground to foster each other's businesses. Lou Gerstner had put IBM on course to restore her lost leadership position. He got lucky betting on C-MOS technology for a new mainframe generation, although the real change he introduced to IBM's culture—true to his former experience—was a new understanding of how to address and gratify customer needs. In the past, IBM's consulting organization had refused to engage in service contracts for non-IBM products. Lou Gerstner modified this long-entrenched policy. If expertise was available, at a price, any and all IT projects were accepted, and customer preferences were no longer questioned. MS's top management listened but could not believe that IBM had turned vendor neutral and was, from now on, willing to treat our products on equal footings. Had the ongoing war clouded our judgment?

We knew the Internet would demand opening up and embracing non-MS standards. But our primary goal was still to defend and cement success for generations of Windows to come. This centric view was causing a bunker mentality, hindering Steve and Bill to trust IBM's promise. In a nutshell, you could say MS was still tapping bricks into her wall of defenses. IBM had started to dismantle hers to win customers back.

Examining the wide gap of practical and philosophical differences between two leaders allowed me to understand why no common ground was found. Bill, fond of his iconic image, presented himself as an esteemed visionary. Lou Gerstner gave the impression of a down-to-earth pragmatist. He was fully respected and at times feared by his employees when making unpopular decisions. On the other hand, Bill always looked for admiration and avoided making the tough decisions, leaving it to his subordinates.

Gerstner once told journalists that IBM did not need yet another strategy, which the press totally misunderstood and therefore ridiculed. In his book, he explained the need for strategies to be "short on vision and long on details." As a consequence, solid strategies "start with a massive amount of quantitative analysis." An anointed vision and gut feelings, to my consternation, were often the centerpieces of ours. In my opinion, Lou Gerstner was correct when stating "vision statements can create a sense of confidence." They "are for the most part inspirational and play a role in creating community and excitement among an institution's employees." Bill and Steve reveled in them with convincing conviction. Lou played them down, considering them a necessary evil.

Observing their exchanges during our legendary meeting, I concluded the core reason why they misunderstood each other: they lived on two different planets. Both were running a successful business, though with dissimilar philosophies and principles and with antithetical disciplines. Lesson learned, mission not accomplished! The two limos idled out from the alleyway entrance and drove off into the late afternoon Chicago sunlight back to the same airport we had both arrived from and still in two vastly different directions.

Shortly thereafter, IBM appointed a new leader for the PC group. After the failed attempt to bring the big boys together a couple of months earlier, I made absolutely certain to meet Sam Palmisano at once and explore how we could improve our working climate. The rumor mill had it that Lou G. was preparing Sam to succeed him. I was full of anticipation and not at all disappointed. Sam was a class act of a manager, and we understood each other immediately. He was polite, to the point, and wanted to explore possible cross-company synergies on a small scale. Sam cracked a couple of jokes about the darker past while, at the same time, obviously wanting us to bury the hatchets. We touched briefly on the failed meeting, and he flat-out told me that Lou had considered it a waste of time. There was no reason to believe a grand alliance would come together in the near future. I appreciated his point-blank honesty. He appeared rather close to Lou, and perhaps the murmurings about Sam as his successor were right on. In retrospect, they were. We agreed to leave the past behind and make sure IBM and MS could at least work constructively on new product introductions.

The first opportunity on the horizon was the launch of Windows 98, at that point less than a year away. I gave Sam my word to fully support him. As I left, he enlightened me on how IBM had survived her antitrust battles with the Feds in the '80s, and he wished us luck accomplishing the same.

As congenial as our meeting was, it did not prevent IBM from supporting competitive products like Linux and Java or aiding to unite the industry against MS and help the Feds in any way possible to condemn us. I couldn't care less; I wanted IBM to preinstall more Windows as long as she manufactured PCs making life as easy as possible for her PC division. I wrote Bill a short e-mail about my meeting, and he responded favorably, saying, "Make Sam your friend. If he indeed becomes CEO one day, I will have to meet him, and it would be nice to finally find some common ground." He never gave up!

IN THE SHADOW OF THE FEDS

THE ONLY MAN BILL GATES MAY FEAR

This was how, in August of '97, the partial headline in *Wired* magazine read. The article that followed was not the only one I had read about the underground conspiracy coalition led by the lone ranger of antitrust justice, Gary Reback, generously funded by our competitors. I understood the word *conspiracy* to mean "a secret coalition of individuals with a common goal." Theirs was to influence the DOJ leadership and sympathetic politicians to relentlessly pursue MS and find an avenue to classify our vigorously competitive behavior violating antitrust laws. In barrister Reback's view, we were a danger to society. As quoted in the *Wired* article, "It's not just our lunch, but our carcasses that MS wants to eat. There are grave societal consequences to that strategy and a whole new wave of opportunity that MS wants to suck into its OSs to maintain its monopoly." Pretty strong and outrageous allegations if you asked me—particularly for a lawyer.

His first actions in his fight against MS had been to deliver a brief to Judge Sporkin's court, rejecting the 1994/95 consent decree as inadequate. When the judge's opinion was reversed and he himself was removed from the case, the Reback group did not give up lobbying the Feds. She gained further momentum when Netscape funded her to investigate our Internet strategy. On June 23, 1995, Mr. Reback fired off a letter to the DOJ, asserting the Internet was a new platform rendering OSs useless. He cited statements made by MS executives and referred as proof to meeting notes he had obtained from Netscape. Concurring with the lofty opinions of his clients, he then demanded MS, the OS powerhouse, should be forbidden to compete in the Internet arena with an integrated Windows browser.

His arguments and actions were straight along the lines of Senator Hatch. We need to protect start-ups like Netscape, introducing disruptive technology. The DOJ's initial response to his letters was lukewarm, encouraging him to lead his group into a guerilla war as he explained in his book *Free the*

Market!. According to Netscape's general counsel, he was sent "to be the government's worst enemy. Criticize them everywhere. . . . Until they sue MS. Then you will be the government's best friend." Certainly a thought-provoking tactic—one reminiscent of the magnanimous ethical standards so many lawyers publicly bear forth into the world!

Let's listen to him again to fully understand his provocative aim: "Microsoft's enemy is not Netscape. It's the consumer choice." Obstructions of consumer choice—a nice, catchy accusation and line for the sensationalist media market, but did reality truly support this? Think about the number of reasonably priced Windows applications available by then, the millions of dollars MS actually poured into supporting competitors to develop them, how OEMs could add browsers to their heart's content to the Windows desktop, and let's not forget all the other vast information choices Internet-connected Windows PC users had.

Yet Reback's relentless quest consigning the MS case to the DOJ continued. By now, the antitrust division had a new boss in Joel Klein. Enlisting the help of Susan Creighton, the conspirators dug out an old Supreme Court ruling, enlisted Robert Bork—a judge with controversial reputation after not making it to the Supreme Court—and got the Texas AG to start an investigation. The group used public statements by MS officials and depositions from the Texan investigations to further beef up their alleged monopolization case. Most of their accusations were boiled down into a still-secret 220-page document funded predominantly by Netscape. The DOJ staffers quaffed the poisonous brew with gusto and issued a document demand to Netscape and MS. A new investigation was on!

Giving IE away at no charge was the next contention. The conspirators cavalierly labeled this predatory. According to Professor Frank Easterbrook, an economic expert whose windy opinion the group adhered to, setting a zero price for a product was irrational business behavior. Mr. Reback described this in his book: "Predatory pricing inevitably harms consumers by eliminating competition." To further prove his allegation, he quoted the opinion of an executive from Silicon Valley Company, saying, "Regardless of how good your product is, you can't compete against free." Think about Google, Facebook, Twitter, or Linux and think again. In the end, the nefarious mercenary lawyers backed away from the predatory pricing theory. The phrase, while catchy, was hard to prove!

Even Steve Jobs got engaged in the conspiracy by inviting Joel Klein to visit his headquarters and telling him to keep MS tied up in litigation, enabling him to gain share with the new iMac PC. Bill had mocked his new baby in public, so he joined the witch hunt and not for the last time.

Tallying it all up, I estimate that the Feds got a nearly $10 million worth of indirect legal freebies from the conspirators paid for by our competitors. Sanctioning their work, the DOJ eventually built a labyrinthine case against us while Garry Reback gained considerable praise—and not to forget fees—for sinking his thirsty mandibles into the warm host hide of evil pioneer MS.

Back in Jackson's court

Sam Palmisano had foresight. Shortly after we met, the Feds, buttressed by Reback's briefs and their own investigation, filed a motion in Jackson's court, accusing MS of violating the 1994/95 consent decree. Most employees, in particular people working for me, could not believe what they read. "Microsoft unlawfully maintained its monopoly by using exclusionary and anticompetitive contracts to market its PC OSs." Allegations and accusations—none of them proven—fueled the fire of public opinion and made my cohorts feel like criminals.

Our contracts were approved by our attorneys, and if they had been exclusionary, they would have never passed muster. On the other hand, being labeled anticompetitive sounded honorable to me. Any company struggling to win in a free market is, by nature, anticompetitive. This held true throughout the world of free commerce, from plumbers to architects and to auto manufacturers.

The Feds' main accusation was that MS was illegally tying IE to Windows. Straight out of Reback's secret play script came the following: "By forcing OEMs to license and distribute Internet Explorer on every PC they ship with Windows 95, Microsoft is not only violating the final judgment,[33] but in so doing is seeking to thwart this incipient competition and thereby protect its OSs monopoly." Therefore, the Feds asked the judge to find MS in contempt of court and order her to cease her tying practice or pay a one-million-dollar daily fine.

The Feds based their petition on the following clause in the consent decree: "Microsoft shall not enter into any license agreement (with any OEM) in which the terms of that license agreement are expressly or impliedly conditioned upon: (i) the licensing of any Covered Product, OSs Software or other product (*provided however, that this provision in and of itself*

---

33 The '94/'95 consent decree

shall not be construed to prohibit Microsoft from developing integrated products)." In other words, IE was supposedly just bolted onto Windows to swart—according to the Feds—competitor Netscape without deriving any valid consumer benefits.

Condemning the integration of IE into Windows ran completely against common practice in our industry. Software products live with perpetually integrating meaningful features to increase product values over time. Otherwise they die. Reading the above quotation correctly, in particular the section emphasized in italics, what we had done seemed expressively allowed under the decree. Were the Feds going back on their own word? And reading it a second time, the reader might agree with the opinion of a clever analyst once quipping humorously, "There is no limit of what MS is allowed to integrate into Windows—even a ham sandwich."

The lawsuit came as no surprise. The Feds had contacted us a few months earlier demanding the separation of IE from Windows. Our legal team had made our interpretation of the '95 decree unwaveringly clear and refused to relent. The DOJ's action was a direct result of MS's refusal to deal.

After both sides made their points in court, Jackson ruled that MS was not in contempt of court. Nonetheless, he issued a preliminary injunction ordering us to deliver an IE-free version of Windows 95 to OEMs. Bill Neukom, our general council, commented about his ruling that same day: "The court denied the Justice Department's petition for contempt; the case should have ended there. . . . It is inappropriate for the court to unilaterally expand the case beyond the scope of the government's petition."

We had no choice but to comply and went ahead separating IE from Windows, with the predictable result of breaking it. With discomfort and a diminutive smirk, I signed a letter accompanying its deliverance, warning OEMs of the separated code being defunct. Not what the judge had had in mind! Steve, dead set against poking Jackson in the eye, was overruled by Bill, who gave the go-ahead to cripple Windows—as he thought it was ordered.

The baffled Feds, reacting at once and with fury, hauled MS back into court. Agitated and visibly angry over our audacity of following his order literally, Jackson let it out on David Cole, then the Windows product manager. He absorbed his rap with stoic grace. MS had told Jackson in no uncertain terms beforehand that Windows ex IE would not and could not work because they shared common code. Not good enough for him. He had to try himself. In open court, one of his clerks removed the IE icon from the Windows desktop, which every user could do. In Jackson's belief, he had just removed IE from Windows—a full-on charade. A mouse click restored it. Jackson's courtroom theatrics only served to emphasize his complete misunderstanding of software-integration technicalities. It took another month to work out a preliminary compromise. Eventually we agreed to produce a functional and partially IE-disabled version by permanently removing only its icon.

I had no desire to deliver such a crippled version to OEMs simply to please a disgruntled judge and the now-victorious and self-congratulatory Feds. The compromise made no sense to me, and OEMs confirmed my doubts: none of them ever installed it. The Internet had taken hold, and a Windows-powered PC without its native browser was no longer considered sellable. Thus, market realities made the agreed-upon compromise meaningless.

To convince the judge to make his preliminary ruling final, the Feds tried to prove that we had bolted IE onto to Windows just to swart Netscape. Their investigation had unearthed a humungous number of memos and e-mails, which they believed substantiated their accusation. At the end of '96, Jim Alchin, our SVP of Windows development, had written, "I do not believe we can win on our current path. Even if we get Internet Explorer totally competitive with Navigator, why would we be chosen? They have 80 percent market share. My conclusion is we have to leverage Windows more." He added, "We need something more: Windows integration."

With Netscape gunning for 100 percent share according to public statements made by Mark Andreessen, a cofounder, the Feds should have applauded MS for taking on the browser juggernaut. Jim Alchin's skepticism and anxiety about the potential unpopularity of his browser baby was typical for a software engineer. A bit of an introvert, like most of them, he was nevertheless humble enough to question the success of his own design. Bravo! If Jim further meant leveraging Windows equaled tighter integration, nobody should have been surprised. First of all, this was the plan we were following since '93, and second, Jim knew no better! Any software developer, responding to cutthroat competition and, in particular, when coming from behind, intuitively knows integrating more features increases value and ultimately benefits customers.

After reading Jim's e-mail snippet, the Feds perceived it differently and concluded, "Everybody saw it; they [sic: MS] really captured the essence of what we believed." Mr. Malone, the chief prosecutor in the case, added, "They [sic: MS] had that power." Power for him translated into MS having monopoly power. Our method of improving products through integration was therefore deemed illegal. What would he attack and forbid next?

In early May of '98, Jackson made his preliminary ruling permanent and extended it to 95's successor, Windows 98. MS appealed. The clock was ticking. With the launch of 98 scheduled for the end of June and no contingency in place to deliver it sans IE, we were in big trouble. My customers needed four to six weeks' prep time to install it and get backup CDs and documentation produced. Not knowing if and when to expect a ruling and how favorable it might be, it felt like we were playing Russian roulette.

Fortunately, the court responded with surprising alacrity and, on May 12, lifted Jackson's injunction for Windows 98. Studying the ruling meticulously, you could easily conclude it would probably hold up for 95 as well because the appellate court reasoned that courts were ill prepared to take upon themselves entrance into the arena of software design. A clear signpost for the Feds to back off? Not really when reading in the same ruling: "The parties [sic: MS and Feds] agree that the consent decree does not bar a challenge under the Sherman act."[34] And that was the comment the otherwise-disappointed Feds unfortunately took to heart most.

While we were jubilant, we nevertheless harbored deep fears regarding their next move. Joel Klein, the DOJ's antitrust division head, immediately recognized that winning the pending 95 case had only a very slim chance. Conferring with Janet Reno, his boss, he received authorization to broaden the litigation by accusing MS of wide-ranging antitrust violations. The Reback gang had finally hit pay dirt. Informing MS's legal team of the developing situation, Joel Klein expressed his desire to settle the case only if we would make further-reaching concessions. Once again, Bill, together with his legal team, engaged in transcontinental shuttle diplomacy to avoid a costly and prolonged antitrust battle.

On May 17, 1998, I received an early morning phone call from Bill at home. Calling from DC, he detailed how he and his legal team had convened with the government attorneys several times since the court had lifted the Windows 98 injunction. In return for not filing the next suit, the Fed's demanded that MS drop IE from Windows and open up certain nondocumented Windows application programming interfaces to all independent software vendors. Our ensuing phone conversation centered on eliminating IE from Windows. The other issue was for Bill to decide. The request to rid Windows of IE meant that the DOJ was slipping its long prying fingers right through a crack in the workshop door to mingle with our product design. My key question to Bill: what would be the next feature on their list of objections? He didn't know but that agreed caving in would be only the beginning of further government interferences. Or as Churchill once said, *the end of the beginning*—for our ability to compete!

---

34 The antitrust laws congress had passed in 1890

Resolve and Fortitude

Bill went on to describe that he had solicited Warren Buffet's advice, who had asked him to yield and compromise. Remembering my German-army education, I disagreed. There comes a time for soldiers in combat to understand the vital importance of taking a hill, regardless of the consequences for their own lives. It's a tactical issue, though often a moral one—testing a person's own courage and commitment to the cause alike. You see the hill, it's the right thing, you summon your resolution and reserves, and you move on to conquer it—bullets or not. I considered this an equivalent situation. We undoubtedly had a noble cause! So I just said to Bill, you cannot let the government determine what underwear you can't sport. You will have to stand up for your right to design products as you see fit. Good products are winning products; government-compromised ones suck. "Take that hill, Bill!" He answered mumbling, "I hear you," and left me guessing what he would finally decide. I had given him my perspective, though I remained unsure if he had the resolve and fortitude to follow through.

I had observed Bill making pretty tough statements in a number of situations, only to compromise hours later. There is nothing wrong with such behavior; I have no doubt executed similar reversals in my business career. My People's Republic of China MS-DOS license negotiation comes to mind. The current situation the company found herself in was different, and not just a simple business negotiation over a couple of dollars here and there. We had to defend an ironclad principle MS had been built upon: the freedom to design successful products. This was our most noble cause, a hill we had to take at all cost. I couldn't locate any middle ground Bill could have settled for.

During our phone call, Bill had reiterated his fear of history remembering him as just another Rockefeller. Reflecting on this comment later, I concluded that I had probably given him poor advice. If he harbored fear for his reputation and his legacy, stalwartly standing up to the Feds began looking daring. Bill would need to let go of his personal feelings and trust his integrity and the compelling nature of his cause. I am unclear with how many other people he talked to during that eventful day and just how much he valued my input. In the end, no compromise was offered, and as promised, the DOJ filed a new, broad, sweeping, and historically profound antitrust lawsuit on May 18, 1998. Knowing the outcome today, I would still have given Bill the exact same advice.

The new filing, lastly, affected a number of people in MS the wrong way. Despite feeling greatly enthused about the upcoming 98 launch, the renewed attack left people depressed or uncertain about the outcome of the looming conflict and fearful of how much real peril it posed. Again I was pressed to field numerous questions in the hallways and over water coolers, e-mails, and phone calls. I radiated optimism and confidence that we could defend ourselves successfully, which helped calm people down.

The mood in the company changed significantly just seven days later when the court issued a 2:1 decision in regard to Windows 95 in our favor. In a nutshell, it said, "The preliminary injunction was issued without adequate notice and on an erroneous reading of the consent decree. We accordingly reverse and remand." The court plainly and soundly overruled Jackson in toto. In very few instances had this magnitude of a reversal ever been levied. The tone throughout the twenty-one-page opinion was harsh and embarrassingly cast Jackson in an awkward and incompetent light. But it was not unanimous—reason to worry?

The most promising fact was the court's acknowledgment of MS's integration power. It found "consent decrees are generally interpreted as contracts," and the ambiguous provision under dispute needed to be judged on the parties' original intent. The court explained that the decree was issued because Novell/DRI back then had sternly opposed the integration of MS-DOS and Windows components into Windows 95. By explicitly permitting this integration, the appeals court found that the Feds had sanctioned "any genuine technological integration [sic: by MS]" into 95 and what we had done passed that test. An incredible total victory for us! In a nutshell, Jackson had failed to perceive the case as a contract dispute, obviously eager and motivated to muddle in software design. In this regard, the appeals court reminded him, "Courts are ill equipped to evaluate the benefits of high-tech-product design." Would this ruling, even as it was not unequivocal, strengthen our position in the upcoming trial and make us head for the victory lane again? I was hopeful.

Not appreciating the harsh ruling and its tone, Jackson was deeply hurt. He never quite regained his emotional state of mind, as witnessed by his closest friends and a few journalists. Yet the saga would continue with him in the center. He was down but not out. In a fresh sweep of irony, one week later, he was picked by the DC district court to preside over the freshly filed antitrust case. Either an enormous judicial blunder or good luck, we will see!

Numerous observers claimed that Jackson, appointed by President Reagan, was a conservative judge. No activist on the bench, Jackson professed he loved free markets and fierce competition and was in general opposed to an overreaching government. But he for sure did not appreciate competition in the legal arena, the one he had to bitterly endure when the appellate court relentlessly picked his final judgment apart. The new and complex antitrust case would challenge him further. In fairness, I will say

Judge Jackson is a human being, and his battered emotions about the blunt reversal are understandable. He could have recused himself from the new case. Yet a flickering fire burning inside him made him go on doing his duty as he wanted to understand it, overshadowing the legal journey ahead of us.

Internal Politics

Despite the legal wrangles, my business was exploding. For fiscal year (FY) '96, ending June 30, my group had earned about half of MS's profits. We were riding the Windows 95 and IE 4.0 combo wave. One year later, we got close to surpassing $5 billion in total revenue, an annual growth of over 50 percent, exceeding my projection by $650 million. I had mentioned this possibility to my skeptical bosses once or twice before. They responded by shaking their heads, expressing I was nuts. Close to the end of that FY, by mid-June we were approximately $350 million short of my personally stated goal, with time rapidly running out. The money was there—I was convinced! Salespeople, who beat their FY revenue objectives early on, tended to slip invoices into their drawers and hold them back to sandbag next year's results. They had done this before, and we had always clamped down on them. They tried anyway. I instructed my controller, "Find the money," and within forty-eight hours, she discovered nearly $300 million in delayed invoices. As the deadline approached, more royalty reports ready to be invoiced were revealed, enabling us to finish the year $20 million over target. We celebrated with pride and deep satisfaction, relishing an astonishing achievement.

The response from our CEO was markedly different. Bill immediately proceeded ordering the corporate controller to conduct an audit into what he called "revenue invention by the OEM division" without warning me. This was his prerogative, no question, though another sign his management style was being impacted by outside events. Our relationship over the years had been built on trust, but now that rare and historically effective essence was eroding.

I knew our result was unimpeachable; I shrugged it off and told my controller, "Let him bring it on!" We actually could have done an extra $150 million, which I decided to hold back—call it my personal sandbagging privilege. They came in handy the following FY. As expected, nothing came out of the audit. All numbers checked out, though I personally never received any praise except from my executive colleagues, who wondered why their shared annually bonuses were so large.

The company had always been cautious in projecting business revenues. Our profits had earlier come up at the hostile Senate hearing, and follow-up articles had labeled them excessive and typical for a monopolist. Producing record results was therefore suddenly considered inappropriate—at least for now. What counted for me: we had done it! Delaying revenue would have only worked for one quarter anyway, so why pile it up too much? It was a golden time of great accomplishments, of self-sacrificing teamwork, and extraordinarily dedicated efforts. In our own little moment in time, we were as good as it gets, old team. Thank you!

The internal audit was just another sign of the company growing political. The huge OEM revenue jump aroused jealousy inside the company. Friends of Steve were covertly working on changing the way revenue was being recognized. They'd been trying for a while, but this time their pleading took effect. As a first step, Steve decided to allow the enterprise sales group to include all versions of Windows in her sales portfolio. From now on, these types of customers had a choice where to obtain OSs from directly—from MS's enterprise sales group or through my OEM customers. Steve was convinced his decision would accelerate our Windows NT workstation share drive. In my opinion, it would mainly lead to discounted OS prices for large enterprises. And it did!

Maybe the sudden change was inspired by placating the Feds in showing less revenue and profit for the OEM division! Being pretty annoyed with Steve's unilateral action, I argued my case hard. I showed him the math; he just shrugged, not wanting to lose face. His decision stood. I told him he was the boss and I would not question this in public. He had blatantly taken away part of my business without proper analysis or consulting me beforehand. Politically motivated as I thought and his to justify!

I was not upset enough to quit or, God forbid, challenge him in front of Bill; I knew better. I struggled explaining his decision to my people as we continued to go about our business. Under these new considerations, the hardest task was preparing next year's budget. No one could predict with any degree of certainty how much business would stay in the OEM group and what would wind up over in the enterprise sales group. As a result, both groups sandbagged their revenue predictions. In turn, fewer resources were allocated, and we missed out on valuable sales opportunities. If this was Steve's intent, he truly managed it to shining perfection.

It would have been infinitely more effective had Steve formulated a business objective for reaching a desired goal and let the division leaders hash out a strategy to achieve it. In doing so, we would have worked closely together, shared resources, and coordinated marketing and sales campaigns. Because of his ivory-tower decision, the gap between the two groups widened, became personal, and turned into rivalry. Steve was learning the hard way.

The not sufficiently vetted decision caused real harm in regard to OS piracy. Product deliveries to enterprise customers had traditionally been less guarded than the ones through OEM and retail channels. Enterprise customers were trusted to report product usage based on an honor code. Checks and balances were nearly totally missing. While most reported honestly and secured the unprotected master copies they received appropriately, some abused the system. Unprotected masters found their way into the open market, enabling pirates to replicate products at will. Over time, certain reporting and replication rules were tightened but never sufficiently, leaving us vulnerable.

The business must go on

Windows 98, finally

Launched on June 28, with the slogan "Works better, plays better," 98 was unmistakably stable and consumer-centric. Allowing for faster program switching, it added DVD, USB drivers, and two-plus gigabyte disk partitioning as core features. Its most controversial aspect remained the inclusion of a nonremovable version of Internet Explorer 4.01. Emboldened by the appeals court ruling, we felt free to take even this last amount of freedom away from end users. (I never understood why!) Adding insult to injury for our loathing enemies, it left the Feds speechless. While the channel bar I had lobbied for earlier soon developed into a minor flop, 98's dynamic HTML[35] engine allowing independent software vendors to easily integrate Web-browsing abilities into their applications was widely praised. A preview in the *PC Pro* journal applauded the updated browser enthusiastically. IE now allowed easy to launch and customizable files and folders to act as hyperlinks on the Web, making them totally transparent and seamless to use for information gathering. Netscape Navigator had lost its edge, and with Apple deciding to exclusively bundle IE with all her Macs, Netscape's last main bastion fell.

Despite the seams-bursting dot-com bubble showing early signs of peaking, my business continued to grow nicely at an impressive rate of 20-plus percent—though considerably less than the impressive above 45 percent, which had catapulted my division through the mid '90s. Exceeding revenue and profit projections, my group continued to show a nearly flawless performance.

We had beaten DRI and IBM and had backed traditional versions of UNIX into a niche position. Apple had remained a formidable competitor over the

---

35 Hypertext markup language, the most used programming language for Internet applications

years but was hampered by the never-ending turmoil in her executive suite and her pricing philosophy. After Jobs had been ousted in '85, the company had gone through three CEOs before Jobs got reinstated in '97. By then Apple was licensing her Mac technology—the exact move we had always feared. This was authorized by Mike Spindler when he was Apple's CEO and was continued by his successor as a mean to gain market share.

Years too late and fortunately for us, the program failed miserably. Most analysts believed our Windows platform was too mature and entrenched, presenting Apple only with a slim chance to gain at our expense. The real reason the late attack failed? First, Macs were still overpriced compared to IBM PC clones regardless of who built and sold them. Second, Apple picked small, underfunded companies to build Mac clones. Her partners were further disadvantaged when Apple kept essential information close to her chest and provided only scant lip service supporting them. The Mac had a formidable application software base by then. In enlarging its manufacturer, base share it should have been rapidly won increased market share. Instead, all of Apple's licensees painted a dreary picture as I visited them, trying to license our Mac apps. As a company steeped in proprietary principles, Apple never valued the profound advantages her licensees could have brought to the table.

After Steve Jobs was reinstated as Apple's CEO, he simply discontinued the program by not extending the deal to the new iMac architecture he invented. Bill was very skeptical about its spiffy design. More aggressively priced than before ($1,300 per unit), it nevertheless sold eight hundred thousand units in the first twelve months and took market share away from Windows. While this hurt some and started a still ongoing trend, the greater danger for us loomed surprisingly in the UNIX corner.

SOFTWARE MINUTEMEN

"Linux is a cancer."

Because it attaches itself in an intellectual property sense to everything it touches, as Steve Ballmer once formulated. Until now my OEM business wasn't much affected by that emerging phenomenon. But in late '98, its threat to our desktop business took a dangerous turn. There were indications that the free-of-charge Linux OS was being bundled with recycled $500 PCs in developing countries. Then plans emerged to preinstall it on low-powered notebooks[36] and Oracle's planned N-cube system priced under $200. Scary scenarios for us, with Windows average price being $55 per unit.

Let's quickly examine why Linux attracted so much speculative attention and support from top-notch companies. As the name and letter combination suggests, it derived from UNIX. A misleading assumption, but in itself the right evocation! Its inventor was Linus Torvalds, a genius of a Finnish programmer who had in '91, while studying computer sciences at the University of Helsinki, succeeded in creating a UNIX OS clone from scratch. Written mostly in the C programming language, he adapted his clone to the Intel 80386 CPU. However, he was not the first person to achieve such a feat; Dutch professor Andrew Tanenbaum had preceded him, naming his cloned version MINIX. While the name MINIX suggested a small system (mini-UNIX), the name Linux derived from Torvalds's first name, Linus, and as such is a nice play on both his and the UNIX name.

Originally, both contained only essential bare-bones components to make them simply functional. They were operated with a command-line language, comparable but harder to use than the ones found in MS's or DRI's DOS versions. Both releases were distributed in source code, unlike the distribution of Windows using binary code. This offered students and hackers around the world an opportunity to study them in-depth and modify

---

36 Often called netbooks or thin clients

or enhance them con gusto. Torvalds's implementation had a monolithic architecture allowing multithreading and contained a task-switching feature. The only criticism of his implementation: tying his 10,200 lines of code so specifically to Intel's 386 did not bode well for portability to other CPUs. MINIX, at roughly the same size, was a more modular system, making a later attempt at portability easier. After exploiting both, the hacker community judged Torvalds's implementation as more complete, refined, and state-of-the-art. Therefore, Linux eventually found its way into everyday use while MINIX remained a classroom tool.

Torvalds had offered his code royalty-free. Could developers redistributing Linux, after making enhancements, at least make money from the handiwork they contributed? Enter the world of Richard Stallman, another genius of a programmer and probably the world's most outspoken free-software ideologist. When working at MIT's[37] artificial intelligence laboratory, he became fed up with the stringent nondisclosure agreements he encountered—similar to what the rest of industry was using—and decided to free the world from such restrictions. In '83 he published a Karl Marx–style manifesto declaring software to be free merchandise. Free as in freedom. Stirring the American soul as a revolutionary thinker, he subsequently went beyond his original idea of unchaining software from commercial constraints. Advocating any software code should be allowed to be modified by everybody, he effectively outlawed the sole distribution of binary code. Going beyond, he radically demanded all resulting code from derivative work on royalty-free software to be dispersed in the same spirit and terms—without dreaded royalty shackles.

By '91 Stallman had published a second version of his General Public License (GNU[38] GPL). It put his intent into crisper legal terms. Torvalds, hesitant at first, followed Stallman's philosophy and agreed to license Linux under the newest GPL. Later, industry observers considered this move essential for the breakthrough his OS kernel eventually achieved. For profit, MS was not amused. Steve labeled Linux a cancer, because all later improvements were in the public domain and could not be explored monetarily.

Linux and its add-ons, freed from royalties and with no other strings attached, woke up dormant talent and created novel opportunities for fame. Think about developers getting hold of it and in their spare time adding significant improvements. Acceptance accolades and fame within their community occurred at Internet speed. Observing the trend and hungry for the same recognition, hundreds of thousands followed trying

---

37 Massachusetts Institute of Technology
38 Stands for "Gnu's Not UNIX," a complete and freely distributed UNIX-compatible OS

to outdo and outimpress each other. Humanity's competitive spirit at its best! Suddenly, working on Linux was fashionable and cool and its minions prolific! Soon the merry army of authors gleefully rebelled against the established fortresses built by the likes of Bill Gates, Larry Ellison, Rob Glaser, and other billionaires of the software industry. No need to throw a pie in their faces anymore just to gain attention; here was an effective, exciting, and peaceful method for actually dethroning them.

In a short amount of time, the legions of volunteer worker bees attained astonishing breakthroughs and software stardom in progressing Linux protuberantly faster than Windows. When the first complete version was released in March of '94, the system had grown to 176,000 lines of code. A year later, it doubled in size, and by January of '99, the system contained nearly two million instructions. Today, as I write this, the system comprises of ten times as many. Developing it commercially from scratch is estimated to cost nearly 1 billion US dollars. The volunteer movement initiated by Stallman and fueled by Torvalds's release had been a truly amazing—hats off—ride for the software minutemen.

Smartly and stringently, Torvalds, now considered both a hero and a bottleneck, managed to maintain control over his various core releases. The gurus surrounding him restlessly wanted to move faster. Torvalds's demand for quality and his scrutiny for robust design held them back. Keeping careful control of the process, his fastidiousness succeeded, rewarding him not only with community recognition but also with speedy client feedback.

By '96, Linux had become an established and reliable server product. In '99, when IBM jumped on the bandwagon, releasing mainframe patches, Linux gained enterprise-class computing honors. Sidelined, our own server group envied Linux's success! Against initial odds, it had been ported to a variety of other architectures. My own group, with no urgent focus on selling server software, so far had encountered limited Linux competition on desktop PCs or notebooks. There were consistent rumors of Linux finally coming of age and becoming easier to operate. Installation and tuning it nevertheless remained no job for novices. The tools available were impressive yet harder to use than Linux itself. Suddenly, though, a change was on the horizon, as commercial outfits jumped into the fray alongside the volunteer network. Hewlett-Packard, IBM, Intel, and the Linux distributors were eager to expand its use into my group's core territory, making me mischievously wonder how to compete with royalty-free software.

From its first stirrings, we closely observed the free-software movement. Bill could not stop himself from exchanging barbs with Stallman in public and vice versa. How could a bunch of hackers deliver acceptable, quality code? MS was proud of her keenly managed developers, her smart testers, and her adopted quality standards. Initially, we regarded the volunteer workforce as a fleeting anachronism. Before long, though, these deceptively disorganized gunslingers were tossing out consistent surprises. As human resource managers know, money is not the greatest or most enduring motivator; I actually believe it is none at all. Reaching guru status within the community was a much more powerful incentive—similarly observed inside MS. With Torvalds's iron enforcer fist in the background, which nobody dared to challenge, this archaic system produced excellent software—though for the moment only for servers.

To make Linux successful on consumer PCs, the ease-of-use aspect needed addressing. Oddly, and luckily for us, the volunteering experts were so far not terribly keen about spending time remedying that obstacle. Loving an intellectual challenge, they had no problem operating Linux effectively. Nobody had ever made it easy for them; why couldn't you assume a certain skill level at the consumer level? Or, quoting David Eisenberg, "The typical end-user sees the computer as a means to an end . . . A typical Linux user, on the other hand, sees the computer as an end to itself." This made ease of use a lesser priority for the volunteering gurus.

The danger of a mind change remained. It would only take one serious software company to produce a consumer-friendly version of Linux. We estimated the effort to be in the neighborhood of thirty to fifty man-years. Cloning Windows completely would take more effort and manpower, but there was no reason to believe that it couldn't be accomplished in less than two years. If a Linux-based Windows clone managed to hit the market, its impact would be severe and, according to several experts, take significant business away from us. We had to win two battles: Limit the damage Linux was doing to our server business, and make sure the Windows desktop franchise would not suffer irreparable harm. Don't rest; move Windows forward by adding complexity through innovative features. Making it a progressing and therefore harder-to-attack target was exactly what our carefully considered counterstrategy defined by Bill called for.

In November of '98, we published a comprehensive Linux analysis for internal MS considerations. It managed, though, to wend its way to the Linux community. Its spies were everywhere! The volunteers eagerly added their criticisms and published the research paper on the Internet, naming it the "Halloween document." The author classified Torvalds's release policy as a hindrance to Linux's success. Linus only controlled a small portion of the code, the so-called kernel. The rest, added later by other entities, varied and therefore led to several versions of Linux being marketed. Companies sinning in that fashion were Linux's main distributors Caldera, Red Hat, and

Apache. As they tried to differentiate their offerings that way, they kept ISVs guessing if their applications would run on all published versions.

Along with their varying core offerings, most Linux distributers bundled application packages like office productivity applications and databases in addition to browsers and a variety of interfaces. All of them at no charge! Major software and hardware companies were now involved, helping Linux to succeed. Helping the enemy as we labeled it! By '98 the estimated user base of Linux exceeded 10 million systems, with more than 98 percent of them deployed as servers and the rest as developer workstations. The report found feature and execution speed parity between Linux and other UNIX systems—amazing. Successfully chasing taillights in the fog of the server and workstation market, Linux had become the undisputed UNIX-class volume leader.

Could this success be repeated for the consumer desktop market? There was no doubt whatsoever it could, especially with the complicity of companies like IBM or Hewlett-Packard. The hypothetically devastating impact kept Bill and me up at night. The substantial support the esoteric but expanding community was now enjoying made MS its prime target. A more sovereign and efficient tactic, perhaps, than trotting through the mouse maze of the courts and filing expensive and ill-founded antitrust claims! Inasmuch as the minutemen rose up against the English, we had to expect that the software revolutionaries were out to obliterate Windows, our crown jewel.

On the face of it, Linux was free of charge. A misnomer. Distributing and supporting it, however, required funding. Windows came with free support supplied by us, our OEMs, or both. The nexus of Linux support services was where its distributors derived their revenues from. Like MS and other for profit software companies, Linux distributors were listed on the stock market. It was the same story on a different slice of bread; free lunch had always been more expensive than advertised. For a simple desktop system, its support fee, including the right to use an information database, today ranges from $40–$50 annually. For servers the fee can reach several thousand dollars. Assuming a desktop PC system lasted three years, a user would then typically—over its lifetime—pay close to $150 for support. Assuming OEM prices for Windows remained stable, this compared unfavorably to—support included—$55 for consumer Windows and $100 for enterprise Windows NT. Stallman's grand vision of free software for non-OS-literate consumers was not gifted to humanity as touted. It's utility and zero sum pricing remains deceptive!

Maintaining such a system was yet another less-than-frictionless task.

Installing device drivers for new devices and making them operable often required expert knowledge. While the Linux community delivered them in their dependent fashion, their quality and reliability were not as solid as Torvalds's own code. Compared to Windows, Linux was also hard to install. There was no plug-and-play technology in Linux yet, another prerequisite for the successful launch of a consumer-friendly version, but again an easily developable feature.

By frequently copying elements from other OSs, Linux developers were prone to violate software patents, making the distribution and usage of Linux vulnerable to court challenges and injunctions. Aspects free-software advocates needed to take seriously. Principally, MS used her patent portfolio defensively while others like Apple and IBM did not. There was no implicit guarantee we wouldn't change our mind. So the patent sword hung vividly over the heads of Linux developers, distributors, and users alike with the ability to be unleashed at any time. While such a patent challenge could not easily be brought against individuals, the opposite was true considering larger and established entities like enterprises or distributors. If sued successfully, a resulting injunction would bar them from running Linux applications or from distributing the OS. Enterprises on the other side were using MS products in parallel. Taking your own customers to court? We knew better. I therefore considered the patent sword a pretty dull weapon and probably ill-suited to defend our turf except against Linux's distributors.

The greater danger from the not-totally-for-free software industry derived from computer science students being trained nearly 100 percent on Linux source code. Not knowing how Windows worked internally and taking the liberal attitude of most educators into account, students therefore may get brainwashed early on in their careers. To limit their admiration of the revolutionaries and discourage them from ultimately joining them, we responded. Opening our kimono, we licensed Windows source code to universities and funded the design of Windows-centric computer science classes. Linux did not dry up or blow away in academia, but we at least created another option for learning modern OS technology.

What counted for Windows in general counted for Linux: innovation on the core platform was an essential, ongoing requirement. Linux experienced much faster release cycles than Windows. The release changes were smaller, constituting a small number of bug repairs. MS responded by issuing "service packs" in twelve- to fifteen-month intervals to keep releases fresh and automated the update process to not require any operator intervention except a couple of mouse clicks. Linux could have responded similarly, but none of these conveniences was rendered important enough by the volunteer movement.

In summary, Linux on consumer PCs was a true threat to my OEM business in the late '90s, and I believed we were ill prepared for battle. The volunteer community enjoyed tremendous public support; we did not. Love for the underdog always surpasses the one for the lead dog, especially one under antitrust investigations, enduring the dark aspersions of public condemnation! Linux's tremendous momentum appeared unbreakable, and we felt extremely vulnerable. When a Hewlett-Packard executive wrote me a letter during that time—asserting that if there would be an alternative, he would stop buying Windows from us—I could not believe my eyes. He had a choice! HP was a huge software company mainly focused on UNIX. She possessed the talent to develop Linux into a consumer-friendly Windows competitor and clone its interfaces. She could have licensed the necessary patents from neighbor Apple as IBM did and produced with her vast resources a resounding, consumer-friendly Windows-like Linux clone—in less than two years. Sold it for free and leaving us in the cold!

In a memo, I warned Bill about the industry ganging up on us in such fashion. Linux was *ante portas*; the threat was real. I suggested that Intel could be behind such an effort. They had the required software expertise. Yes, CPUs are hardware components, but the underlying embedded software makes them tick. Intel knew Windows inside and out, and a free OS would help sell extra PCs and therefore CPUs—no doubt there was a motive.

We were fortunate and extremely lucky that none of our customers or competitors, especially IBM, seized the opportunity and mounted an attack, catching us off-balance. It took until 2006 for a consumer-friendly Linux system to show up. A distributor development with not enough muscle behind, it failed after patent challenges were mounted by MS. In 2007, Dell begun offering a Linux consumer version called Ubuntu[39] preinstalled on her PCs. The most recent danger has arrived from Google's Android and Chrome systems. Based on Linux and first developed for cell phones, Android found its way into computer tablets in 2010 and has by now been adapted for Intel-powered PCs. Only time will tell where this development will eventually lead to. Dethroning Windows becomes markedly harder with every added feature, but I would never completely rule it out.

---

39 Translates into "Humanity towards others"

Price Wars

By 1998, the trend toward radically lower PC prices became reality. Consequences for Windows? A passionately debated topic inside the company. We received $55 on average per Windows-powered PCs, selling typically between $500–$1,500. Our percentage of the total system price therefore varied from 3 percent on the high end to over 10 percent on the low end. For hardware companies, not exactly devoted software fans, paying higher percentages for software was worth resisting. They took to victimized posturing. I had to defend my current prices inside and outside the company. Bill and Steve argued for reductions, potentially eroding shareholders' values overnight. The most favored and intensely debated idea was to tie royalties to PC prices. Good in theory only! I could not find any correlation between hardware and software prices! If anything, producing new versions of Windows with added features resulting in expansive code needed extra programmers paid for with inflation-driven salaries. There were production efficiencies and gains through integration available for hardware design, not so for software code. Any correlation you ever find will probably be negative, meaning as hardware cost comes down, software cost goes up. The only factor offsetting this trend was a substantial increase in PC units—yet not always enough, as we extended ever larger discounts to the most successful shippers.

What my bosses did not know: we had tried a similar pricing model once with Philips, the large Netherlands-based consumer electronics company. She had dabbled in the PC business early on but eventually shut down as the margins grew thinner. When she tried a second time, opening a manufacturing plant in Canada, we negotiated a royalty agreement with a sliding scale correlating with the suggested retail prices of her PCs. We entered into the deal with the objective of learning if such a model could work. Philips was the right experimental partner for us: her volume was low, and being located in Canada, we did not expect the news to travel fast and far.

I was skeptical, but we took the risk with a DRI threat looming. The deal rapidly spiraled into the abyss. Customarily, PCs sold together with monitors. After the crafty Dutch separated them from their systems, the average price dropped 15 percent to 20 percent at once. As we cycled merrily along, old inventory was sold off at huge discounts, making room for newer technology, temporarily reducing royalties as much as 30 percent. Philips then decided to drop her high-end lines, and the bottom fell out of our calculated average assumption. Stripping keyboards from PCs gained another reduction. As the one-year deal ran out, Philips came up with the idea of basing royalties on production cost. This time I did not bite.

Remembering our brief flirtation with a scalable model, I firmly stood my ground. In my opinion, Windows was doing basically an identical job on low-end or high-end PCs. There was no need to charge a substantially different price beyond what we had once introduced when slightly differentiating its price according to the CPU used. The only valid rationale for changing our pricing would have derived from offering functionally different Windows versions. We had done this with Windows NT versus Windows 95/98. For low-end PCs, a single-application-running Windows version was yet another idea popping up in our internal discussions. For the time being, nobody wanted to develop, in one line of thinking, such a crippled variety. But eventually, the company followed a similar path by producing a home, an advanced and a professional version, as Windows moved to a unified NT code base early in the next millennium. Until then we stuck to our guns.

Piracy in China

By '96/'97 our business with PC manufacturers in the People's Republic of China (PRC) had taken off nicely. While they by now paid for most of their OS software, local screwdriver shops did not. Finding this highly unfair, the larger manufacturers asked us to intervene and eliminate what they considered an unfair advantage caused by still-rampant piracy. How did violators circumvent our now much-tighter security measures? They obtained unsecured product copies through pirates, who used enterprise customers or hackers as their source. Unable to convince our enterprise group to tighten security requirements sufficiently or enable to catch the hackers, we set out to attack the issue by working closely with the largest PRC OEM: a company called Legend.

As she rapidly expanded, her name was changed to Lenovo, thus avoiding potential trademark conflicts outside PRC. Her first successful product was a Chinese character-set add-on board for PCs. Together with her nifty software, it enabled PRC end users to display native characters on DOS driven monitors. This earned her a prestigious award just a year before I visited Beijing in '89. Seven years later, Lenovo had established herself as the number one PC manufacturer in PRC and Hong Kong, all without exporting her PCs to the West. Her chairman, Lui Chuanzhi, an engineer by profession, was now navigating the unpredictable political waters while overseeing the company he had founded. Lenovo's status as the best-selling brand in the PRC could be credited to her younger and most dynamic CEO, Yang Yuanqing. By now, the company was listed on the Hong Kong stock exchange with the PRC government owning one-quarter of her shares.

My local OEM crew, still operating out of Hong Kong, had worked diligently and established a friendly and productive relationship with Lenovo's executive team. In '97 we managed to finalize an IP[40] agreement with the

---
40 Intellectual property

goal to reduce piracy. As in all developing countries, most PCs sold were low-end. Non-PRC companies had gained a foothold in China, though with mixed and limited success. AST Research had been an early winner. Going down fast, she was replaced by Compaq and Dell. PCs from the West were just too expensive for consumers. Most were bought for reasons of prestige by enterprises, leaving the consumer market firmly in control of local companies. The Chinese government, strong-arming the market, advised the public to buy hardware that was built locally, thus limiting the growth perspectives of foreign intruders. Ever heard about antitrust? No need for such a law existed. As experienced earlier, the government regulated unchallenged.

We estimated that at least 60 percent of the market was still in the hands of local screwdriver outfits at a nearly 100 percent piracy rate. The alliance with the market leader Lenovo intended to inspire change; other PRC OEMs watched closely. Like them, Lenovo wanted a larger piece of the pie while we desired to sell more legit OS copies. Combining forces might just achieve this for both of us. We therefore offered Lenovo a generous amount of marketing funds for special ad campaigns conditioned on containing strong antipiracy messages. A bit of a risky move without knowing if her patronage would influence consumers to take the honest road and ask for genuine PC software. Our hope: with Levono behind the campaign, the public might just assume that her main shareholder, the PRC government, had sanctioned the move.

We supplemented our push by instituting an annual crusade focused on large population centers. Here we addressed the screwdriver community in road-show-style gatherings. Whenever I visited, I spoke at them. We drew huge though not always welcoming crowds. As a prerequisite, we had beefed up our distribution system by engaging our Hong Kong–based replicators and expanding our PRC-based distribution network. We soon discovered that requiring US currency for our products presented a formidable hindrance to success. I made the decision to deviate from our long-standing policy of accepting only US dollars. Revenues, from now on, would arrive in nonconvertible local currency and remain in PRC, helping to develop our steadily growing local subsidiary. This last coup together with the effective Lenovo ads, which we complemented with our own, saw our business with the novitiate screwdriver community grow beyond expectation. Maybe buyers had gone superstitious as they suddenly demanded a genuine OS for their PCs. The PRC government not objecting worked—plain and simple. As a result, I estimated in 2001 that at least 70 percent of all PCs sold in PRC had a legit OS installed. Not bad considering we had started at ground zero back in '89. Confucius was turning over in his grave!

PRC's struggle with honoring IP rights as mentioned before reached back for centuries. The current laws on the books were murky, difficult to interpret, and hardly ever enforced. We had to get to the policy makers and enlighten them about the opportunities China was missing by disregarding IP enforcement. In '95, then first lady Hillary Clinton attended a conference in Beijing concerning women's rights and gave a pretty caustic speech addressing, loud and clear, the lack of them in PRC. Only Hillary!

Something clicked when I read about her provocative assertions. I began mulling the idea of conducting a similar incendiary conference, this time about IP, aimed at politicians, judges, and policy makers. Our local people liked my idea but were not well-enough connected to lead such an effort. I found receptive ears in an organization called the National Bureau of Asian Research (NBR), a local Seattle think tank. I had been elected to its board after I began running the Southern Hemisphere portion of MS's retail and enterprise business in '95/'96.

NBR's roots go back to Senator Henry M. Jackson's goal to provide research and information to US policy makers on key trade and political issues affecting US relations with Asia. Through MS I had helped the think tank by sponsoring database developments, donating software, and financing portions of its computer equipment. I liked the work the board was doing and appreciated the nonpartisan atmosphere of our quarterly meetings. Initially my idea was greeted with understandable skepticism. But with the backing of its astute and forward-looking leadership team of Larry Clarkson and George F. Russell Jr. and its agile and whip-smart president Richard J. Ellings, the project was approved as a worthwhile adventure. Brigitte Gort-Allen, the good soul of the organization, gets the kudos for developing and relentlessly pursuing our plan, doing an extraordinary fund-raising job and organizing the events to perfection. MS contributed around one-third of the needed funds; nearly half of them were donated by our competitors sharing similar interests, and the rest came from the Jackson foundation and my OEM customers.

Our first conference was held in '98 in Chongqing, and I was pleased to greet Lenovo's CEO as one of its participants. The second of the three big conferences was scheduled to kick off in Shanghai on May 9, 1999. Two days earlier, during the Yugoslavian conflict, US airplanes mistakenly bombed the Chinese embassy in Belgrade, killing nearly forty people, most of whom were Chinese citizens. In response, Chinese demonstrators attacked the US embassy in Beijing, and for security reasons, all US-China activities in the country were hurriedly cancelled—except one. While chaos spread and many Americans headed for the airports, former Shanghai mayor Wang Daohan, whose protégé Jiang Zemin was then China's president and party leader, met with our project director, Professor Mike Oksenberg. Fortunately for us, Mike was a well-known and respected historical figure in PRC's political circles. They had never forgotten the influential role he had played in helping to negotiate the normalization of US-China relations during the Carter administration. Wang and Mike, the two old friends, decided that this one bilateral meeting would go on as planned to symbolize that the bilateral relationship could weather an extraordinary crisis. Underlining his unwavering support, Wang Daohan demonstratively sat in the front row the morning of the ninth as our second conference commenced. Mike Oksenberg's commanding figure strode up to the podium and led a long moment of silence before launching in his welcome address spoken in fluent Mandarin. IP rights, at least for several days, took center stage. I spoke at the first and the last conference in 2000 in Beijing and came away with the adamant conviction that all of these conferences truly contributed to a better understanding by Chinese policy makers in regard to IP protection as a valuable business proposition.

I recognized our efforts were only a small flanking maneuver, and IP issues would still remain high on the list with the PRC government for years to come, even after the PRC was accepted into the WTO.[41] Damages caused by PRC companies taking advantage of Western companies' IP are today estimated to be $50 billion annually. Approximately the same amount is estimated to be the damage software pirates, whom I sometimes refer to as IP terrorists, are causing the WW software industry, with MS alone suffering north of $10 billion losses annually. Both are nontrivial amounts causing an estimated loss of five hundred thousand jobs in several industries and demonstrating that severe efforts will be needed to turn the unfortunate and undiminished state of affairs around.

---

41 World Trade Organization

NOBODY IS IMMUNE TO FAILURE

Right after launching Windows 98 in June of '98, MS reorganized again—on the top. The dreaded executive committee was abandoned and replaced after only eighteen months by a single president. You guessed it: Steve Ballmer finally got the job he had long yearned for, making a nice clear path for him to eventually become CEO. With Steve's appointment, the power play inside the company had once and for all been decided and called to order.

Bill was showing early signs of withdrawal. The antitrust trial, now set to start three months later, was casting a protracted and menacing shadow. The Feds had continued their drive to discredit him personally. The public sentiment, whipped to pitchfork-and-torch-bearing frenzy by the hostile and headline-hungry press, had turned derogatory. In public, Bill was no longer the celebrated genius and recognized visionary. By putting Ballmer in charge, he believed he could back off from the day-to-day business and focus his energies on trial preparation and strategic investment opportunities. He loved directing the lawyers, convinced he knew more about law books and legal tactics than most of them. I was astonished by the change I discovered. I could not get to him as often as I was accustomed to despite the fact that my ex-admin now worked for him. When we eventually talked, the business at hand interested him less. I concluded that being labeled a monopolist had finally gotten to him, even with accusations remaining merely allegations. The annoying redundancies had done the job. Bill had slowly grown less confident about his once-gleaming status, the company's well-being, and his legacy in general. He began spending enjoyable time with his growing family and was finding solace with his charitable foundation.

Steve, on the other hand, completely and boldly seized the opportunity with a weakening CEO at hand, forcefully taking the reins. With Bill less energized and buoyant, Steve articulated just the opposite. As far back as '89, he had been paying for advice from the McKinsey consulting company and had put their recommendations to use when reorganizing those facets of the organization he was responsible for. In his new role, McKinsey's engagement was further enlarged. Having heard too many less-than-successful stories of the fruits of her advice, I was no fan of this. Steve, on the other hand, appreciated the prescriptive input he was getting and went so far as to hire several McKinsey consultants for key management positions. Most of them failed miserably in their new assignments, and I was left scratching my head over how quickly they had rocketed up through the ranks. The whiz-kid equivalents of the bourgeoned McNamara era had found their way into MS. They needed to be young and smart; experience would count less. Steve was trying to mold the company to his liking and was energetically imitating the colorful tactics applied over at GE by Jack Welch, whom he held in highest esteem. He had gotten to Jack via Jeff Immelt,[42] whom he knew from his Procter and Gamble days while sharing an office with him. Still in learning mode, Steve had not fully found his leadership style. At times I felt certain changes he commanded were done simply for the sake of introducing change, and the promotion to president had fostered in him a bolder incarnation.

Bill's image within the company was undergoing profound change. People expected him to lead, not delegate visionary leadership to his loyal knight who was still learning the ropes. Great leaders, as I knew them, mastered their most imposing challenges, subsequent breakthroughs, and victories by confronting harsh realities with outermost logic, determination, and pragmatism—gaining respect. Admiration could wait. At this crucial junction, Bill failed to take that introspective path, dig deeper, and question a few of his own traits like his overly assertive tactics. The government was deviling us and the means by which the Feds pursued their quest was maddening and painful. A true leader would have gutsily leveraged this unfairness into an opportunity and reviewed business policies and procedures. One would have pulled heads out of the sand and demonstrated that he was able to cause change in places where the company had genuinely erred or had been over the line, as Gerstner had done when he struggled solving the IBM crisis. There was a call for urgency and a demand for honesty coming from all corners of the IT universe. Unexplainably, the company's commander retreated or answered with more of the same.

The push for paranoia continued unabated. The newest target was the DOJ. Paranoia, though, is not a smart way to win the end game. Changing behavior is. Openly irritating executives from Apple, Intel, AOL, and Netscape not only made enemies but also demonstrated unnecessary macho behavior. The barely veiled threats tendered by over-the-line employees emotionally poisoned the well where business flowed from. If

---

42 Today he is CEO and chairman of GE.

Bill would have jumped over his own shadow, like jumping over chairs—which he sometimes demonstrated and did astonishingly well—he would have become the agent of change the company needed. The empty void would have been filled up with character.

The genius in him, however, pleaded for love and admiration and responded with intellectual righteousness, signaling defiance. As a response, the Feds and our conspiring competitors accelerated their attacks by taking extra shots directly at him—the weakest point in the net he had concocted. He could not stomach it. His befuddlement signaled to the world that he was no longer willing or able to steer the ship. Crushingly disappointing to all employees working for the company—a sad moment of reckoning!

Sea changes were occurring not only inside MS. A number of founders and pioneers had made serious money by now and were becoming distracted. The charitable foundation race was on. Bill had established his in '94 and was promptly followed by Michael Dell and Gateway's Ted Waitt, as if they were competing even here. As much as these foundations were doing wonderful things for the world of the needy, they nevertheless stole valuable executive time. In the case of Bill Gates, our public relations people considered his charitable giving a great new way to portray him favorably in the public eye. Bill took their advice to heart, and under the guidance of his wife and father, he transferred most of his wealth into his foundation. Animated by their enthusiasm, Bill threw himself increasingly into guiding its strategies and began traveling extensively—now in his own private jet—to lead his newly minted philanthropic missions. Steve tended the shop at home.

The sudden surge of interest in the world of charity signaled an escapist flight from reality into a distracting new realm where he hoped to reestablish a permanent and reaffirming legacy. The public relations people loved the heartwarming drama his unprecedented charitable activities created and zealously depicted him as an incredibly caring man—the polar opposite of the now-out-of-fashion, shrewd, and scheming business Buffalo Bill. As I saw it, he had not changed one iota. His razor-sharp intellect combined with his desire for world domination remained unaltered. Through family and outside influences, he merely redirected his ambitions, as people working within his foundation confirmed. A fascinating image transformation was underway, a fresh path into an on-the–surface, meaningful, and altruistic life combined with a steady retreat from MS's day-to-day business affairs.

In addition to Bill's highly visible makeover, the company stepped up her efforts to do what most of her competitors had been doing effectively all along: proactively shaping the political landscape. In the United States, most companies affect legislative outcomes by tendering politicians direct campaign donations and funding lobbyists, promoting company-specific agendas. Like pouring legislative octane over political wildfires—not always successful, but without trying, prospects look even dimmer. In my mind, these direct donations made within legal parameters constituted nothing short of bold-faced bribery. Lobbying activities, to me, constituted a more ethical attempt at shaping political outcomes. Naturally, the recipients of the former would find ample reasons to disagree. Corrupt as the system seems, it remains legal.

Until '94, MS had been totally naive about how to buy favors from politicians. The consent decree increased awareness, and as a result, a political action committee (PAC) was formed. Four years later, it had increased annual donations eightfold from the $30,000 in '94. When I left the company in '02, that amount had climbed to over $1.3 million. Derived from employee contributions, the money was dedicated to specific causes or politicians by the donors, though most were simply contributions to a general fund. A closer analysis reveals that the contribution pendulum was having wide swings between Republicans and Democrats. This lack of loyalty to a particular party was both due to our collective desire for a favorable outcome of the upcoming trial and our proclivity to support the party who was more likely to lean toward a tax-friendly legislative climate. You can't buy political influence in the US in the open market?

What bugged me was how the PAC contributions were collected—supposedly a voluntary affair. But each year I got phone calls asking for donations to the PAC by people who claimed they were directed by my boss to do so. Steve and I never talked about these solicitations. I resisted the pressure or gave only to a specified candidate of my choice. The one year giving $5,000 to the general fund, I was deeply disappointed on learning how the PAC leadership had allotted contributions. Believing the wrong people or causes had gotten them gave me a perfect excuse to never contribute again. The trend into politicking represented a troubling shift in our business matrix—from the purist pursuit of serving PC users to serving the makers of laws with their vapid and purchasable attention spans.

Changes were not limited to MS. After fifteen years, the PC clone industry stars from the past were overtaken by newcomers. One of the early victims was Beny Alagem, Packard Bell's (PB) CEO. I saw him a last time in his Sacramento, California, headquarters. An old army depot that he had leased for a pittance back in '95 after an earthquake had struck and partially destroyed his old site at Northridge. My first visit to the new location late that year impressed me, as he proudly toured me through his 370-acre kingdom. During the same year, he had sold a 20 percent stake in PB to Nippon Electric Company (NEC)—the beginning of the end.

On my next trip to Tokyo, our friends at NEC took me aside, telling me they were watching Beny carefully and, if I should observe anything odd, I should feel free to inform them in strictest confidence. This was a clear first sign of suspicion and discontent between the new partners! I never made use of their invitation. The cautious and circumspect Japanese would no doubt have issues with the boldly aggressive former tank commander. All of Japan was watching their move. This substantial investment was a first for any Japanese company outside her homeland. If not, successful heads would roll and NEC's management would lose considerable face.

When the investment came about, PB was in peak form, having just been named the highest-volume PC seller in the United States by *Dataquest*. The company that the Compaqs, Hewlett-Packards, and IBMs of this world had long waved off as being too low-end to be taken seriously had beaten them all. With his industry-shocking success, Beny was now the undisputed king of retail, and in order to finance his ambitious expansion plans, he had let NEC's capital in the door. The Windows 95 launch gave him another jolt as he successfully expanded into Europe, filling the gap Amstrad had left to be explored.

Even after the initial NEC infusion, PB continued having difficulties financing Beny's expansionist's dreams. A '96 merger with Zenith, owned by the French company Groupe Bull, had brought an additional investor into the company. Now my friend had to maneuver between both of them. Beny, a shrewd negotiator, was by no means a diplomat. Quite accustomed to running the show as he saw fit, receiving a 650-million dollar capital infusion nevertheless meant losing part of his cherished freedom. A year later, he became disengaged and out of sorts. He had to devote considerable time and energy to cope with his new pseudomasters who were watching him closely, particularly as PB inched into unprofitability. As the losses mounted, so did their disagreements. Retail marketing in '98 was getting tougher. To maintain his market share, he had to counter companies like Hewlett-Packard, Compaq, and IBM, strafing PB with shock-and-awe price skirmishes. The newest entry, eMachines, a Korean-investor-backed company, created a tsunami atmosphere when offering PCs at record-setting low-price points.

The last time I met Beny as CEO of PB, he was a changed man, though still with a roaring fire in his belly. He told me he was no longer willing to put up with the prescriptive BS the board was throwing at him and would take his money and leave. The old warrior exercising Auftragstaktik expected the same from his board members. Living in an ancient world of inflexible mandates and egocentric myopia amid the ossified ruins of lost empires and forgotten armies, they simply could not find it in themselves to oblige. Shortly thereafter in '99, he tried a comeback after buying the assets of AST, another PC manufacturer who had not survived the onslaught. Unfortunately, he did not succeed a second time and decided to venture into other areas of interest. I missed wrestling with the shrewd tank commander and relentless warrior; he was an impressive character—smart, inventive, and to the bitter end ever true to his own best interests.

With the Feds attacking us in court over the legality of protecting the copyright for Windows, our OEMs got increasingly bolder in asking for total configuration freedom. Solid relationships began to erode as a result—best explained by exploring the Gateway (GW) conundrum. In early '98, CEO and cofounder Ted Waitt moved GW's headquarters from the wind-scoured plains of South Dakota to San Diego, California. As I remember it, from down-to-earth prairie life to southern California glitz! The proffered reason: he could no longer attract the appropriate level of management to the less-than-thrilling setting despite a by-now paved parking lot. The local community was at first little impacted, with most manufacturing operations remaining in place. My suspicion was that his new wife was behind the move, unenthused about life in the chilly, dreary, flyover country of the noncosmopolitan Dakotas and having to tolerate the stench of a nearby meatpacking facility. This may explain the purchase and remodel of an expensive new homestead in La Jolla, competing with Bill's $60 million mansion on Lake Washington.

Simultaneous with the move to the glamour coast, tensions between the two companies accelerated sharply. I experienced it myself when we renegotiated a new contract for bundling our Office applications. To smooth feathers, I had to personally get engaged and meet Ted at Denver International Airport. The mile-high oxygen-deprived atmosphere was at once outright hostile. He asserted that our original agreement to sell Office application bundles was solely responsible for the dominance we had gained in that product category. Then he imploringly complained that we should have been far more grateful to him after giving us that opportunity. I could only shake my head as he refused to acknowledge that GW already had the best deal between OEMs in regard to Office. I personally was his staunchest fan. In the end, though, none of my arguments convinced. He looked like he possessed a kind of sad, delusory logic, not trusting me any longer. No wonder it took hours to arrive at a reasonable agreement; in this case, it was accomplished by me personally breaking the deadlock and making an essential price concession after talking to him at length one-on-one—a rare occurrence indeed.

Hotly contested issues developed over the Windows desktop configurations between me and Jim Collas, his man in charge of product procurements. Being an ex-Apple employee, he probably didn't have too much sympathy for MS from the get-go. Initially, our relationship was cordial at best; now he veered sharply into the anti-MS camp and made an unending flurry of demands. My way of responding to the most outrageous ones had grown simple. "The matter is before the courts, and I'm not willing to relent until the case is decided upon." I considered using the current legal dispute to leverage truly far-reaching entreats nothing short of blackmail.

More irritations were caused when GW bought all of Amiga's patents, making our patent-assertion clause harder to negotiate. We never understood why Ted acted upon this specific patent portfolio, though we learned that he later sold it to Escom in Germany. Neither company ever developed plans to produce Amiga PCs.

Influenced by advice from insiders and outsiders, Ted slowly but surely refocused his allegiance and realigned his stance toward MS. I blamed it on the brilliant SoCal sunlight. Had the down-to-earth boy from the Dakota plains gotten too much of it, or had his obvious success finally gotten into his head? After briefing Steve about the changing relationship, he asked if I had heard anything about Ted secretly dealing with the Reback group or the Feds. Nothing direct, I replied, just rumblings through the grapevine. To get to the bottom of these rumors and dispelling or confirming them, we invited him to meet with us. Emphasizing his status as chairman and CEO, he made his attendance conditional to having a solo audience with Bill.

I let him have his twenty minutes in eternity. Not surprisingly, Bill told me afterward that they mostly talked about their foundations. Then he joined us. At the end of our talk, Steve asked him point-blank if the alleged rumors had any foundation. Friend or foe? Reacting with extreme evasiveness, Ted was not about to be forthcoming, avoiding eye contact and denying any behind-the-back dealings. After accompanying him to his waiting limo, I went back to recap our examination. Steve told me straight off the bat that he felt Ted had deceived us. I just nodded. Taking a deep breath, he said, "I never want to see this guy again!" Ted missed a golden opportunity of redemption by airing any grievances openly. Steve, having an admirable capacity for compromise and forgiveness when confronted directly by customers, was always ready to make amends.

Later that year, Ted groomed Jeffry Weitzen for the CEO post and stepped away from day-to-day business while retaining the chairman title. Having grown tired of running the place, he wanted to spend time with his family, enjoy his fortune, and tend to his foundation. Cloning Bill? By '01 he was back in the saddle, running the company after the appointment hadn't worked out. The rest of the GW story is not a pretty one. After buying a low-end supplier (eMachines) in '04, the once-phenomenal company went into a sudden and colorful financial tailspin. In '07, the Taiwan-based Acer acquired what was left. Having obtained substantial wealth in the process, Ted resigned. I wish him all the best in his new endeavors, and I hope he remembers a lesson learned!

The other company going through a lot of change was behemoth Compaq. In early '97, her CEO Eckhard Pfeiffer finally made an attempt to explore the mail-order channel by trying to purchase Gateway. As Bill and I departed a meeting with Compaq's upper brass, we observed a line of stretch limos idling up to her headquarters. Despite the gleaming smoked-glass windows, I spotted a Gateway VP in one of them. Bill and I exchanged puzzled looks. The Compaq crew had been quiet about the happenings, but a few days later, Eckhard called Bill before he could read about it in the papers. However, the deal never closed despite Compaq's generous $3.5 billion offer. Friends in Compaq later told me that Ted had gotten cold feet after recognizing the changes in store, particularly for his top management team and the South Dakota facilities. Ever the country boy, he later confirmed to me that he had no desire to put his existing hometown employees in danger of being laid off. While the deal did not materialize, its due diligence did, and soon thereafter, I took calls from Compaq execs asking for the same Office deal we had extended to their acquisition target. Confidentiality had again gone out the window. True to form, my answer was swift: "Give us an identical commitment, and I will give you the same treatment." The bitter truth was understood though totally underappreciated.

Instead, Compaq bought Tandem, the NonStop computing server company—the last manufacturer to ever launch a computer system with a proprietary OS. Gary Stimac, Compaq's SVP, had been on her board since March of '97, just three months before the takeover offer was extended. Loving her technology, he was the main driver behind the acquisition. The Tandem purchase was a clear sign of Compaq renewing her enterprise's customer focus. Her obvious plan was to take on IBM.

For that to be successful, her CEO needed a consulting arm offering IT-integration services. In a surprise move, he made an offer for ailing Digital Equipment Corporation (DEC). My old friend John Rose, having joined Compaq several years earlier, was the driving force behind that deal—stunning me. The '92 departure of Ken Olsen, DEC's admired founder, caused by her first quarterly loss, had left that company polarized and with no clear direction. Her new CEO, the capable Robert Palmer, one of the founders of Mostek, had since then struggled to turn the ailing ship around. Despite successfully launching an advanced minicomputer line and executing massive layoffs, he failed. After consuming the deal, Compaq's management had a tough task ahead. DEC was a wallowing, wasteful company, and her reputation with customers had begun to break down. A situation comparable or worse to what Gerstner had found when taking over IBM.

Evaluating Compaq's acquisitions, Bill and I thought Pfeiffer had bought anchors, not sails. By early '99, merging the three companies ran into tremendous difficulties. When an earning crisis developed, dropping Compaq's stock price 20 percent, Pfeiffer got ousted. Chairman Ben Rosen took the title of interim CEO. Replacing him four months later was a lackluster insider, Michael Capellas, who tried his luck to fuse the floundering wreck into a coherent and meaningful competitor to IBM. During this turbulent time, Steve Jobs—still mad at Bill for not liking the iMac design—attempted to upset the MS-Compaq relationship by offering Compaq a license to the Mac OS. Fortunately, her management rejected the idea and stayed loyal, though it again showed how vulnerable MS remained.

Eventually, Hewlett-Packard (HP) came to the rescue by buying Compaq through a stock swap worth around $25 billion. The deal was completed in '02 against the objections of Walter Hewlett, HP's major shareholders and director. From thereon out, the combined company was the undisputed PC king. Carly Fiorina, HP's CEO and chairwoman, tried for another long four years to restore and revitalize the shipwreck, forgetting that in a wickedly competitive environment, the scent of blood travels fast in the high-speed winds aloft. This time to the east! Carly's failure allowed Dell to seize the opportunity and erode HP's market share deeply enough to lead to her ouster. The appointment of Mike Hurd managed to put HP back on track, restoring some of her old glory. A genuinely astonishing achievement! By '04, the consolidation in the industry left HP and Dell with a combined 40 percent of WW PC volume, a concentration of power never experienced when I was still running the OEM group.

Forging Ahead

My last remarks have ranged well beyond my time with MS, so let's get back to early '99. Despite his promotion to president, my boss's relationship with Compaq never completely recovered from earlier mishaps. Whenever Steve visited the company's execs, Gary Stimac, Compaq's SVP of engineering, stood a life-size cardboard cutout in the conference room, in lieu of him attending personally—an insult poorly taken by Steve. Observing the struggle Compaq meanwhile experienced was the last motivation he needed to embark on a mission to acquire new friends by cultivating her competitors. Concerned about a promising pursuit of the enterprise server and workstation business, he fostered a rapport with Dell's president Kevin Rollins, future CEO and Michael Dell's right hand. The marketing money to seal the envisioned pact, greasing the tracks, and forging a cooperative spirit came directly out of my budget. Unbudgeted. Looking back, it turned out to be a good decision. I did not resent striking the deal, though I was deeply hurt by the way it was forced on me impulsively in a meeting with no advanced warning or prior analysis.

With Steve firmly in command, I recognized another change in the company, one I personally had never wanted to witness. The meetings where decisions were supposed to be made swelled into teeming crowds involving all manners of logistical impossibilities. Anyone who remotely had something to learn, and most of the time nothing to contribute, would now be invited. The presentation folders distributed grew into Manhattan phone books. The PowerPoint presentations labored on ad infinitum. The review meetings I now had to attend often ran for days, at twelve hours per. Ironically labeled team-building exercises! McKinsey and the whiz kids were casting their long dark shadows. Nobody was allowed to step on anybody's toes any longer or display the normal MS style—raucous at times. Drawing strength out of healthy disagreements was suddenly a lost art. Right or wrong, one company, one team! As practiced, the resulting charade suppressed true opinions, maverick passions, and controversy—attributes that in the past had led to impressive progress and breakthroughs fanning a healthy competitive spirit. The earlier culture suffered. Forgiveness was favored over accountability. Being a friend of Steve was becoming important. He was making an effort to keep his temper under control; I never thought that day would ever arrive. The company was drifting into a decision-by-committee organization. I often did not possess enough *Sitz Fleisch*[43] to endure these never-ending sessions and began finding excuses.

As the decision-making style in the company shifted, I refused to be party to spreading the cancer. Once, after calling for an operational review with my logistic group, nearly fifty people showed up for a meeting scheduled for three hours. I looked at my business manager with astonishment and said, "I am going back to my office, and when I return, I want four people in this room. If there are any more, I will start firing the rest until we get down to exactly that number." I barely finished issuing my instructions as a mad scramble for the door drowned out the last part of my final phrase. After we were down to the quintessential group, we finished the agenda in less than fifteen minutes. I had to repeat the same exercise again during a marketing review. From then on, decision-contributing gatherings never exceeded six to eight people; all others needing to know were later informed by e-mail.

I was nearly at the end of the rainbow. Steve must had sensed this when he expressed his desire to build his own leadership team and wanted to know if he could count on me. What inspired me to stay was twofold: first, the desire for an orderly and smooth transition, and second, my loyalty to the nearly five hundred people working for me. The trial was not over yet, and I firmly believed that an experienced hand was needed to steer the group through the turmoil of the aftermath.

Assuming I would join the exodus in progress was not far-fetched. Early employees like me had created enough wealth to explore other venues or simply retire. As the stock price drifted further and further south and stock option packages became less valuable, a lot of talent left, absence felt; there were ghosts in the hallways. Paul Maritz grabbed me one day and, looked straight into my eyes, imploring, "I am not sure why the two of us are still around!" He departed before I did.

As Steve detailed his ideas further, he expressed his desire to build a team of executives in their late thirties or early forties. Why mention age? Did experience not count any longer? Was loyal service and years of stellar performance no longer rewarded, honored, or appreciated? He made me think harder with several not-so-off-the-cuff remarks. I had not made up my mind completely, but I had a capable person in my division to take over for me. Owning the conviction that age and performance are not necessarily

---

43   German word expressing "not having the patience to sit it out on one's flesh or behind"

correlated, I ignored his anti-age comments and carried on—yet I felt freshly motivated to seriously consider my life after MS.

In a follow-up meeting, I told him I would be happy to stick around until the end of FY '01[44] before wanting to move on, suiting my remaining stock option vesting schedule. After thinking it over, he agreed to my proposal, and from then on, we were both actively on the lookout for a successor. Knowing that my time was coming to an end had no impact on my group's all-star performance. Steve, writing a personal note on my bonus notification in August of '99, saying, "JK, another huge year. Superb!" must have had the same impression.

---

44 Meaning June 30 of 2001

JUSTICE ON THE BLOCK

TRIAL SETUP AND STRATEGIES

The MS antitrust trial of '98 shook the IT industry at its foundation and was soon dubbed the trial of the century. Attempting to bring down the most influential software and largest capitalized company in the world, the Feds caused lusting minions to circle like vultures above the veld. Not far behind, the press venally manipulated the truth to drive an appetite for extra copy, sell more newspapers, or attract a larger TV audience.

My research shows at least 150-plus voluminous white papers or books have been published to postanalyze that historic event. Why write another long chapter following the path of no return. In its place, I will share with you how I experienced the trial as a participant and observer and reveal my own feelings on where justice, finally, came to reside within the process—focused mainly on my OEM business. Its practices were inarguably its centerpiece, and my chronicling stage lights will therefore track and illuminate and scrutinize those facets, without neglecting to capture the emotions of the key players as I witnessed them from the inside out.

After Anne Bingaman's departure as chief of antitrust enforcement, Joel Klein took over. Skeptical of winning the ongoing consent-decree-violation complaint, he hired Jeffrey Blattner, the talented but rough-hewn former chief counsel for the Senate Judiciary Committee, to covertly broaden the never-halted investigation. His true role was a well-kept secret: to dig deeper into the documents already in the DOJ's possession and to distill and condense e-mail snippets and other evidence into a deadly antitrust Molotov cocktail. Just in case Jackson's ruling wouldn't survive the appellate court. Somebody was thinking ahead.

Blattner was assisted by top-notch gunslinger David Boies. The DOJ, engaging him first as legal consultant and after hiring him fulltime, made him chief prosecutor in the newly filed case. Boies came loaded with marquee credentials as a most celebrated and successful trial lawyer. In 2000, *Time* magazine dubbed him lawyer of the year, stating, "Mr. Boies's memory is one of the first things when people discuss his strengths. What's most impressive about that gift—focused as it may be by the intensified concentration that his dyslexia demands—is Boies' uncanny ability to recall a key fact, legal citation or piece of contradictory testimony at moments of the most intense pressure." He had garnered firsthand experience with antitrust laws when defending IBM against the Feds. The historic trial lasted from 1969 to '92, and Mr. Boies had ever since been recognized for devising IBM's winning strategy.

A victory coinciding with an all-new Regan Republican administration noted for having the least active antitrust enforcement, motive, and record in modern history. Yes, politics do play a dreadfully active role in antitrust cases. By now we were under Bill Clinton's rule—his administration certainly not known for welcoming or practicing Reagan-friendly enterprise policies as we were experiencing firsthand.

In the spring of '98, the joint Blattner/Boies mining operation burrowed deeper and deeper into the never-receding cordillera of documents. Digging for the mother lode and carefully considering the latest appellate court ruling, they waxed confident of having their case wrapped up. Soon thereafter, the Reno DOJ filed the aforementioned sweeping shock-and-awe antitrust suit, accusing us of having illegally obtained and maintained a monopoly in violation of the infamous Sherman Act. A bloody and protracted battle was guaranteed. The gist of the accusations was still based on tying IE features illegally into Windows. A lot of others, variations to the theme of anticompetitive behavior, had been added. Twenty states had joined forces, with every single one of their AGs dreaming of headlines and inaugural balls in the governor's mansion. They were coming at us in droves—sabers raised. Success surely breeds contempt! Political activists were milling and murmuring in the restless night, pitchforks in hand and torches ablaze. The home state of MS, Washington, declined to join them.

Judge Jackson's top priority for the upcoming trial was to avoid a prolonged battle and complete the proceedings before the Clinton era drew to its close. According to later-revealed interviews, the judge claimed he was still open-minded toward MS despite our earlier eye-poking performance. He made no secret whatsoever of having been hurt and humiliated by the harsh ruling of the appellate court. A bad omen? I heard the phrase "Humiliation is a license to hate"; let's see how humiliated he felt.

Jackson had a lot of discretion and elbow room conducting the trial. He used it boldly by fast-tracking the trial for September of '98—curtailing our prep time and slightly advantaging the Feds. Next he limited the number of main trial witnesses to just twelve and a scant three for the rebuttal phase. As a multifaceted company with the action of one branch of MS not necessarily reflecting the company as a whole, we felt straight jacketed. Hostile or not, the company's legal forces went into overdrive. The Feds kept adding to their workload by requesting ever more information, further pinching valuable prep time.

Total documents delivered: approximately three million, barely fitting inside two 18-wheelers. To capture them, our attorneys, on behalf of the Feds, raided employees' offices with no warning and regardless of whether you were in or out. Whatever they deemed relevant got shipped off. They harassed everybody, and many took the intrusions personally. A pest and a nagging distraction from running the day-to-day business for sure!

As the Feds began conducting depositions, we, in return, deposed government witnesses and collaborators. Another time-intensive task for our extraordinarily committed but equally stretched and overworked legal team as the clock approached deadline. In the end, Jackson relented and agreed to delay the trial until October 19, 1998. My friends on the legal staff, totally exhausted, were barely ready.

I was named one of the twelve main MS witnesses. No surprise—this was all about my business. Paul Maritz and most of his direct reports found themselves on the same list, while Steve's and Bill's names were absent. MS employees were shocked, recognizing that the top two honchos we had worked and sacrificed for, followed, trusted, and obeyed were dodging the trial. The news spread like wildfire on campus. People were dismayed and in stunned disbelief.

Upon questioning Bill Neukom, he explained that Bill had indeed volunteered to defend the company, but our legal team had deemed him a bad client. With all his hyperintellectual smartness and word-mincing abilities, he apparently lacked the coachability required to transform him into a sensibly forthcoming, judge-placating witness. Or, as the NY Bar Association defines one aspect of a bad client, "He has tunnel vision about the matter and does not want to listen to new ideas, is not interested in other options, or can't face reality about his role in creating the situation." Not privy to how Bill N. had concluded this, I concurred after watching a video snippet of Bill's own deposition. Churlish, thin-skinned, and agitated, he reminded me of President Clinton's depositions in the Paula Jones and Monica Lewinsky cases, where he questioned the meaning of the word *is*.

Our policies and actions were the direct and accretive result of Bill's and Steve's visions, plans, and business acumens. Like Siamese twins, they were joined at the hips as they outlined key strategies and instilled a take-no-prisoner operating style into everybody. Then and now, they were the architects of the MS house of cards, which could collapse anytime from the pressure the DOJ was putting on it. Them baling out compares to refusing to be at the hospital the day your wife has your baby.

I suffered through a lot of explanations with my people and, against my belief, excused our leadership. I never talked with Bill about the vacuum he created, one we all felt, but I mentioned it to Steve, who just confirmed he had left the witness selection solely to the lawyers. I harbored profound reservations on the appropriateness of such a delegation but dug no deeper into the man whose enduring passion was rallying and rousing his troops. Not being selected to be on his beloved stage this time needed no further insult. How he could maintain his proud, upright bearing and justify his no-show to himself was inexplicable to me!

Decision made, my life went in a different direction. I was not eager but certainly proud to be a witness and defend our profoundly successful actions, well-considered policies, and our people's tireless work. Emphatically, they all wished me good luck. I thanked them for their support with all my heart! I would need it!

In September of '98, I spent one week with our attorneys scrutinizing documents. I was extremely thankful for their in-depth preparations for yet another deposition and hoped my former experience would help me to get through this one with flying colors.

Mr. Malone, the prosecutor in the earlier contempt case, deposed me, and no smoking gun was revealed. Compared to the first DOJ attorney I experienced back in the early '90s, he was well organized and understood the vagaries of the OEM business reasonably well. He was cordial, polite, and stayed on his agenda. I was forthcoming enough in my answers, but once in a while, I had to resort to the worn-out Oliver North phrase employed during his public grilling by congress: "I do not recall." Small wonder, as several events he questioned me about had happened six and seven years earlier. He focused on the OEM business along with relevant e-mail exchanges or documents I had written or received. His style was diligent and persistent but not gratuitously mean. As experienced before, he tried to trick me by asking what other people had meant when they wrote their opinions down. When I told him "Ask them," he countered by asking me what I had understood as the receiver. Several years later, who can truly remember all connotations? So we struggled along, Malone laboring by means of inference and innuendo to prove evil intent of the authors or the recipients and me grappling to deny him that same pleasure.

OEM pricing was a central issue he pursued. He failed to appreciate a number of my answers as he pushed hard to contrive how we were overcharging customers. He relished hearing how we had arrived at our royalty prices. Whatever the method, he always implied we had pricing power. He was loath to accept the truth of having only scant competitive information to base them on. As a hired skeptic and deputized enforcer of the law, he therefore cavalierly concluded that he had found yet another way to prove our alleged monopoly power—not accepting an obvious information gap. He confronted me with the letter Hewlett-Packard's VP Romano had sent me, decrying Windows as the only game in town. I pointed out other options close to what I have explained earlier here. For Malone, the opinion of the Hewlett-Packard guy neatly trumped my own deep experience and considered logic.

Last but not least, in one of my e-mails to Bill, I had mentioned an application barrier of entry for our competitors when competing with Windows. I reasoned that the Windows standard we had created luckily protected us—until now—while pointing out our luck could run out soon. For Malone, the barrier he had in mind was absolute, as his economists had told him, enabling us to defend our alleged monopoly forever. The trial was casting a shadow.

I should mention one last item that had captured the Feds' attention. In a separate mail to Bill back in '94, I suggested sending IBM a hit team in case they failed to comply with one of our requests. An ill-chosen word for sure. But in any high-powered sales organization, *hit team* is a colloquialism implying no physical harm, meaning "Let's be sure to send the highest-caliber, toughest negotiators possible." For Mr. Malone, it expressed that we had power to harass and coerce customers. As long as I could remember, IBM never seemed intimidated by us. But for a government attorney like Mr. Malone, such language was not permissible and proved malignant intent or harm punishable by law.

Naturally, his other interest regarded the by-now-famous Windows boot-sequence restrictions we had imposed on our OEM customers. Again probing for anticompetitive behavior, his main line of questioning was about the reasoning for our regulations. Why did we restrict OEMs from removing our icons from the Windows desktop? He for sure did not like my answers, but our copyright existed, and our restrictions defended it. After two days of intense cross-examination, he left—not terribly satisfied. My attorneys praised me for a reasonable performance. I had survived another day.

The DOJ strategy for winning the trial had become crystal clear as we studied the substance and shape of arguments our witnesses had been entangled in during their depositions. In all antitrust trials, one of the key elements consists of defining and sizing the relevant market for the judge. The Feds wanted the market definition as narrow as possible, with us insisting on the opposite. A judge could not easily condemn a company of having monopoly power if her competitors had 85 percent market share. Judge Posner described the struggle to define the relevant market in his book *Antitrust Law* as an always present dilemma: "The importance that antitrust law attaches to defining a market is another consequence of the law's failure to have developed an approach at once genuinely economic and operational to the problem of monopoly." No science here as I understood his remark—the subjective beliefs of a trial judge sufficed condemning you. This aspect of paramount importance would resonate throughout the trial delineating utter consequence.

The other obvious part of the Feds' strategy was to discredit as many of our witnesses as possible. That tactic had priority above all on Boies' courtroom menu. He is quoted in *Fortune* magazine as saying, "What you do is set it up so you get the jury [sic: in this case a judge] to think the witness is lying—before you even suggest the witness is lying." Most trial lawyers attempt this, though Boies arrived with the reputation of having mastered it. If he accomplished that feat with our witnesses, as he had already done in Bill's deposition, our cards would look really bad. Boies profusely understood that Jackson, when issuing his final verdict, would be disinclined to base his reasoning on witnesses he failed to believe in. If Boies, in his henchman role, could trick our witnesses into tying their own hangman's noose, no need to escort them to the gallows himself. The judge would just sit back and watch them swing in the breeze.

The third strategy was to convince the judge of the validity of a highly abstract and unconventional economic model. In a classic antitrust case—and in all cases to date so far—an offending monopolist, after arriving at its unique position, restricted output and raised prices above reasonable levels, maximizing profits to the detriment of consumers. Promising to never charge for IE and keeping our Windows prices pretty much flat since '95 and unadjusted for inflation without ever restricting output could hardly qualify as monopolistic behavior.

To overcome that hurdle, expert witness Franklin Fisher, on behalf of the DOJ's economist Daniel Rubinfeld, planned to present an esoteric abstraction of an economic model based on path-dependent network effects. The underlying theory is derived from what mathematicians label chaos theory. Despite its name, it describes a serious statistical model used to study highly dynamic systems dependent on, or being predominantly sensitive to, initial conditions. Good examples for these are the big bang, quantum movements, and weather patterns. A few avant-garde economists, like Stanford's econ scion Mr. G. Saloner, found this self-same statistical model helpful to describe and predict anomalistic market behaviors. Their main deduction: as in physics, history can determine the ultimate outcome of a network-driven market economy and lead to irreversible results. He therefore postulated that our success with Windows could sufficiently and scientifically be explained within the means of this model. Exact science or ambitious thinking, who knew?

Accepting this as a highly plausible theory, sufficiently describing how we had achieved our dominance, the DOJ's economist concluded that we could therefore never be overcome by competitive forces. The destiny of being top gun forever was for them path-dependently predetermined (by God?) and manifested by networking effects working in our favor. As dubious as this sounded to me, the Feds' economic team was dead set on putting this freshly baked model to a legal acceptance test.

A gamble in the face of missing traditional—Chicago School of Antitrust—evidence for classic monopolistic behavior better described as clinging to a desperate last straw. Nevertheless, the Feds planned to suggest that the large-scale Windows demand we had created rewarded us with unstoppable returns to the detriment of competitors and consumers alike. Buying into this colorful sci-fi depiction of our undiminishable dictatorship would give the judge hopefully enough reasons to topple a nasty despot and save the industry and mankind from disaster.

We, on the other hand, characterized customers buying and staying within the Windows environment a genuine selection process, by volition. Potentially, a natural network had been created by an avalanche of entities joining up, comparable to a phone network or the Internet. Economists understood how a popular and growing network could naturally shape the structure of a demand curve and lead to high returns. For the Feds, the extraordinary steepness of that curve in regard to Windows distorted the proper workings of the industry, unfairly tipping the playing field to the advantage of, and wrongly rewarding, a monopolist.

Historical events contradicted this. In the early '80s, Apple had reigned supreme with her Apple II platform, and the myriad of application programs written for it served as a barrier of entry for competitors. Attacked by the emerging DOS-powered IBM PCs, it eroded and eventually faltered. The barrier of entry was overcome by free market forces: risk-taking entrepreneurs (PC manufactures and ISVs alike) and customers opting for emerging computing platforms!

Three years later, Apple attacked the by-then dominant IBM PC platform with her superior Mac PCs—unsuccessfully this time. The Mac went on and blossomed as a niche product until Windows 95 and its successors stole more of its limelight. As with VCR standards before, an inferior product (VHS) won over a better contender (Beta) because consumers—speaking with their pocketbooks—favored the less expensive choice. . Much like what happened to Apple's Mac sales when consumers preferred the more affordable DOS-based IBM clones and later the ones based on Windows. Early on Steve Jobs and the ones who followed him at Apple did not get it right!

In all these cases, the market had spoken, applying its own common sense realism—not needing government interventions. Leaving only one conclusion: as slick as the Feds' economic model looked academically, it did not pass the reality test of the PC revolution.

Our strategy to win the trial was a much simpler one. Based on our experience with the judge, most members of our legal team did not trust his ability to comprehend the intrinsic matters before him. Others believed he was tainted by our eye-poking behavior in round one. Both had a point.

Our legal team believed that all our actions were defendable and that the market we operated in was much larger than the Feds alleged. Admitting to owning a monopoly was therefore not in the cards. So let Jackson do his work, and if needed, appeal his ruling. It worked once; it should work again. Banking on winning on appeal was a delicate strategy to execute and could easily backfire. We needed to assume the judge would do everything within his discretionary power to avoid another stern rebuke from the higher court. He was an old hand and knew the procedural rules well enough to navigate them to his advantage.

Unbeknownst to us, he had already decided to issue his finding of facts separately from his final verdict, hoping that even if his verdict got overturned, the finding of facts would survive. What he, like any other judge, would issue would not be facts in a true sense. They represented in a condensed form what Thomas Penfield Jackson, as a subjective human being, had discovered. As long as the trial record even vaguely supported his findings and bias could not be explicitly proven, it would be hard to disregard them. In talking to our attorneys, I got the impression that they knew full well the risks of betting on such a simple yet narrow path for victory and believing in the appeals process as the savior for their client. It struck me as odd, but I was in no position to recommend an alternate path.

WITNESS PARADE

On the pretense of shortening the trial, all witnesses had been asked to prepare written statements so testimonies could be obtained in advance. Several added videotapes explaining technical details with graphical images and compilations of event sequences. I never composed a single word of my written testimony. Our attorneys did. We reviewed the draft, and with the exception of modest changes, I approved what they had produced. My submitted videotape was a different story. The prosecutor had found mistakes in two of them, presented by MS witnesses preceding me. To avoid another disaster, we therefore remade my tape. It was comprised mostly of screenshots of Windows desktops as delivered by OEMs. The revised version showed a growing number of OEMs placing competing browsers and other competitive icons on Windows desktops. This would not sit well with the prosecution wanting to prove the opposite.

At the opening of the trial, the government paraded her witnesses first: not a single OEM customer under them, making the Feds claim they were afraid of us. Seven of the prosecution witnesses were from competitors, two were economists, and the three others were software experts. Our legal team had a chance to cross-examine them first and to soften up their written testimony. The government used its time trying to prove MS had systematically harassed these firms so she could erect and defend a competitor eliminating barrier to entry of her alleged monopoly.

The DOJ's two economic experts attempted to convince the judge to accept the narrowest of market definitions. Subsequently, they tried hard to get their theory of a network-effects-driven economic world order in software markets validated by him. An attempt was made to make our zero-dollar pricing for IE look predatory. Windows pricing, on the other hand, was labeled excessive and typical of a monopolist without providing sufficient empirical data for the allegation. To prove this, the DOJ's experts compared our unvarying Windows royalties with ever-falling PC prices. There is no positive correlation between the two. A correlation between the price of a car and the price of steel or aluminum certainly existed. Yet the price of a car and the price of gasoline, which you for sure needed in '99 to actually drive one—did not.

When it came to our alleged monopoly power, the DOJ economic expert Franklin Fisher concluded in his direct testimony, "It is difficult to imagine that in an open society such as this one with multiple information sources, a single company could seize sufficient control of information transmission so as to constitute a threat to the underpinning of a free society. But such scenario is a realistic (and perhaps probable) outcome." The company who had wanted every human to have a PC on her or his desk—a threat to mankind?

What both of the Feds' gentrified economists did not provide was sufficient empirical evidence for their broad-stroke wisdom. Bill Neukom confirmed my observation in a press conference: "This is the fundamental flaw in the government's case: they apparently never took the time or brought in objective analysis to the question of whether there's any consumer harm to our actions." When Fisher was cross-examined in court and asked directly if we had harmed consumers, he had to concede: "On balance, I'd think that the answer is no, up to this point." So why were we a danger to society? When Boies reexamined him later, he tried to change his story by saying that the consumer harm had been done by depriving them of choices through restrictive OEM contracts. I should have known!

The Feds' next argument centered on raising Windows prices. For six years we had kept them unchanged. But in '98, I reduced some rewards OEMs could achieve by a small amount for the simple reason that they had made no attempt to gain them. Reducing incentives was therefore presented by Boies as a price increase. Yes, he could argue along these lines. Without boring the reader with details, rest assured that the increase did not even compensate us for increased development cost or inflation. Neither was the higher-value proposition for the improved 98 version ever considered or analyzed. Therefore, my math-driven argument that we had kept Windows prices basically flat during the time in question still stands. Mr. Boies nevertheless scored brownie points with journalists, exhorting, "It's clear that Microsoft has been raising its prices at a time when everyone else in the computer industry has been lowering them." His noise made great headlines without being close to economic truth or hard math. Again there was and never will be any proven correlation between software and hardware prices in any industry except a negative one.

An executive from Apple testifying for the Feds was no economist but insisted MS had a monopoly based on his understanding of "both the industry and the technology therein." He referenced our high market share and the symbioses between Windows and its applications. The guy was speculating wildly. Neither high market share nor the natural symbioses between any OS and its applications proved his point. All the same, his amateurish-appearing evidence was allowed to stand.

Richard Schmalensee, our own economist, testified quite differently. He, like Fisher, got into a pickle early on when attorneys were able to show that his written testimony contradicted with what he had written in earlier print publications. He went on, however, presenting a detailed and thorough analysis, using formulas well accepted by economists to prove at no time had we ever charged a monopolistic price. Without waiting for the prosecutor, the judge freely opined that a monopolist looking for long-term gains did not necessarily have to charge a high price. Was he arguing for the prosecution? While Schmalensee did not make the greatest impression on the witness stand, the government econ experts were equally lusterless. One mistake committed by our defense team was having Richard Urowsky examine Schmalensee. Richard was the self-same advocate who had won our first case arguing our appeal, thus helping Jackson to more than one black eye. Who knows how lucid his memory was in this respect?

In summary on the economic front, the Feds were hard at work trying to prove how we had harmed consumers. Expert commentary was abundant. A typical distillation of critiques appeared when William Koviac, a former attorney for the FTC, was interviewed by the *Washington Post*: "The government has not introduced that much evidence to demonstrate that consumers are suffering grievous harm today at the hand of MS. It's one of the weakest elements of the government's case and one of the strongest elements of the MS defense." An interesting comment came from Mark Patterson, a Fordham law professor: "Unless it's very clear, the courts worry that the potential of the long term harm may never materialize." Reiterating the conclusion that courts, not possessing a certified crystal ball, normally bore a skeptical view of any company, causing long-term hurt.

The Feds' technical experts told the judge that they could find little if any consumer benefit in the integration of Windows and IE. Judge Posner, in his book, *Antitrust Law,* made a wise comment about such experts: "There are very few genuine neutrals in the technical field judges can rely on." And this one was not one of them. The benefits were obvious for consumers and software programmers alike! The judge himself did not ask too many questions. He fell asleep on the bench now and again—a bit detached.

After the Feds had presented their witnesses, the onlookers were eager to glimpse at ours next. Our legal team decided to call me as witness eleven, with just one to go after me. Playing excerpts from Bill's videotaped deposition the day the first MS witness showed up, Boies stunned the courtroom and alarmed our legal team. At once our lawyers contested the legality of his exploit; Jackson allowed it. Bill had done a lousy job during his interrogation, fighting David Boies tooth and nail every step of the way. Not picking the winnable fights, he looked unreasonable, had numerous lapses of memory, was evasive, and stubbornly refused to explain the most obvious details. A certain degree of complicity is necessary to a witness' credibility. Bill wanted none of that.

For the rest of the trial, David Boies mischievously introduced supplementary snippets of Bill's video footage to convincingly manifest that our dear leader could not be trusted. It worked. The court of public opinion was turning sharply against him and consequently all of MS when played on TV nightly. True to form, Boies had achieved what he set out to do without putting Bill on the stand.

Jackson expressed this in a comment made to a journalist later: "Here is the guy who is the head of the organization and his testimony is inherently without credibility. At the start, it makes you skeptical about the rest of the trial. You are saying if you can't believe this guy, who else could you believe?" In Boies's own words, "The chairman of the company doesn't have any credible explanation for what she did, even though he was intimately involved. If he doesn't have an explanation, then how can you credit the explanations of his underlings?" The strikes were mounting. Any company executive following this spectacle would face a nearly insurmountable task of convincing Jackson of being trustworthy. Brilliant lawyering by Boies, which Jackson later dubbed an "ingenious strategy," had achieved its first objective.

Next, our intent to crush competitors while gaining and defending our market position was on public trial. Franklin Fisher, the DOJ's expert witness, had formerly written about this subject in an academic publication: "The subject of intent of a company is difficult to determine and will usually reflect nothing more than a determination to win all possible business from rivals—a determination consistent with competition." He expanded his thinking into "On retrospect evaluation of whether there were more desirable alternative actions that could have been chosen, would be to elevate competitors above competition and threaten the entire competitive process." His former opinion no longer counted. Was intent alone enough reason to condemn us on moral grounds? All eyes were on Jackson: would he boldly opt to carefully weigh the tangible impact we had actually caused or blindly believe Boies's version of the truth?

The court battle raged on as I was preparing to enter the tribunal in February of '99. To be certain I was thoroughly prepared, I took some days off and went to Hawaii,[45] carrying two suitcases. My personal belongings in one and several thousand pages of documents in the other. The handle broke from the latter's weight as I checked it through. I spent the next ten days mostly in a hotel room, examining the material. The few beach walks and swims I took were well deserved and helped me to stay focused.

My next stop: Washington DC and the law offices of Sullivan & Cromwell. Here I met with a portion of our legal team. Steven Holley, the lawyer who had been assigned to me, was present along with Dave Heiner, Bill Neukom, and much of the time, Richard Urowsky. Compared to my deposition prep, the one for the trial was brutal and exhausting. We worked ten to twelve hours daily, combing through relevant documents and discussing the potential issues from all possible angles. Even during dinners and lunches, we stayed on the subject. I made notes during the day and, before going to bed, typed them into my notebook to better memorize them. Two days before my scheduled court appearance, I snuck in under radar, familiarizing myself with the courtroom surroundings, the judge's imposing style, and the crafty vagaries of the prosecution.

During the last day of preparation, my nerves went into overdrive. The legal team had decided to involve me in a series of merciless mock trials. In no time they tricked me and tied me up in knots. We tried again and again, and eventually I got used to the mind-set and methodology a prosecutor was likely to employ. The team effectively used the very words and tactics Boies had used with former witnesses—one benefit of being next to last. The team had studied him with careful precision. I felt better than in the morning, but my confidence was waning, so I skipped dinner and spent much of the night browsing through my notes, reading depositions and former testimonies over and over again. I am glad I did—my memory, now refreshed, helped tremendously to overcome most traps and snares David Boies had waiting for me.

Next morning, on February 24, '99, the legal team accompanied me to the US district court at 333 Constitution Avenue in Washington, DC. The air was sharp, clear, and chilly. I felt tension heating up inside me. Major national TV networks had their communications trucks fitted with huge satellite antennas parked around the building. To reach the courthouse, I wended a maze of press and a crush of onlookers—a pulsing mob of dozens of bulb-flashing photographers as well as dissonant platoons of journalists shouting provocative questions, trying to elicit a careless response. The fella granting no interviews for over ten years, MS's secret power broker,

---

45   I later learned Paul Maritz had done the same.

had arrived for his day in court. Bill Gates's alleged enforcer supposedly wielding the pricing sword for MS, no longer hiding—open game.

The zoo-like atmosphere created by the paparazzi was startling and nerve-racking. TV cameras all aimed directly at me, the flashes of the Nikons stinging and nearly blinding my eyes. Inside the courthouse, the atmosphere was equally hectic as many people tried to get close, hoping for a printable snippet. I quickly disappeared into a small witness prep room where I patiently waited my turn. The judge had not finished with the former witness. Instead of being called at 10:00 a.m., I endured another long and restless ninety minutes before my chance to testify.

Once I was up on the stand, I could no longer consult my attorneys until my testimony was in the record in its entirety. Legal or not, I considered Jackson's order unfair. I found out later that Mr. Norris, an IBM rebuttal witness, was indeed allowed to talk to his IBM attorney between his days on the stand. How could the judge justify such inconsistency? I left my briefcase containing my handwritten notes and a notebook behind as I was finally called upon.

By now, I knew Jackson was in a rush to wrap up the first leg of the trial by Friday afternoon, two and a half days away, allowing him to leave on time for the weekend. There was one additional witness after me, so I assumed that I would be on the stand for two days at most, not the three originally planned. Plotting in my head how to best take advantage of the rush to justice, I assumed the prosecution would now have to squeeze all questions into the much shortened time span. Something I hoped to capitalize on. Long-winded answers without annoying the judge came to mind!

The first ninety minutes were occupied by Steven Holley, presenting the videotape we had prepared. Waiting on the stand settled my nerves and let me study a courtroom filled to the last seat by the press and the public. Our legal team was positioned in the back, well in sight of the judge, not being able to signal me as long as he stayed awake. We always had up to five attorneys in the courtroom. The government had a platoon. Piles of papers were stacked up on benches close to the witness stand. I rightfully assumed that these government lawyers had a narrowed focus. While I was responsible for the full spectrum, the prosecution confronted me with the laser-tight expertise of boutique lawyers, each one with their own niche. They had ample time to burrow into any vital detail the capacious recesses of Mr. Boies's prodigious memory failed to instantly summon and stack the odds against me.

If Boies was truly as good as advertised, why did he need such an elaborate backup system? Maybe in the end he was beatable! My inner voice issued a stern warning: "Don't get too confident, Joachim. Be on guard." Two artists were present, scratching away at sketches of me and the prosecutor. With no cameras allowed in the courtroom, these were the only real-time pictures the public would see the following day. I tried to buy one of them but considered the price too steep. I still regret walking away from that purchase.

After Steven finished, he handed me over for cross-examination. Here he was, the man my legal advisors had cautioned me to respect. Mr. David Boies, a lion of a litigator. Brimming with subtleties and unseen currents! With his penchant for intentional drama, his tall gleaming cranium, and his Midwestern patois, he had me finally close to his fangs. In an ordinary-looking, baggy suit, curly gray mane, darting eyes, and a quick tongue as I would soon find out! Supposedly a beast in court—a striking and disarming presence with no need to wear expensive suits! In his trademark black sneakers and his slight, mobile stature, he was an unsettling admix of Philip Spitzer and Bill Murray. "Intuitive and resolute." For the next two days, I would be at his mercy—a word not to be found in his dictionary. His reputation of extracting from our witnesses precisely what the Feds wanted the judge to hear preceded him. I was armed with the discomfiting foreknowledge that he'd earlier eviscerated a couple of my colleagues.

My lawyers had instructed me to be extremely careful with him, no playing of games and cautiously picking only fights I could win. But how do you triumph against a legendary talent and unforgiving monster like this? Concentrate, never drop your guard, and beat him at his own game. Have a better memory than his, and be super context aware. My inner voice called out: "Joachim, he has never run your business. You know more about the IT industry than he could ever dream of. Don't let him intimidate you. You stood up to Bill and Steve and other imposing figures. You can stand up to a *Panthera Leo*." With sixteen grueling courtroom hours ahead of me, forbidden to talk to my legal team and ominously warned of having little chance of escaping the stalker in a courtroom-cage setting, I did not despair. On the contrary, game on; who was being hunted?

In the Lion's Cage

He introduced himself in a warm and unassuming voice, piercing eyes glistering and a body language coiled for action. A prosecutor's role was not necessarily to discover the truth and nothing else, but the truth as people falsely assumed. His job was to artfully manipulate and massage it to his advantage. All being relative, my job was to offer my version and make the judge believe it! Him being a pro and with me being a courtroom novice, I had my work cut out.

The battle heated up quickly as he quizzed me why the content of my remade tape had been significantly altered, contrasting the first one. Very simple: OEMs had gotten bounty greedier. Not appreciating my answer, Boies suddenly asked if I knew what OEMs were allowed or not allowed to do when shipping Windows 95. After four years, I remembered several, yet I decided to make him work for the answer: "If I would keep this all in my head, I would really need to add some memory sometimes." His response, sinister and swift, revealed how much I had annoyed him: "And it's harder to add memory to some people than to computers." Hearty laughter rippled through the courtroom, dispelling tension. My provocation forced him to dig in and remind me of the specific liberties or restrictions he wanted me to address—wasting precious court time. Right away, I had slowed him down; eliciting a sweeping response had not worked.

Not getting too far, he ventured into the freedom OEMs had in altering Windows code and why we had restricted that. The desired answer: to hurt Netscape. He came back to this topic several times during testimony in a kind of blurry concurrency. My countertactic was to address the subject as vaguely as possible, denying him what he was looking for and complaining to the judge whenever possible about the speculative style of his nebulous pursuit. It irritated him visibly.

Another tactic I frequently used early on was to break eye contact with him, turning to the judge when providing answers. Avoiding his piercing eyes diminished his high self-regard. Feeling snubbed, he circled me desperately. My goal in this cat and mouse game was to make Jackson my friend and ally, like him or not. Ever aware, this prowling and hypnotic palavering lion, with his disarming charm, was my true enemy.

As soon as I mentioned that we did not allow OEMs to "tamper" with our core Windows code, the lion took the bait. In the following tense exchange, I turned to the judge, explaining in careful detail what I meant with the word *tampering*. Annoyed with my long-winded explanation, Boies admonished me with "to the extent that you are trying to be helpful, I would like you to wait until counsel examines you just in the process to get this through quickly." Yes, I was tapping the brakes to moderate the speed of his little joyride. To be sure, I asked the judge if my explanation had been acceptable. He told me to delay my helpfulness until my own attorney examined me. I thanked him respectfully, mentioning this was my first time as a witness and, in my naïveté, I simply did not know better. The lion moaned and groaned, and I envisioned his tail sweeping the courtroom floor, restraining his feline anger.

Suddenly he leaped forward, asking me if we restricted Netscape from being a featured Internet service provider (ISP) in advertising sequences OEMs had inserted into Windows. My prompt answer: "I am unaware Netscape is an ISP." Wasting a lot of time on technicalities, he exhausted, not for the last time, Jackson's patience, who interrupted him by saying, "I think we are having a semantic problem here," prompting me to reply, "That is exactly what I believe too. I am sorry." The whole Q&A had stalled. But persistent and clever as a pursuing lion is, he simply decided on another path from a differently angled direction to pursue his game. Still not able to extract what he wanted, my accuser tried to trip me up by pointing out that I had answered the same-self question differently in my earlier deposition. The following exchange grew pretty heated as I was able to prove the lion's prodigious memory had failed him. In response, his fangs let go of the document he was holding, angrily hurling it onto the courtroom floor.

From here on, he grew thirstier for bloodlusting revenge and, without hesitation, deployed a pretty dirty trick. He showed me a Gateway (GW) document that I had never seen before. This at once confirmed that GW had been actively helping the Feds to plot the case against us. The referenced document was under seal, meaning only attorneys knew about its content and existence. Looking at the large stamp Sealed on Top, I asked if we could even discuss this in open court. Jackson's prompt answer: "It's my order, and we can do it." Unjust for two reasons: Without prior access, I was absolutely unprepared in regard to its content. Secondly, I could not ask for legal advice. Jackson then prodded me to answer the pending question without reading the sixty-page document completely. Considering his actions unfair, I took my time, painstakingly delaying my answer to the long-suffering Boies.

He then engaged me in a discussion about why the latest version of IE was no longer uninstallable by end users. He was asking the wrong guy. Neither had I made that decision, nor was I privy to any technical reason why it had been disabled. I had my opinion about how unfair this was. No one asked about it. Spending a lot of time on the subject without getting the answers he was hoping for, the judge eventually asked Boies to watch the time, and as soon as he relented, we adjourned for the day.

A friendly securities analyst I ran into when leaving the courthouse stopped me. I told him I was forbidden to talk about my testimony, and he politely refrained at once. In parting, he gave me a thumps-up and said if any witness would have done as well, MS would have had a better chance of winning. Just nodding, I said, "Thank you." Returning to my hotel room, I had a lonely sequestered dinner.

Before drifting off to sleep, I reflected on what I had learned during the afternoon. As a debutante witness, I felt I had done pretty well. Now I was unconsciously analyzing Boies's interrogation style and identified one technique he persistently deployed. He would ask a broad and leading question, purportedly referring to a document he was presenting. After careful inspection, its content was often a nuance apart from how he had described it. Catching this early had protected me from being lured into the misty, alluring realm of speculation or volunteering information. Was this questioning technique correlated to his handicap or just his way of seducing a witness to be less guarded, loose, and vulnerable? I concluded that only staying alert and vigilant would prevent him from leading me down a dangerous path onto a craftily prepared trapdoor. I needed to keep him honest and his scope narrow. My legal team had warned me about this being one of his dangerous tactics and had helped me by watching him and interrupting the proceedings when he misquoted. So far, together, we had done well catching his mischief. With this assuring thought, Morpheus took me into peaceful arms.

If I'd anticipated the second day on the stand being less nerve-racking, I was soon to be proven wrong. I had slept well, gotten up early, reviewed my notes again, and entered the courthouse through the usual carnival atmosphere, this time with a bit more confidence. As soon as I was back on the stand, animated and eager to score a big point, Boies launched back into detailing browser choices in boot sequences. I could smell something was up.

All of these superstar prowlers are first and foremost absorbed in gratifying their egos; truth comes later, if at all. Boies was no exception, and his afternoon press conferences demonstrated that trait to the hilt. As a natural dramatist, he took great satisfaction in posing as victor and reveling in providing useful snippets journalists could march out into print. Yesterday afternoon, he had nothing spectacular to show off—truly disappointing for the brawny king of the animal dominion. Not having accumulated the de facto bragging rights he was accustomed to when devouring yet another victim rekindled his vigor. As we continued turning around and around on the by-now well-worn subject of where Netscape's advertising was allowed or disallowed, I briefly faltered—losing my concentration. I had already answered the self-same question thoroughly, at least twice, and his insistence on clinging to and hammering onto the moot subject riled me with annoyance. I dropped my guard, supplying him with an untrue answer—the one he wanted in the record.

The gallery groaned. Boies abruptly and skillfully changed topic. By now he had a huge grin in his face—finally he had scored! I interrupted him forcefully and turned to the judge, whom I told flat-out that I had misspoken, not having understood Boies's question properly. For ten long minutes, the lion refused to accept my argument, clawing back at me again and again. When I adamantly insisted on a correct answer for the record, he accepted it under ferocious protest. I ignored him and took the opportunity to make additional points, addressing the judge directly, infuriating my opponent further. I had recovered, through persistence and stubborn perseverance, but I was not happy with my performance. Smelling a faint of blood had excited the beast in him; would I survive the next fight? I longed for more than the tepid glass of water in front of me to regain my strength. Bourbon to the rescue!

I was then asked if we had ever incented OEMs to make our browser the default one. Still shaken from the prior incident, I couldn't recall this right away. But as Boies pawed deeper through the subject, the details reemerged, and I offered him a long explanation for why we had done so. He exploded with impatient anger. By all inference, he derided my memory, which wasn't supposed to recover so fast. Unable to shake me and score, he eventually accepted my answer under protest. After this last bout, my nervousness suddenly disappeared.

In the next round, I paused during a longer explanation, something he for sure did not desire to hear. As he impatiently chimed in, I told him "I am not finished" and just pushed ahead with my answer. When I paused for the last time, he angrily roared back at me, "Are you finished?" I was getting on his nerves, and he was getting on mine. In a game my attorneys had told me I could not easily win if it annoyed the judge and especially against such a seasoned and battle-savvy brute.

Analyzing during recess what had just occurred, I decided to exercise extra caution and restraint. Back in session, before I could collect myself, he struck with retaliatory precision and speed. He showed me an e-mail Bill had written about the importance of winning browser share, and when I battled with him over how to interpret its content, he questioned my ability to understand and communicate in English. Another opportunity to advance the clock! As I calmly and deliberately led him down the path that I can speak and write in several languages, he tried to establish that my correspondence with Bill had always been in English. A straightforward question would have produced a quick answer; instead, valuable minutes slipped by. Irritated, the judge was looking once again at his watch.

So the jousting clattered on, Boies unhappy, interrupting my answers constantly, and my lawyers asking the judge to reprimand him. A nice interplay of courtroom drama developed, ending with Jackson issuing the detached admonishment "Let him finish," with the lion snarling, "I will." Boies's forehead was now glistening with sweat. He understood after having hurt me once that it would get much harder to gash me a second time.

I gave him no inch when he showed me a second e-mail from Bill expressing concern about OEMs displaying other browsers more prominently than IE. For Boies this meant Bill was definitively telling me to stop OEMs from loading Netscape Navigator onto their systems. Bill never mentioned Navigator specifically. Boies would not accept my correction; for him, only one competitive browser existed, so again, in detail, I listed all the ones I knew, squandering valuable time on a hairsplitting issue.

My fear and respect for the lion's ability to successfully stalk me began to wane. No firearm needed escaping him in finality, just careful attention to detail combined with dogged persistence. Maybe he was tired and less prepared after having been in court for such a long time, or perhaps his support system was functioning less thoroughly than expected. My internal voice came back sharply: "Careful, he nearly got you down once. Don't become overconfident."

His ego wounded and his desperation visibly hungrier for a kill, Boies continued playing word games, prompting me to turn to the judge and voice my concern about his defamatory tactics. Boies' response: "The record will have to show that! The court will have to decide that. I am not permitted to comment on that. All I can do is ask questions." I was amused and amazed when the prosecutor felt obliged to justify his slippery tactics!

Back after recess and more alert than ever, Boies switched subjects: just another clever obfuscation to test my short-term memory. Behind him, once Boies's busy minions recognized I had not proffered the desired answer, they burrowed deeper into their mountains of documents, locating supplementary confrontational evidence. The paper shuffling never stopped. Still probing the motivation for implementing our restrictions, letters from OEMs were produced, outright condemning them. Way too theatric to be taken seriously, nor ever quantified by their authors. These were opinions of a scant few citing higher support costs and user confusion as reasons for challenging our zero alteration rules. For the prosecutor, they proved we caused consumers harm. I argued that we had a perfectly valid copyright and distributors, such as OEMs, were not supposed to change Windows without consent. I used Melville's *Moby Dick* as an example. Imploring that as a reseller of books, you can't just tear the last chapter out if you don't like the ending! You can't replace the introduction either, can't change the names of characters or the color of the whale. I insisted the changes OEMs had demanded were equivalent to "butchering" Windows. Our software pros did not want their masterpieces altered by amateurs. The judge listened.

During the rest of the morning, I repeated dozens of times that I considered IE part of Windows. Boies again paraded out e-mails from technical personnel naming IE a standalone Web browser. I could not help on this issue; I had only sold and delivered Windows and IE as a whole. Boies had to find other witnesses to claim differently.

Soon I was off to lunch. After finishing my meal, I rated my performance. On the first day, using a scale from one to ten, I had probably come close to a nine, but not today. I had to focus and improve. Going for the lion's throat had to be replaced by entangling him in his own net.

Despite my adamant refusal to acknowledge Windows and IE as separate products, the prosecutor just as obstinately refused to give up on eventually tricking a complicit statement out of me. So, immediately following lunch recess, he continued with his useless and futile exercise. He tried every possible angle, from the simple fact that we called IE a browser (purely a name for one of Windows's many features, like Media Center) to IE being delivered on a separate CD. He knew very well that the CD always included more than just IE code; nonetheless, he pressed me to say the opposite. Eventually he moved on, asking me if I had influenced integrating IE code deeper into Windows. Not a smart question, since when does a sales exec make intrinsic tech decisions?

The dispute about the browser ended in telling him there were other competitive responses available to Netscape to respond to our deeper integration efforts. In a nutshell, that company did not need to sell an OS to win the battle of the browsers—just a better browser. Now the judge was listening and interrupted, offering me an opportunity to explain why our browser was the better product. I explained my impression to him from a PC user's perspective. "I believe we really simplified basically the operation and usage of the system for the consumer." The lion was pacing impatiently as I embarked on a lengthier explanation; his growing agitation and desperation showed in a deeply creased forehead.

He abruptly switched the topic to browser distribution and was now eager to elicit from me the OEM channel was the most important one. He was wrong from the start; there were way too many other options. Boies was trying to prove we foreclosed Netscape in the OEM channel despite the fact that OEMs were shipping plenty of other browsers on their PCs, as my tape had shown. He tried to get me to guess how browser-distribution market share was split between the various options. I would have none of it.

To illustrate his persistent interrogation tactic, let's listen to one of the dialogues as Boies asks me if I knew the percentage of PCs being shipped with Netscape's browser. My answer: "I personally have never tracked this. I do not know." Boies: "Just give me an approximation." "I do not know." Boies again: "Can you give me a range at all?" And this goes on and on. "Give me a wild guess" ends in "Give me any range or approximation," and so on. So to shut him up, I said maybe 20 or 30 percent of the systems came with Netscape's browser, expressing I absolutely did not know any precise number and adding this was merely my gut feeling. Jackson allowed it.[46] Next question: "Now, all I am trying to do is probe what the basis of that gut feeling is. Has somebody told you information that you are relying on?" This continued for a long time without me confirming any exact data. He was truly badgering me—on a fishing expedition, in waters where no fish existed. Even desperately hungry lions normally don't venture there.

Never missing an opportunity to incriminate a witness, he launched into the testimony of John Rose, a Compaq executive, given earlier to help our case. When I told him I had read that particular section, he accused me of lying, claiming I had denied this in the morning. He was dead wrong and painfully had to admit he probably misremembered and immediately thereafter apologized.

Afterward, he stirred up another debate over why we restricted OEMs' shells not to boot directly into Windows, and got nowhere. Erratically, he switched back to prove OEMs were the key distribution channel for browsers. An assumption opined in another memo written by an MS

---

46   The true number later discovered in a document validated by Netscape was 22 percent.

product-line marketer. He had written "Users follow OEM's lead into the Internet." I again disagreed. When Jackson interfered, I told him maybe 20 percent of the users would do this because of obvious convenience and carefully defined my estimate as purely anecdotal. Not good enough for Boies. He continued prodding and probing into the subject further. Telling him I had no statistical evidence for my opinion, he finally let go.

A short recess allowed me to cool down a bit and step back from the heated exchanges I had endured. Back on the stand, Boies insisted on continuing this already fruitless topic of OEMs' role in browser distribution. I was getting frustrated—maybe the effect he sought to create. Next I was shown a document Paul Maritz had written defining "browser share as job one." Paul seemed to have feared once that programmers would abandon Windows and write only Netscape Navigator–centric applications. Not likely but remotely possible. This went to the heart of the government's case. In the David Boies's hall-of-mirrors fun house of logic, this was the reason why we had thwarted Netscape! Not believing this in the first place, I was the wrong guy to testify on this subject.

After more paper shuffling, he probed yet again for answers as to why we had integrated IE into Windows. I told Boies integration was our way of building a better mousetrap and shifting the ground rules on a competitor who had nearly 100 percent browser market share. Disliking my answers, he tried to lure me into technical details of the actual code integration. One subject I could not at all help him on, though he fecklessly and desperately clung to it at length. The judge was shifting around in his seat with growing impatience and asked Boies how long he wished to continue with the topic.

Agitated, the lion requested a bench conference, roaring out his frustration! I listened carefully to the dialogue between him, Jackson, and Holley. Boies complained that after asking me a question, he would "get a dissertation or a speech or statement." True, I wanted to slow him down! He was "reluctant to let that stuff sit there on the record without cross-examining the very statements" I had made. Then he accused me of saying I was just a salesperson and knew nothing about technology yet freely offered unexpected speeches about the benefits of integration. He was painting me as a chameleon able to camouflage the nature and dimension of my knowledge so the majestic lion could not catch me and knock me over. Claiming he was running out of time, having to reexamine things he didn't expect me to say and "that [sic: my answers] aren't really responsive" or relevant to the questions asked. Prompting my attorney to respond, "Your honor, I obviously take issue with this," asserting that after Boies had shown me technically oriented documents, I had responded with my best familiarity. Jackson agreed, adding, "His scope is rather limited in terms of what he can testify to knowledgeably." So the trio agreed to classify me as a salesperson. Hallelujah, I had achieved my goal, fencing in the beast. The reason for Boies's frustration rested squarely with him. He was drifting back and forth with a feral and misguided priority of chasing after opportunities to corner his prey, inviting—demanding—unqualified testimony in regard to modern OS technology I simply did not possess.

Revenge, though, was being craftily plotted, with questions getting longer and windier as their substance became less precise. I simply refused to answer them, forcing him to break them down and clarify them. Not showing such terrific focus anymore, and being thrown off script during the bench conference, he reinvestigated old issues already covered ad infinitum until we arrived back at the subject of copyright. I had supplied a copy of our copyright applications for both Windows versions with my written testimony. Even here he wasted time by flailing away over whether or not we had a copyright for the different versions of IE as well. Yes, we did as part of Windows, but not separately!

After another sealed document was unzipped on the fly, Boies then went ahead and asked questions about prices Dell paid in comparison to Gateway (GW). The judge cautioned him because the document he was referring to was sealed. He knew full well this was confidential information and pricing issues were not to be discussed in open court. After my attorney intervened, I assumed Boies would no longer delve any further into similar topics. Stubbornly he persevered, asking the same question about Compaq's Windows prices in relation to GW's. Neither Jackson nor my attorneys objected. I was flabbergasted, but with the damage already partially done, I proceeded to answer truthfully, in public. The TV stations in the evening and the newspapers the next morning made this part of my testimony the big story. Boies openly basked in the fact he seemed to have proven GW had been disadvantaged—as he phrased it in his afternoon press conference—because "they did not play by the Microsoft rules." The truth was different. GW paid slightly more for Windows because she shipped less than half as many PCs than the competitors mentioned!

At last we reached the IBM relationship. Weirdly enough, Boies arrived upon it by probing into whether we had ever tried to talk IBM out of supporting Sun's Java product. He was trying to prove a pattern whenever an OEM supported a competitive environment we would intervene. IBM's support for Sun, a competitor of hers, did not appear to be smart. Pointing this out to Sam Palmisano in our first meeting, he had told me this wasn't his decision to make and he could care less—semihonest as I thought. True or not, it mattered little to me; I had made my point, and that remark could not be counted as coercion.

The background was a document written by Bill voicing that if IBM was really in the Sun Java camp, it could affect our relationship. We hardly had any. Mr. Boies, however, worked feverishly to have this comment translated as just another evil attempt to coerce IBM. A part of a pattern he was trying to prove convincingly. Since when couldn't we express our dislike of IBM supporting one of our competitors' products? A heated exchange developed when I refused to second-guess Bill, letting Boies say, "I'm entitled to know whether you understood this meant that Mr. Gates was saying . . ." Since when was Boies entitled to me reading Bill's mind years later? Asking me in the same context if Bill was "being primarily concerned with IBM's welfare or MS's," I answered truthfully, "With both," hoping Jackson would believe that Bill wanted peace with Big Blue.

After a brief recess, I was confronted with yet another sealed document, this time generated by Compaq. It stated that one of my former employees, Jan Claesson, had told one of Compaq's VPs four years earlier that if Compaq would ever opt for just a per-copy license, this would be a "big issue." Boies's idea was to demonstrate we had coerced Compaq into a per-processor license with this not-even-vaguely retaliatory comment. When Jan's discussion had taken place, Compaq operated under a highly advantageous per-processor license, which still had another three years to run. A per-copy license would have no doubt yielded higher royalties. I expressed to the judge how wildly illogical that statement was considering the circumstances.

The most serious issue Boies brought to my attention was a discussion Jan C. supposedly had with Mike Clark, one of Compaq's VPs. He had whispered in his ear what other OEMs were paying for Windows 95. Employees working for me knew such information was not to be disclosed. Jan, being a supervisor, made his breach of confidentially extremely severe. When Boies asked me point-blank how I would have reacted if I would have known this earlier, I told him flat out I would have fired the guy. In a painful moment of truth, we all agreed.

Wanting to demonstrate how badly we had damaged Netscape when bundling IE with Windows, Boies probed the details of a competitive review Steve had once initiated. A long-winded question followed, hard to understand. My response: "That's a long question"; his answer: "I will break it down." Unsatisfied with my answer, another of his long, loopy, run-on questions followed. Boies was obviously getting tired, and his questioning style reminded me of Thomas Mann, a German novelist of the nineteenth century, who wrote the largest longest German sentences I have ever read.[47] I noted: "Mr. Boies, it is very late. Please break it down so I can

---

47 It extended over more than three pages in his famous 1901 book *Die Buddenbrooks*.

understand it." He answered: "It was a long question and awkward and I apologize. It is late." Jackson nodded.

The following exchange was one of the funniest. Obviously frustrated, Boies asked if I knew how many people were downloading browsers from the Internet. I answered: "I do not know, but I believe it is substantial." He then tried to lecture me: "Would you be surprised, sir, that for the first nine months of 1998, it was negative?" I promptly responded: "Meaning they sent their browsers back?" By now we were closing in on 5:30 p.m.—the courtroom, including the judge, exploded with tension-dispelling laughter. Boies was wearying. It took him considerable time to recover at last, asking the right question a few minutes later. What he was referring to was a survey showing a decrease in the number of Internet browser downloads.

He had a point. In '98 we were already shipping IE version 4.0, a winning product neatly proving exactly the opposite of what he was trying to ram down the judge's throat. Not until we had such a superior product did users stop downloading Netscape's browser. As could be expected, improved features and quality had won!

The next session happened in camera, meaning without the public. It took half an hour and focused on OEM pricing and policies in intimate details. There were no surprises, though Boies made certain Jackson understood we had pricing power; we had for sure, but this in itself meant nothing. I disputed his argument of not having competition, but he insisted we operated only in the Intel-based PC market and no other computing environment was relevant. This was for the judge to accept or reject. We finished by 6:10 p.m. It had been a long, grueling day, and we had gone an hour beyond the normal close of court—a first for Jackson. Another lonely meal completed my day.

The next morning found me fresh, optimistic, and energized, knowing in a matter of hours my ordeal was coming to an end. Back on the stand, Jackson reminded me—with a sarcastic undertone—I now could be helpful. As his rebuttal began, Holley grinned at me: "Don't mess it up on me!" He did an excellent job asking precise questions and, as a result, got exact answers. Holley was narrow and focused and not ambiguous like Boies had been. I only hoped the judge would take this part of my testimony to heart. Another part of Holley's rebuttal made another of Boies's many tricks transparent, showing me a document without all of its attachments. As Holley pulled them entirely out, the formerly suppressed piece contained exactly the opposite of what Boies had alleged. By now the atmosphere in the courtroom had palpably shifted from tense to pleasant. When he asked me about the current piracy rate on IBM PCs, I estimated it at about

15 percent. Jackson, not understanding my answer, wanted the number confirmed: "Was it 50 percent?" I corrected him, and Holley responded, "Luckily." And I added, "If it would have been 50 percent, I think I would have not done my job," with Jackson chiming in, "You would be out of a job!" Inspiring another round of healthy laughter.

The lion rose for a last time—for one last charge. He posed a meaningless question to me in final desperation or as if putatively filling in the last lines of his inquiry. Making it easy for him, I answered promptly. He turned with a sigh, indicating he was finished with me—no more growling. I was free to go and talk to my attorneys, delighted the ordeal was over at last.

Preparing this book, I reread the court records, reviewing my testimonial battle with the clarity of hindsight. Boies's tactics appeared to be predictable and transparent. Well-prepared and circumspect witnesses should see right through his inner schematics laced within the moment rhetoric. Painstaking preparations combined with steady nerves, the will to hold his feet to the fire as he holds yours, and never giving an inch will wear him down. My legal team gave me an 8.5 performance rating. I would have done better if it hadn't been my first showing in court. I had no quarrel with Jackson except when he unsealed secret documents at gusto. I found him attentive and more supportive than anticipated.

The critical responses to my performance were mixed. *Business Week* reported I "took the court room on a wild semantic roller coaster ride." Yes, the businessperson being grilled employed a different vocabulary than the prosecutor and differed with him in the definitions and inferences within the vernacular of that language. The papers reported correctly, "For Boies, proving a point in court was like pulling a tooth." More painful for him than for me because it made him work for answers and shed time. The same article stated that Boies "was clearly annoyed at Kempin's performance," as he should have been. Nevertheless, he mentioned to journalists in one of his daily press conferences that I was "one of the smartest men working at Microsoft." Coming from him, and probably meant sarcastically, he was obviously reflecting on his own prowess.

BETRAYAL OF JUSTICE

REBUTTAL

After a recess of thirteen weeks, the trial continued with three rebuttal witnesses from each side. Ted Waitt was rumored as being one of them. At the end, he was not called upon or didn't want to show up. A bit of a tactical chess game developed. Who would name witnesses first? I believe MS made a mistake in doing just that. The early rebuttal phase was little more than a rehash of what was presented earlier, as each side tried shaking and stirring former statements from witnesses who had been called a second time. From behind the scenes, I understood that Bill had been beseeching the attorneys to lift their ban. They however determined that his showing would no longer affect the outcome. Exposing him to this—in their mind—extremely hostile judge would do more harm than good for his soul and the case. Steve's name never came up. For a second time, a lot of employees were mystified. A no-show for the first round and in a lack-of-leadership redux, not present for the second one either.

We scored some points when a hostile witness had to admit Netscape had distributed 160 million browsers before being taken over by AOL—representing more than nearly two browsers for each existing Internet user. Franklin Fisher finally determined that the monopoly price for Windows could be calculated at just above $200. He did this with a generous elasticity factor as Schmalensee, our expert, later pointed out, but at least we had a number. Numbers speak loud and clear, and with our average OEM price at $55, we were looking forward to how Jackson would take such a crucial fact into account.

The last witness was a big surprise for me and our legal team. The Feds had finally coerced—sorry, convinced—IBM to allow one of her employees, Mr. Garry Norris, to testify against us. Quite possible that Boies had used his old IBM connections to accomplish this. The man sent by IBM was on a mission, and being last, nobody could refute his testimony, meaning the ultimate monopoly of knowledge had been bestowed on him. I knew him personally, and when I read what he boldly asserted, I was outraged. My lawyers got the rap. Mark Baber, an ex-employee who had been directly negotiating with him, came out of retirement; and when asked by a journalist, Barber stated Norris's testimony was nothing other than "total fabrication."

Jackson was unmoved. When one of our attorneys suggested he should call me back on the stand, Jackson deflected his request, insisting, "I would have to answer for a lot more," indicating he would be wasting his time. By which he openly implied that he was unlikely to believe me anyway. Here, finally, is my side of the story.

Now came before the court what the Feds would later call their star witness, on loan from Big Blue. Norris's personal motivation to tell the whole truth and nothing but the truth could be found in his résumé. Plying his trade from '93 to '95 for IBM's personal software division licensing OS/2 to OEMs was surely a less-than-fulfilling career, being consistently pummeled by MS and never making quota. Leaving just a bitter taste in his mouth or lust for revenge? Something had consciously or unconsciously clicked when, in March of '95, he changed divisions, receiving a promotion to director of software strategy for IBM's PC Company. He did not sell hardware but oversaw licensing negotiations with MS and Lotus. In the case of Lotus, his engagement lasted less than four months, coinciding with IBM buying the company outright.

Mr. Norris had a saleable witness résumé: an accounting background and an MBA degree embellished by an aborted law school stint. Coached by IBM's legal team, the onetime wannabe attorney came with an advantage on the stand. He gave the impression he had been quickly accepted by IBM's hierarchy. Inspiring much trust from above, he indicated he knew exactly what was being planned for the good of Big Blue. The prosecution made the judge—who later claimed he liked him—believe he was close to the decision makers who trusted him to be a reliable mouthpiece for a victimized company and its poor, defiled past.

He was transferred and promoted to his new job four months after his old division had started the nasty "IBM First" campaign at a time when our infamous IBM audit was already in its seventh month. Admittedly, he was immediately and thoroughly briefed by his predecessor on what had transpired during '94 before he arrived. Whatever he had been briefed on then definitely got refreshed by IBM's legal team before his trip to DC. Analyzing his testimony in details indicated neither his refresher course nor the on-the-job briefings had left a sufficiently reliable and lasting footprint in his memory.

The judge therefore had to differentiate carefully between how much the man from Big Blue had really witnessed and what was plain old hearsay.

As an example, let me cite what happened when Norris was asked about the IBM audit of 1994/95. First of all, he claimed the audit was primarily concerned with LAN-Manager, by then a passé product. Yes, IBM had licensed it, and therefore we did include it into the auditable product list. However, the witness did not seem to know what the audit's main focus had been: OS/2 triggered by IBM's marketing group publishing aggrandized numbers. Did our truth-outing mission bruise Norris's ego, or had he even lent a hand to create the scam when working in that group?

He never admitted how much IBM had tried to block the progression and completion of our audit. Not knowing these details puzzled me. His IBM lawyer knew for sure. Norris's remarks about the audit related mainly to a June '95 meeting where IBM's upper management finally came embarrassingly clean, announcing the requested data could not be compiled in time. My immediate response to IBM's late confession was asking her negotiators to pay up and move on signing the 95 license at hand. Norris portrayed my proposal as a sudden and out-of-the-blue retaliation because IBM competed with us with OS/2. He adamantly repeated this numerous times, and not only in this particular context. I am convinced that my reaction—several times expressed before—was no surprise for the decision makers in IBM. He suitably forgot to remember. The inconvenient truth he refused to admit: equal and fair treatment of all OEM customers was what I always had on my mind. IBM's obsession with OS/2, while disturbing, never determined how we in OEM dealt with her.

Let me further touch on how Norris testified about the '94 alliance proposal we had worked out with IBM. He informed the judge he was clueless about the content of our final, generous offer, though nevertheless characterized our earlier ones as distorted and unbalanced. His memory lapses just following a coaching session by his IBM lawyer were a bit too fitting. Yet with no one else testifying on the subject, his testimony stood unchallenged and oddly profound. He never acknowledged how the nasty "IBM First" campaign, initiated in December of '94, left no room for friendly maneuvers and resulted in both companies pursuing a no-cooperation-with-the-enemy policy for years to come.

Mr. Norris's testimony allows me no choice but to classify him as an unqualified storyteller, considering his inexplicable memory lapses and his absence on so many occasions when vital decisions and promises were actually made. Hearsay trumped, and I would have loved to clean up his mess. With him as the last witness, the trial simply marched right along.

After he left the stand, Jackson encouraged both parties to settle the case. Obstinate state AGs[48] made sure the talks went nowhere. Jackson, observing from the courthouse, acted by publishing on November 5, 1999 as finding of facts the impressions he had garnered during the trial. By then he had started giving background interviews to a select number of journalists to "demystify" the methods employed by a judge in arriving at what he called his findings of facts. Billed as a noble cause to educate the public, these interviews will later give us useful insights in his motives and character.

Before deepening my analysis of his ruling, let me point out Jackson's handling of hearsay evidence during the trial, defined as "a statement not made by a witness testifying in court and offered to prove the truth of the matter stated." If allowed, it deprives the opposing side an opportunity to cross-examine the person originally producing the information and to get to the bottom of such evidence. Meaning the so-called truth. Instead of striking hearsay from the record immediately—like most judges will—Jackson left the door open and promised to weigh it appropriately when preparing his finding of facts.

An example of how far his hearsay relaxation went can be demonstrated by the oral testimony of a prosecution witness. One of our attorneys asked him if he had any other basis for what he had opined in his written testimony other than what he had absorbed from press articles. His answer: "I do not have any other basis for the statement related to the provisions in the agreement they [sic: MS] have with their manufacturers"—meaning the OEMs. Holley therefore moved to strike this particular section, reasoning neither the witness nor the journalists mentioned had ever read MS OEM contracts. Jackson denied the request without prejudice—meaning without telling either side if he would use the hearsay info or not—giving him a chance to later secretly roll a dice in his chamber.

---

48 Attorney generals

Monopoly power

Historically, antitrust convictions are based on how much share and market power a company possesses. Believing the government experts, Jackson concluded our OS market share of the IBM PC clone market was 95-plus percent and rising. To ever arrive at such large number, he had to include all pirated copies without leaving enough room for OS/2, Linux, BeOS, and UNIX, to name a few of our competitors. Considering MS and DRI were still selling—not on trial—pure DOS systems, the number strives even less. He nevertheless ruled that we had monopoly power in this narrow segment, surmising that no reason existed to believe any serious contender could emerge to challenge our presumed monopoly in "a few years." His assertion neatly avoided specificity and contained no empirical proof or any other quantifiable substantiation!

Jackson dismissed Apple as a serious contender and never mentioned IBM's PowerPC ambitions. To play down Apple's role, he argued she could never endanger the IBM PC clone segment seriously without relicensing her technology to other companies or significantly lowering her prices. Completely ignoring Apple's ability to change this by her own competitive volition—with the stroke of a pen.

The judge further neglected or dismissed alternative computing devices and the degree to which emerging paradigms could profoundly change the competitive landscape. Server computing was played down (today you would call this cloud computing), the power of handheld devices was not taken into account, and consumer interest in competing information appliances was ignored.[49] Alternative game consoles weren't treated as serious competitive contenders either. The rise of cheap networking computers laden with Linux, later called netbooks, was played down with arguments over existing network latencies and security concerns. They would be handily overcome in less than a year!

---

49 I have several friends who stopped buying PCs when handheld devices became powerful enough and were fitted with rich Internet communication features.

Jackson's constrictive and narrow focus of the computing market made him classify consumer investments in Windows application programs as something they would hold on to forever, right along Franklin Fisher's testimony. After having chosen the Windows PC path, he predicted consumers would consequently behave pitch-perfectly according to chaos theory.

I observed consumers switching from IBM PC clones to Macs and vice versa, freely accepting the costs when they outweighed the potential benefits—sometimes just one killer application or a unique hardware innovation was enough. The judge forgot to mention that most application software licenses are PC specific. Meaning that if their owners bought replacement PCs, they would also have to buy fresh licenses. Considering application programs had a limited shelf life, his reasoning became more peculiar. Every two to three years, when a new majorly improved version appears, users will have to reinvest to stay current. PCs do not last forever either, and instead of paying for expensive repairs and upgrades, people tend to buy new ones to enjoy the latest technological advances—the perfect time to switch platforms. Consumer behavior was consequently not written in stone or determined as much by chaos theory as he had himself convinced of.

Jackson further forgot or did not understand how effortlessly software is cloned. Legally. Cloning Windows was certainly not an insurmountable task, and it actually was accomplished several years later by two small start-ups. IBM, Hewlett-Packard, Sun, and Intel, as I had once feared and expressed in a memo to Bill, for sure had that expertise. By pooling resources or engaging the Linux volunteer army, these high-powered contenders could have accomplished this easily within a year or two at minimal cost, contradicting Jackson's belief that such an undertaking would be prohibitively expensive.

He opined: "The overwhelming majority of consumers will only use a PC OSs for which already exists a large and varied number of high quality full featured applications." I am sorry but the IBM PC entry, according to this principle, should have never made inroads in the already-existing PC market then dominated by Apple, Commodore, and Tandy and its rich bandwidth of application software. He further mentioned that developers tend to write to the computing platform with the largest installation first. Not always true! The IBM PC started with very few. The same holds true for Windows version 1.0 and the Mac OS. All succeeded because developers risked their livelihood writing for unproven platforms, and consumers gambled by giving them a try. Let's further not forget that in spite of frequently incurring exorbitant costs, software vendors often simply hedge their bets.

Calling the phenomenon of an existing barrier to entry a "chicken and eggs" problem, Jackson exclaimed that it "would make it prohibitively expensive for a new Intel-compatible OS to attract enough developers and consumers to become a viable alternative to a dominant incumbent in less than a few years." How on earth did he know how much risk developers were willing to take? The MS trial experts expressing a differentiated view were totally ignored as they looked beyond the narrow Intel-based PC platform. According to them, a thorough and objective big-picture discussion of the topic should have included all extant and semi-interchangeable computing platforms.

Now we have arrived at the barrier of entry and why in, his opinion, it could not be overcome by any new contender at reasonable cost. The government experts did not quantify these, and neither did Jackson. In its place, he blamed positive network effects "by which attractiveness of a product increases with the number of people using it" as the major reason for his ruling. He called this phenomenon a "self-reinforcing cycle" or a positive feedback loop.

Judge Posner, the antitrust expert, defined the common meaning of a barrier to entry in his book as "any obstacle that a new entrant must overcome in order to gain a foothold in the market, such as the capital costs of entering the market on an efficient scale. This is a dubious usage, since the firms already in the market must incur expense to remain there, must continuously overcome, therefore, the same hurdle that faces a new firm." His conclusion goes to the heart of the case as Posner opines that the existence of a barrier of entry is ill suited to be the main reason for a monopoly condemnation.

Contrarily, Jackson characterized that positive feedback was solely responsible for creating a "vicious cycle for the would be competitor" without giving us any credit for stunning innovations, advancements, competitive struggles, and hard work. Adding to this, a new entrant would have to overcome the seventy thousand applications already in existence for Windows. Had he ever heard of quality beating quantity when it came to software? He then went on to claim that recruiting independent software vendors (ISVs) would be prohibitively expensive for any new contender. We spent about $80 million on recruiting and nurturing them annually. Windows had been launched at the end of 1985. By now, nearly fifteen years in the market, what we had invested amounted to less than $1 billion. A drop in the bucket for the larger competitors interested in ousting us!

The most fascinating part of this section was how he judged IBM's OS/2 failure. It was explained as IBM's inability to attract sufficient ISVs. The reader of this book will hopefully characterize this as an unsustainable argument knowing the OS/2 design sins IBM committed and how consumer unfriendly it was. To put Jackson's argument once and for all to rest, let's dive again into Gerstner's book, *Who says Elephants Can't Dance?*

Years before the Feds dragged us back into court, Mr. Gerstner's stated conviction was "the reign of the PC was coming to end"! (Today we know his foresight arrived way too early.) While he would have liked IBM to be a dominant PC vendor, he strongly believed—deviating from his predecessor—"it was no longer strategically vital." Not wanting to chase the hardware lead any longer, he consequently labeled the fight for PC OS dominance an "expensive distraction" and counter to his view of "where the world was headed": to open standards in his opinion! He then continues, "The last gasp was the introduction of a product called OS/2 WARP in 1994, but in my mind the exit strategy was a foregone conclusion. All that remained was to figure out how to withdraw." He then describes the exit strategy Big Blue engineered so her existing customers were not left in the cold and goes on to mention that his decision caused a lot of "emotional stress" inside Big Blue because many employees believed OS/2 was the better product and could, at some point, have prevailed.

Unfortunately, and by judicial design, these facts would not be revealed in the courtroom. IBM's Mr. Norris offered Jackson a different account: MS killed OS/2 with her policies and by intimidating OEMs. Admittedly, Gerstner's book had not been published when Jackson wrote his finding of facts. IBM's witness, who Jackson said he believed and who presented himself as being close to IBM's decision makers, should have known whose decision eventually killed OS/2. "A foregone conclusion" admitted and enforced by IBM's own chairman and CEO! I rest my case.

According to Jackson, our innovation power enabled us to "push the emergence of competition even farther into the future by continuing to innovate even more aggressively . . . While MS may not be able to stave off all potential paradigm shifts through innovation, it can thwart some and delay others by improving its own products to the greater satisfaction of consumers." Are there any other means to compete and win?

The next topic he tackled had to do with Windows 98 upgrade pricing. I remember that the product group was trying to lower the upgrade price substantially from $99 to $49. I intervened at once. In my opinion, the marketing guys had hugely overstated the incremental sales effect of such a price decrease. In the end, they agreed with me. I recommended a price of $89, which was adopted. Establishing $89 as the new price meant for Jackson that we exercised monopoly power because we had "substantial discretion in setting the price ." Yes, we did, and as a net result, we lowered the upgrade price. Is this how a monopolist behaves?

His next condemnation was aimed at the market development agreements we offered OEMs. He cited Gateway and IBM as the two OEMs resisting their adaptation in various ways. He speculated that we offered these agreements "to enlist them [sic: IBM and Gateway] in our effort to preserve the application barrier of entry." In reality they were designed to promote technical improvements, reduce piracy, and spread brand awareness. I argue they made Windows-powered PCs in general more competitive and consumer friendly. Participating OEMs advancing the platform got rewarded accordingly. They not only received discounts but also improved the values of their PCs, ultimately helping them to win against alternative PC platforms. Gateway and IBM independently decided to be less active in this regard. I assumed their decisions were based on well-thought-out reasons and analysis. Where was the evil intent—enticing willing customers to produce more attractive choices?

Had we charged a monopoly price for Windows? "It is not possible with the available data to determine with any level of confidence whether the price that a profit maximizing firm with monopoly power would charge for Windows 98 with the price that MS actually charges." Case closed? He condemned us anyway by reasoning that if we would have charged less, it would not be "probative of a lack of monopoly power" but only a sign of "a low short term price in order to maximize profits in the future." Our time-honored tradition of holding OS prices steady over long periods of time went unnoticed. Nor did he stop there. According to him, we kept Windows prices low to stimulate the IBM PC clone business and attract new users to "intensify the positive network effects that add to the impenetrability of the application barrier to entry." What else would you expect from a company publically stating she wanted a PC—any PC—on every desk and in every home?

According to him, our evil intentions went farther. Not maximizing prices, we spent our monopoly power on restricting OEMs from promoting "software that MS believes could weaken the application barrier." He concluded that incenting them to build and sell NT-powered workstations was solely done to hinder them from developing thin clients (like netbooks). He speculated they would not be based on Windows[50] and therefore could have hurt our business. Our NT incentives had only one goal in mind: increase the sales of NT units and help participating OEMs sell higher-end and more profitable systems. The facts were in the record. There was no evil brainwave in my

---

50 A lot of them eventually were!

head when promoting NT, nor had I ever thought about hindering the development of thin clients—this was totally out of my control!

Jackson, siding again with the Feds, found us guilty of having attempted to crush competing products with prohibitive anticompetitive actions. Several incidents were available for him to potentially arrive at such a conclusion. For him they manifested a pattern of misbehavior by our employees, which a monopolist, according to him, should have refrained from, while our competitors were allowed to use them unabatedly. With this opinion, he contradicted what Judge Posner postulates in his book *Antitrust Law*: a monopolist has the right to compete vigorously!

As an example for such an incident, let's look at how Intel in '95 tried to improve application programs using audio and video signals. Considering a solution we had embedded in Windows 3.1 substandard, one of Intel's subsidiaries developed its own software. Introducing it in the summer of '95 developed into a nasty conflict between the yins and the yangs of the PC industry. By then we were less than three months away from launching Windows 95 with OEMs totally occupied to launch ready their PCs. Not interested in experimenting with Intel's new software, which was not written for and did not work with 95, they told Intel after talking to us to shove it. Intel, though, not convinced we would hit the release date, pushed back, advising OEMs not to believe us.

Behind the scenes, Bill tried to overcome the deadlock and meanwhile public feud. Eventually, Intel's CEO Andy Grove and Bill met to solve the conflict. As a result, Intel promised improved CPU features so we could utilize them easier for audio and video processing. What bugged the Feds was how the two guys arrived at the deal. According to an Intel witness, Bill threatened Andy with not supporting Intel's next CPU model if he didn't let go of competing software activities. Even if this had been put forth, it would have been silly. We had little chance of ever following through. Bill's ego had simply gotten the better of him. There was no alternative supplier to buy from. AMD, Intel's nearest competitor, never possessed the capacity or the financial muscle supplying the total market. To think we could have entered the semiconductor business was nothing other than nuts—we had neither the expertise nor the capital to accomplish such a feat. This so-called threat was an empty, off-the-cuff remark—hollow as an old oak tree. I assumed Bill knew, and so did Andy. After the two top honchos passed the olive branch, Intel prospered, and so did MS. No harm was done. I believe Bill had simply persuaded Andy Grove to stick to his guns and in return offered him increased cooperation where he wanted it. Partners in arms, as we were, do exactly that.

In his opening salvo in regard to Big Blue, Jackson used strong words:

"The IBM PC Company relies heavily on Microsoft Corporation to make a profit, for few customers would buy IBM PC systems if those systems did not work well with Windows, and further, if they did not come with Windows included." I do understand the "works well" comment but not the "must come with Windows included"'—you could always install it later, and millions of users did. Let me add that at no time had anybody in MS ever sabotaged Windows from working properly and efficiently on IBM PCs. Why sabotage our own sales potential?

The judge then went on, detailing that IBM marketed products competing with MS offerings. "This has frustrated the efforts of the IBM PC Company to maintain a cooperative relationship with the firm that controls the product [sic: Windows] without which the PC Company cannot survive." There are at least two erroneous conclusions in this statement. First, we wanted to sell IBM as many Windows copies as we could, and we therefore made all our products—not just Windows—work flawlessly on her PCs. With this in mind, IBM did not need us to survive, then or ever. Windows was freely available in the marketplace, and IBM did not have to license it to sell PCs. Remember, she did not install any OS on their early PCs and did quite well.

We had indisputably helped IBM grow her business by supporting her hardware with our software portfolio and, in the process, created a whole new and meanwhile booming industry. Everything else paled, and the divorce in '91 was definitely not our fault alone. MS in general and Bill especially, with enduring passions, wanted Big Blue as a friendly partner. We tried repeatedly to find ways to forge a new alliance—in 1994 and 1996—proving we were serious in overcoming the challenges of the past. It was IBM who refused!

The challenge within her PC division derived from having to serve two masters: the popular demand we created and Big Blue's own software ambitions dictated by her corporate executive committee. Caught in the middle, her PC sales declined in '94 while the rest of industry was growing. With the tail wagging the dog, the PC company leadership was entrapped in cross fire. Expanding on these distortions, Jackson accused us of leveraging the licensing of Windows 95 "to move its [sic: IBM's] business away from products competing with Windows and MS-Office." When IBM continued her OS/2 pursuit, we punished her with "higher prices, a late license for Windows 95, and the withholding of technical and marketing support." The facts, as I've detailed earlier, were quite different—IBM's stubbornness to comply with the audit caused a stir—but by no means the harm Jackson cited. By believing an unqualified storyteller and not allowing the best witness to come forward, Jackson handicapped himself.

Next he addressed why it was unfair that IBM paid higher royalties than Compaq and Dell. Having a competitive product in OS/2 and therefore diverging goals, he therefore argued that IBM had to refuse being an active market development agreement (MDA) participant. While somehow plausible, he missed the real reason for the higher prices. Both companies he mentioned sold more PCs than IBM. Our volume-sensitive pricing model therefore remained rightfully favorable for the market leaders, and IBM should have swallowed her pride, used a spreadsheet, and gained the available MDA rewards. It was in her best interest to make her PCs more competitive and Windows-ready, something most of her customers demanded.

When IBM complained to us in '94 that we were giving Compaq an advantage, her management was wrong and knew it. IBM had—as Mr. Norris explicitly confirmed in court—the best (and unwarranted sweetheart) deal in the industry. Offering to work out a marketing alliance against our initial conviction was Bill's and my response. We were ready to comprise and tailor a deal by taking her internal restraints into account. It included marketing funds, not royalty reductions as Mr. Norris wanted the judge to believe. The deal further asked IBM to get behind Windows 95, installing it on around 50 percent of her systems, a move her negotiators verified would not endanger OS/2 ambitions. Judge Jackson nevertheless interpreted our proposal as "Of course, in accepting the terms, IBM would have been required to abandon its own OS." There was nothing in the record to support this conclusion. The deal fell through because IBM's management chose war over cooperation and later sent Mr. Norris to shift the blame, camouflaging a self-consigned defeat.

The most problematic statements in Jackson's so-called facts can be found in regard to delivering Windows 95 to IBM. Contrary to what Mr. Norris declared, IBM got the final version of 95 delivered exactly as other OEM customers did—without having a signed license in place, totally against our own policy, which I personally overruled.

At last, an off-the-cuff remark I made one time in Western Montana. I remember well when and where. After one of our clandestine meetings west-bound on I-90, between Clinton and Missoula, I mentioned to Tony Santelli that if IBM would drop Lotus Smart Suite from her PCs, we could shoot the moon together. The way I said it, and the way Tony understood my remark, was indisputably nothing beyond a naughty little joke. We both knew full well that engaging seriously in a *quid for pro* on this subject would result in the darkest of all possible legal consequences for both companies and for us as individuals. We had a chuckle over my flippant humor, and neither of us ever seriously broached the subject again. Contrary to the reckless conjecture of a few historians and gossips, my entire division from top to bottom, including me, was thoroughly trained in antitrust issues by our attorneys. For me or any of my people to engage in any serious discussion about kicking Lotus out for a reduction of Windows royalties is absolutely preposterous.

In his findings of facts, Jackson published an excerpt of a letter I wrote to the head of IBM's PC division. It speaks for itself, and I am grateful he did:

> As long as IBM is working first on her competitive offerings and prefers to fiercely compete with us in critical areas, we should be honest with each other and admit that such priorities will not lead to a most exciting relationship and might not even make IBM feel good when selling solutions based on Microsoft products. . . . You are a valued OEM customer of Microsoft, with whom we will cooperate as much as your self-imposed restraints allow us to do. Please understand this is neither my choice nor preferred way of doing business with an important company like IBM.

And in closing:

> You get measured in selling more hardware and I firmly believe if you had less conflict with IBM's software directions you actually could sell more of it.

Leaving this subject behind, let me explain how Jackson believed we had harmed consumers. Amazingly, he gave us credit for charging zero dollars for IE, therefore reducing costs to consumers and forcing Netscape not only to match our price but also to improve Navigator. In the same breath, he accused us of doing this to defend our turf, meaning the famous and ghostly barrier to entry. Our actions therefore "have also caused less direct, but nevertheless serious and far-reaching, consumer harm by distorting competition." He offered no empirical data to proffer his statement, which, by now, is no longer surprising.

Concluding his findings, he wrote that our "prodigious market power and immense profits" had been used to harm firms insisting on pursuing their own interest in competing with our core products. We had hurt these companies and "stifled innovation. The ultimate result is that some innovations that would truly benefit consumers never occur for the sole reason that they do not coincide with Microsoft's self-interest." A speculative, galaxy-shattering accusation, without naming any specific product, company, or independent market data.

Our legal team had arrived at dark suspicions about Jackson's incompetence much earlier and was less surprised by how he had failed to see how the monopoly case in front of him required sophistication beyond a run-of-the-mill Chicago school specimen. Before him, in my view, was a case for wide-open rumble-tumble competition, potentially leading to a network-driven natural monopoly. In general, not a valid concern for antitrust enforcers, as insightful commentaries in several law books remark.

VERDICT

I spent a great deal on Jackson's finding of facts—a 141-page document—analyzing and rebutting mostly the parts related to my old business. He erred profoundly, but the public damage was impossible to undo. Instead of delivering his verdict pronto, Jackson encouraged the parties to again engage in settlement talks. With the finding of facts hanging like a Damocles's sword over our heads, he introduced Chief Justice Richard Posner of the Chicago appellate court as mediator. Posner was a good choice and, for the moment, a promising sign. He was one of the leading antitrust experts in the country and was in the process of publishing, as mentioned earlier, a highly respected and often-referenced antitrust bible. Yet even this sagely experienced judge could not bridge the rift between the stubborn state AGs and the DOJ. Despite MS's best intentions to settle the case, and after agreeing with the DOJ on several drafts, the mediation fell apart—the overreaching State AGs to blame. A frustrated Posner informed Jackson of his failed attempt, and on April 3, 2000, Jackson swiftly published his conclusion of law, convicting MS of violating the Sherman Act by illegally monopolizing the OS market for Intel-based PCs and maintaining that monopoly illegally with anticompetitive measures. He then ordered the commencement of an expedited penalty phase.

In lieu of rambling on at length about his verdict, allow me to reference just a single issue that will cast a clarifying light on his character. Admonished by the appellate court in regard to the IE/Windows tying argument and strongly cautioned not to engage in tech-product design, he audaciously contested the higher court's earlier verdict. Bluntly, he questioned the legal test the appellate court judges had applied when deciding the prior case. To prove he was the better judicial thinker, Jackson pitted their logic against highly respected, nuanced, and cleanly defined Supreme Court decisions.[51] The appellate court had instructed him not to put "a thumb on the scales" of computer design and innovations. Remembering that he naughtily countered with the childish admonition that MS had indeed put her thumb on the

---

51   Jefferson Parish and Eastman Kodak

scales of competitors' innovations. Therefore, he revisited the validity of MS's plausibility claim, in regard to IE integration, and erroneously applied the Supreme Court's conclusions in the aforementioned cases. According to him, it needed to be judged "upon proof of commercial reality, as opposed to what might appear to be reasonable" benefits users obtained through that integration. Explaining "the commercial reality is that consumers today perceive OSs and browsers as separate products, for which there is different demand." Consequently, he found MS guilty of defending a monopoly through illegal tying, primarily because software can be "comingled in virtually infinite combination." Let's leave this for the appellate court to answer.

MS's reaction to Jackson's verdict was predictably swift, despising his ruling and promising an appeal. By now Jackson was talking intensely on record to reporters while banning them from publishing his remarks until the case had been fully concluded. During several appearances, Bill and Steve boldly insisted on no wrongdoing. Not terribly smart or diplomatic, though certainly understandable. In Jackson's view, these defiant public statements by our top commanders indicated we respected neither his judgment nor his intelligence, wading out into the dark, marshy terrain of character assassination. Touché!

Meanwhile, a new administration had taken over Janet Reno's department of questionable justice. Republican President Bush had been ushered in with the help of the Supreme Court after a much-contested election outcome in Florida and Boies losing contender Al Gore's case in the Supreme Court. With the old staff still mostly in place and no change of heart forthcoming, the Feds marched on, asking Jackson to use the most drastic measure he had in his repertoire to remedy our perceived antitrust violations. Stunning the IT world, he swiftly complied and ordered the breakup of the company, denying us, as required by federal rules, an evidentiary hearing in regard to the harsh antidote.

To evade a regular review by the abhorred appellate court, he made use of his privilege, which allows judges in antitrust cases to send a verdict straight up to the Supreme Court for confirmation. Rarely does this ever happen. With the Supreme Court being able to pick and choose a case, he hoped for the best. Most judges would have known that a case riddled with so much upended detail and blurred uncertainty had no chance in hell to get accepted. Simultaneously, and to our utter surprise, he softened his stance at MS's request and, in last–minute, stay-off execution disbelief, suspended his breakup order. Our relief could be heard from sea to shining sea and certainly reverberated in Washington, DC. We remained, for now, one company. As expected by most legal experts, the Supreme Court declined to take the case, enabling us to file an immediate appeal. Let the final ballet of justice begin.

My last hurrah

A new CEO

With calendar year '99 coming to a close, we were getting ready for Windows Millennium (ME). Its primary new feature, the "system restore" aspect, finally compensated for badly written application programs. NT had further matured and was now called NT 2000 professional workstation. One year later, its underpinning would be used for all Windows versions. After eight years, we had finally unified its code base, naming the two Windows products appropriately: Home Edition and Professional Edition, targeting consumer and business users with different feature sets at different price points.

This next version, called XP for *eXPerience*, was the last Windows version I helped to launch and sell—in itself a stable OS, but vulnerable to rampantly growing security attacks via the Internet. The Linux camp and the hacker community, especially, encroached on and slashed into it at will. We struggled mightily keeping it safe and protecting it from viruses through releasing a multitude of security upgrades. The good news: XP's retail and OEM versions employed nearly bulletproof antipiracy measures through encrypted log-in sequences controlled via the Internet. Except for still less protected enterprise versions, a long arduous journey was winding down to a happy ending as my departure drew closer. Our developers and their management had finally enacted the right measures.

With undiminished intensity and stamina, my group had grown to 550 employees while the company has exploded to around 50,000. Like in all other segments of MS, valuable employees had left, but nearly all of my managers had stayed on. People liked working in my group and being part of what they considered a focused, well-guided, and no-nonsense elite sales team reaching beyond $20 million annual sales revenues per rep. I had rotated my managers through varied assignments over the years, educating them to understand and accept different cultures and their unique business dealings. My management style had created a group of empowered and closely knit leaders who understood how the company wanted her OEM business to succeed. Yes, even then we sometimes had healthy disagreements. They enhanced how we pursued our cause. I wanted managers with strong egos and brimming ambitions, focusing their energies to move the business forward and not fight or blame each other as I had observed in other parts of the company.

By the end of '99 and in the spring of '00, we conducted business as usual during a truly unusual time. After Jackson had issued his finding of facts, the sword of a potentially devastating verdict hung over our heads. Our policies had not changed one iota, except our moves were now watched not only by the outside world but also equally closely by our own legal department. Not a healthy environment to do business in even if you closed your eyes. There was nervousness in all parts of the company, and people showed signs of being palpably afraid. Auftragstaktik was being kissed good-bye! Receding slowly into the archival dustbin of MS's legend. I made it my personal responsibility to ensure that the OEM group did not fall into this yawing pit of fear and loathing. In my group, the pillars of trust and empowerment were there to stay. My people knew how far they could go when they needed to reach, and I made sure they were never undercut. As far as I was concerned and regardless of what the judge had propagated, we were on solid grounds, and the company would eventually find her way out of the crisis.

The only strange altercation I experienced was an internal one. Steve, putting his foot down, instructed me to share revenue recognition for the fast-growing system-builder business with the subs. He no longer wanted to endure their persistent and nagging complaints—another political move no doubt. With Bill distracted and growing moodier by days, there was no way to fight his edict. I swallowed my pride and agreed to the dreaded accounting changes. They did not improve the business one iota and fostered no friendships inside the company apart from saving time arguing. Going one step further, he suggested that by next fiscal year, I should give the subs sole control of that business. My passionate answer: "Over my dead body as long as I am running OEM!" The subs had gotten a hold of a little finger and were now looking for the rest of the hand. My group had done the brunt of the work, and now our fastest-growing segment was to be transferred to a bunch of managers eager to revel in its success, potentially destroying its structure, hard-won stability, and pricing integrity. My fervently presented arguments hit a nerve. Steve then tried to mollify me by promising that pricing would remain under my control. I told him, "Dream on!" I had no desire to play an internal policy enforcer role. In the end, he relented, leaving the business within my domain until I left my post.

Steve could have easily avoided this conflict by identifying an alternative business goal he wanted me to pursue, such as doubling this business within two years. I might have concluded that this could only be achieved by handing the business responsibility over to the subs, and I would have done so immediately. Absent this, he insisted and settled upon an accounting change, which only served to artificially bolster subsidiary revenues. The command-and-control system I liked best was then for sure not his.

Right after the judge published his finding of facts and threatened to break up the company, the board named Steve Ballmer CEO. Bill, still chairman, added the title chief software architect to his name, indicating he was from now on responsible for strategizing and planning the development and delivery of leading-edge software products. Having grown sick and tired of the persistent public humiliation, he was in full retreat. For the first time ever, a rift between him and Steve could be spotted. Not superobvious to outsiders, but the rumor mill had it nailed. In an angry aside, Steve told selected people he did not need Bill any longer to run the company! True to form. I was convinced he could manage her complexities, but could he lead and initiate meaningful change?

From then on, he grew more eager than ever to build his own team, engaging me seriously into succession planning. He proposed promoting Jeff Raikes to head MS's sales force and asked me how I felt about reporting to him. Surprised about his change of heart, I felt obligated to remind him of our understanding. I had not changed my mind about leaving OEM in '01, but in good old Steve Ballmer style, he had gotten impatient, probably emboldened by his newly gained title. Jeff Raikes—one day to assume the reins of the Gates Foundation—had the two stamps I mentioned earlier on his forehead. His intellect and management skills were held in highest regard by Bill and Steve, and they both considered him a personal friend. Credentials in hand, he had methodically scaled the company ladder, performing reasonably well in his assignments. I had no reason to disrespect him but believed he lacked the adequate experience of running a large sales force. At heart he was a marketing executive, knowledgeable of and practiced at running a product group—so I objected.

The concurrent Caldera antitrust lawsuit settlement, with MS coughing up nearly $300 million, did little to help my standing. I suspected the OEM group and maybe me personally were being blamed for the costly outcome. Undermined by the preliminary ruling in the current case, the sweetheart settlement Ray Noorda received was the result of a developing no-balls attitude and a weakening defense team.

I sensed at times, with Bill being moved aside, that I had lost my key mentor and protector. After another pushy session with Steve, I wrote him a well-thought-out e-mail asking him to honor our agreed-upon deal and to respect what the old warrior had done for the company. Telling him point-blank I saw no benefit in reporting to Jeff—implicitly suggesting I was his senior. Underscoring, what Steve had admitted, managing me was not a lot of work for him. I proposed to leave the OEM group untouched. He, in turn, thanked me for my considered and thoughtful summary and patiently refrained from any changes. My trust in him was restored.

My direct reports discovered Steve was from now on showing increased interest in the OEM business. No wonder—he knew I was likely leaving within the next eighteen months. The organization had always followed my directions, but straightaway I spotted a certain behavior shift among my senior managers. I had encouraged them to showcase their talents to Steve to better determine if they could succeed me. As a strange but plausible result, a few of my senior guys now wanted major decisions cleared by both of us. This ground fog of ambivalence appeared to be helped by the general uncertainty taking root in the company. In these tricky and ambiguous times, my crew was looking for additional reassurances. Arresting the trend early and with full authority stopped the nuisance. Even as Steve and I occasionally agreed to disagree, no second-guessing was needed. With few exceptions, I knew where my authority ended and how much elbow room I had. My crew interpreted this as me being bulletproof. I hadn't been before, and I wasn't then; everybody is replaceable, but as long as they thought so, I just laughed it off.

By mid '00, the search for my successor had started in earnest. There had always been a leading candidate in Richard Fade, who had just recently rejoined my group. He was well liked and generally respected by my people and, more importantly, by Steve and Bill. Richard, though, had encountered unfortunate health matters in his family, forcing him to take an extended leave of absence. I wasn't sure about the level of commitment he would be capable of if he stepped into my shoes. Shortly after I took over from my predecessor, I had promoted him to supervisor. He excelled and had been instrumental in resolving our distribution challenges and addressing the tricky piracy issues. Consistently a devoted top performer, he owned a deep familiarity of the OEM business. I entrusted him with larger roles over time, until Steve lured him away to run the Far East region out of Tokyo. With his move came a then-rare VP title. I encouraged to him to jump at the opportunity. Back in the United States, as head of the consumer and desktop division, he had been instrumental in launching the successful and stable Office 97 product. I liked him a great deal as a person but sometimes believed he wasted time when trying a bit too hard to build consensus. I sincerely hoped the home front situation would not take a larger toll on him. I was aware that not all of my direct reports appreciated his management style as much as I did, but that's business life, isn't it? We had to expect and tolerate personal shifts and departures after I was gone. Wisdom and maturity required as much, and my group's structure was healthy and solid enough to easily absorb a hit or two.

For the first time ever, Steve asked me to attend with my crew the company's worldwide sales meeting next year in Atlanta, Georgia, and forgo our distinct annual OEM summit. No friend of participating in such a monstrous event, I nevertheless reluctantly agreed. Expecting ten thousand participants, its format had to be rigidly structured just to move the crowd along. Typically, you could expect three days of training sessions punctuated by geography-specific meetings, reward ceremonies, and several all-attendee rallies, often in a sports stadium. Always featuring Steve as the frothy, cheerleading company whip followed by Bill as the rational and placate keynote speaker. Johnny Carson and Ed McMahon! With Steve reveling in the role of group orchestrator and augmenter, Bill had completely slipped in the role of regal software architect and elder statesman—a touch less aggressive and vibrant than in the past. I had witnessed several similar events before and often left with growing skepticism about the display of rabble-rousing righteousness and choreographed chest-thumping pedantry.

The annual OEM sales summits had evolved quite differently. They were always held in June, allowing us a running start for a fiscal year starting July 1. For the first two days, I brought my sales and marketing team to Redmond for intense and intimate training sessions. Speakers from relevant business groups, customers, logistic experts, and as highlight, Steve and Bill addressed the assembled crew. Every speaker's performance was measured on meaningful content and less on showmanship. Afterward, my troops precisely understood our product road map, our sales goals and marketing campaign objectives for the upcoming year, and our previous year's failures and accomplishments. As a special bonus, we then embarked on the fun part of the gathering, each year full of surprises, adventures, and team-building exercises. The crescendo was the official naming of winners from last year's contests during a gala dinner. Thanks to our fabulous admins and their extraordinary inventiveness and hard work, they grew into happenings highly envied by other parts of the company. People talked about them for years. So going to Atlanta in the hot, humid summer of '00 in its place was not welcomed, and it sucked. Steve kept his word, letting me devise next year's gathering in our proven and memorable format. My last one as I already knew it would be.

The reason we organized our annual events differently from the rest of the company could be blamed squarely on me. The first two, after the one in Hood Canal in '89, had transpired at cushy resorts, featuring extended leisure and camaraderie. I hated them. I was managing a group of vital young professionals, yet we behaved like retirees lying on the beach and enjoying the easy nightlife. Too many people got drunk, and except for soaking up sunshine, there was nothing too terrific to report about. The third year I surprised my crew. We went into the Canadian Rockies, slept in unheated tents at below freezing, endured wild canoe races, and engaged in competitive games exploring the untamed wilderness. The spirit of the organization changed on a dime. No one, except my fellow German female sales reps, was the least disturbed when the hot water for their morning shower ran out before commencing on a tough sixteen-mile hike. People were exhausted yet inspired by the exhilaration of outdoor adventures combined with team-building fun. The traditional last-evening poker game did not materialize either—the day had been too grueling.

Morale had never been better and from there on; expectations set, we geared up for extra adventures and team-building opportunities, inducing motivation and creating long-lasting memories. The outcome, my team felt rewarded for what it had accomplished and was less afraid of whatever lay ahead. In good old MS style, no words were minced—camaraderie triumphed. When the enthused and exhausted crew rejoined their coworkers in the subsidiaries around the globe, they truly had lofty adventures and lessons to extol.

Unfortunately, for the first time since I had taken the reins, results for the FY year had been slightly off, missing our revenue budget by a whopping 2 percent. A considerably improved profit margin had still enabled the division to deliver the anticipated contribution to the corporation. Our growth was down to a mere 15 percent. The dot-com bubble had burst, and overall PC volume was gravely disappointing. The enterprise group had sold more Windows NT workstation copies than forecasted, resulting in lost revenue for us. Our absolute growth was nevertheless fabulous. Compared to the previous year, we had once again increased our revenue by approximately one billion dollars. No one was depressed, and no doom-and-gloom feelings set in despite our slight underperformance.

Xbox

By 1998/99, planning for an MS entertainment console had gone into overdrive. MS, the ultimate software company, was boldly venturing into the sharply different dimensions of being a hardware vendor. Yes, we had produced computer mice and ergo keyboards before, but a fully functioning computer system had never been on our list of offerings since '86, when we had abandoned an early game console in Japan.

Since then, the competitive landscape for game consoles had changed radically. With Nintendo in the lead, Sony had launched PlayStation, immediately making life difficult for all competitors. Keeping a wary eye on her progress, Bill was concerned that this alternative computing platform could conquer the living room. Sony's division was run by Ken Kutaragi, an energetic and impressive entrepreneur and business executive. He had equipped his console with an advanced, lightning-quick proprietary graphic processor that was far beyond anything found in PCs, attracting a huge number of leading-edge game developers. Nintendo with her GameCube and Sega with her Dreamcast system felt pummeled as gamers left opting for the newcomer.

We landed a deal with Sega, making Windows CE her OS of choice. It got us a foot in the door. Yet Bill insisted on gaining a larger piece of the action by launching a full-blown game console of our own, stopping Sony's march. To justify his ambitions to the board and the shareholders, he wondered if we could offset predictable console losses with game-software profits. Ed Fries, a MS veteran and the genius behind MS's immensely popular flight simulator program, made a convincing case confirming Bill's idea, and soon the race was on. Promising to attract and recruit a critical mass of independent software vendors for the new platform, Ed sealed the deal. Short on detail and long on vision, MS style.

I was stunned to hear about this new endeavor. My main concern: how much would our OEMs appreciate our move into producing a complete computer system potentially endangering their PC business? We had enough daunting challenges already; why take on yet another where initially huge losses were expected? I further questioned our ability of sourcing parts cheaply enough and managing production aspects in a cost-effective and timely manner. The idea of still-to-be-created game software coming to the rescue for this hardware adventure did not strike me as a solid business foundation. Questions regarding the software being delivered on time and us being able to create hot-enough titles crossed my mind. To accomplish our profitability goal, we first of all had to sell lots of consoles at probably substantial losses for years to come.[52]

Despite my customer-related concerns, Bill was dead set, brimming with optimism and perfectly willing to gamble with shareholder's money. The ambitious project soon went one step further with Nathan Myhrvold, our chief technology officer and resident propeller-head, insinuating we could eventually transform our game console into a full-blown PC. I shook my head. Fortunately, cooler heads prevailed. At least for the moment.

In the end, Bill listened and offered me a chance to approach several OEMs, swaying them to build and launch a game console on their own account using our blueprint. We committed to supplying a special Windows edition along with a promise to develop game software exclusively for the new platform. We talked to Dell, Compaq, Hewlett-Packard, Acer, and Nippon Electric Company (NEC). Thanks but no thanks! They had studied the Sony and Nintendo business models in detail and didn't believe they had the ability to compete profitably. Leaving us to stick our neck out, alone! We should have listened closely. After this setback, there was talk of buying Sega or Nintendo, neither faring too well, but no deal was ever seriously attempted.

Just after I left the OEM post in 2001, Xbox was launched. The project had gone through several iterations and numerous delays. MS had a hard time designing the box properly and sourcing the components at acceptable prices, shocking AMD when deciding to go with a special low-priced Intel Pentium CPU. Ed Fries did a super job recruiting ISVs, but even then, games for the new platform were coming along slowly. ISVs were outright skeptical if we would pull it off and were busy writing PlayStation games. Buying one of them enabled us to publish one big hit, *Halo*, a supersuccessful war game, yet not enough to compensate for mounting losses. In the following ten years, the company lost well over $ 10 billion on Xbox before breaking barely even. Not the type of success I would have aimed for.

---

52 Judge Jackson, in his logic, would have labeled our plan suicide because of the famous barrier to entry, which in this case Sony and Nintendo were guarding.

One last aspect to our hardware story: Rick Thomson, who had been in charge of building all MS hardware products, joined the Xbox team in '99 to build the new platform. As he left the company a year later, frustrated by internal politics, our hardware group experienced compounding difficulties. New management came in and decided to abandon the OEM mouse business I'd been nurturing all along. I was deeply disappointed but no longer had the passion to fight for its revival. The little critters would from now on be supplied by other manufacturers while MS would focus entirely on the supposedly more profitable retail sector. Opportunity lost!

Late challenges

My travel in 2000 was mainly to developing countries such as South Africa, China, and India. I was not satisfied with our progress in India in particular and decided to visit the country for the first time ever. The leading IT Company to visit was Hindustan Computer. Founded in '76 and still mostly focused on India but with plans to branch out into Africa, she had moved beyond selling hardware and transformed herself into a top-notch IT consulting and engineering company. Bill recommended I talk to Jai Chowdhry, her CEO and founder, calling him one of the smartest people he had ever met. After my arrival, he graciously gave me an educational overview of India's fast-growing IT industry and the local PC market in particular. Typical of a developing country, India's PC business was firmly in the hands of small system-builder shops. Among the relatively few larger manufacturers, HCL was the most eminent. To stay competitive with the screwdriver outfits, the smaller manufacturers sold their PCs mostly without any software and let the pirates find a way to take over at the point of purchase. HCL, with her established brand name and her primary target of serving enterprise customers, was a noble exception to the rule.

I visited India a couple of weeks after Bill Clinton had toured the country in a formal head-of-state visit. My local team had planned the visit following his route, including some of the same hotels and restaurants. You do live well as president of the United States. By contrast, the country's poverty rate was staggering. The other much-too-apparent observation I'm compelled to mention was the tremendous pollution I observed. Mexico City was bad, and Beijing may be close behind, but nearly any large Indian city tops them in spades.

I found the business climate in India a touch aggressive and tense, though the people I met were sharp and highly educated with advanced business experience. As a result of my visit, we teamed up with several OEMs, incenting them to promote genuine software bundles. It helped move the needle in the right direction for a while, but not as much as in the People's

Republic of China. The main reasons as I saw it: the government had no stakes in these companies and was not supportive of anybody who wanted people to pay for software.

South Africa (SA) was my next stop. I had first visited the country in '94, shortly after the apartheid regime had been removed from power. I had gotten to love its wild primal aspects, which still existed in an ethereal and natural state as long as you were willing to travel into the veld. Unfortunately, her poverty rate had remained way too high even as her new government had made long strides in easing some of that pain.

The local PC market was dominated by a company called MUSTEK, with 40 percent market share. Ran by her founder, the Taiwan-born and Pittsburg State–educated David Kan, her huge share was a result of the apartheid boycott, which had prevented overseas PC manufacturers from branching out into SA. David arrived there from Taiwan, incorporating in '89 and later floating his company on the local stock market. His Taiwanese connections enabled him to import enough PC components to make an impressive run. I had met this brilliant and energetic businessman several times in the past, and we developed a close friendship. David had no ambitions to compete on a global scale, but he had cleverly branched out into other parts of Africa proper. To shore up his core business, he established a profitable component-distribution business including Windows, helping us to reduce piracy in this remote part of the world.

One reason for my current visit was to review the progress of a special project we had launched exclusively in SA. The idea was the brainchild of our MBA whiz kids. I had predicted failure but was interested enough— along with a skeptical Steve—to let the experiment proceed. Despite of only a slim chance for success, we opted for gaining valuable experience in exploring an alternative way of selling MS Office.

Priced at about $300 down there, a lot of buyers in this developing country were hesitant to shell out that much money for personal productivity software. Instead of requiring the full price up front, the whiz kids proposed to allow customers to pay in monthly installments. Their plan called for OEMs to conveniently preinstall the product like an OS, and if end users agreed to purchase it permanently, a low monthly fee would be charged to their credit cards. To limit our risk and exposure, the small and remote SA market had been selected for a test run. My reason to be skeptical about the idea was twofold. First and foremost, credit card ownership was not exactly widespread in SA, the Internet usage was low, and last but not least, how would we police the customer's failure to pay up?

David Kan, together with other locals, had agreed to participate in the trial. The initial sales had been slow but promising. The OEMs had gotten a small sales bounty, and as predicted, payments beyond the initial ones had been hard to obtain. Not being eager to deal with public outcries when cutting off people's right to use MS Office, local management was weary. Still in SA, I sent e-mails back home, asking to abandon the project and extend amnesty to current participants. In retrospect, we were ahead of our times; most likely we'd simply chosen the wrong country to launch the experiment.

For quite some time Bill and I had been talking about trying something similar with Windows. Exploring the concept in detail, we soon discovered how hard it was to implement and enforce payments. In theory, we could have charged OEMs zero dollars for preinstalling Windows and given PC purchasers thirty days of free usage. Agreeing to pay us an annual fee would have kept their PCs functioning. Once modeling such an annuity concept, I discovered, despite an obvious first-year revenue shortfall, the gain derived from no longer having to extend volume discounts to OEMs, and adding a small finance charge could boost per-unit revenue.

What spoke against the idea? Not every end user had Internet access, and any other way of implementation would have been awkward, inefficient, and probably impossible. Nor did all Windows customers possess a credit card. The final issue breaking the camel's back concerned enforcing the annuity deal. Windows, like any OS, remained essential to keep a computer running, unlike an antivirus program, for example. Would we put our money where our mouth was and enforce nonpayments by making PCs dysfunctional? Instead of disabling them completely, implementing an automatic slowdown and reducing the number of applications to run simultaneously[53] were other technical possibilities. Undoubtedly, any of these measures meant creating a lot of unhappy PC owners. Last but not least, hackers could go to work and repair the inflicted damages; in the end, it was just software, and even if encrypted, it could have been cracked. We even considered selling an encrypted hardware dangle with every Windows copy, which turned out to be too expensive to manufacture but might be a way to implement a stronger protection and enforcement scheme today. Our final conclusion then: wrong thing to do, dream on about OS annuities, end of discussion!

Back home I had to attend to an unexpected crisis stemming from the antitrust trial. Documents the Feds had confiscated during their investigation had been posted on the Internet for public perusal. Naturally, they were a treasure trove for people trying to gain insights into our internal discourses.

---

53   Some of these implementations later got us awarded a joint patent.

Best compared to WikiLeaks. The ones published had been scrutinized by our attorneys, ensuring confidentiality. Redacting three million documents, nevertheless, was prone to errors.

One occurred when a report from a negotiation with Compaq was not well-enough guarded. The lowball Windows price mentioned in the erroneously published document immediately caught the eyes of Kelly Guest, Dell's corporate counsel. Freely but mistakenly assuming it reflected Compaq's actual royalty, he immediately apprised Michael Dell and Kevin Rolling, Dell's president, of his findings, which at once raised a howl of despair and disfavor. Constituting about half of what Dell was actually paying, the news made the execs understandably furious, signaling a de facto huge disadvantage.

Welcome to the world of most favored nation clauses where trouble is sure to abound. Their origins stemmed from treaties between nations in which signatories agreed to grant each other the same favorable terms as offered to others. Similar clauses had found their way into commercial contracts. I hated them with all my heart and soul. Yet in supertough negotiations, we had deigned to extend them but so far to only three customers. (Proving we had way less pricing power than the judge was made to believe.)

Under a most favored nation clause, we guaranteed customers the absolute best royalty. Guaranteeing this to Compaq, the top volume shipper had for now no apparent risk. Making sure we could honor such a promise, I instructed my attorneys that anytime a similar clause was headed toward another agreement, they needed to obtain my advanced approval. A couple of years later, Dell got wind of her neighbor's favorable contract term, and Kelly Guest—why not?—asked for reciprocal treatment. With Steve trying to climb into bed with Dell's execs, there was no denying.

With Compaq outselling Dell by 30 percent, I was adamantly against extending her prices to Dell and asked our attorneys to find a way to soften the absolutely-best-price guarantee. Under the modified clause, Dell could only obtain Compaq's price when she bought the same number of Windows copies. Dell understood the finesse but nevertheless called us on the carpet. A painful meeting with Steve commenced. Between Kelly Guest arguing eloquently, Michael Dell playing hurt, and Kevin Rollins seeming pissed, we were in a grand and dramatic pickle. Wanting their cooperation, Steve had no desire to stand up to them. I passionately wanted our contract language to prevail. In the end, none of my arguments got through. I got plainly overruled. We left with me sporting bruises and bad feelings but nevertheless agreeing to apply the reduced royalty rate going forward. Kelly Guest's political gamesmanship triumphed, having judged the situation correctly. I had met him several times before, got to know him quite well, and had a great deal of respect for him. He had pulled a fast one. I swallowed my pride.

When Steve and I recapped the situation shortly thereafter, I angrily told him to consider himself from now on Dell's chief account manager. The sales reps were better off dealing with him directly when micromanaging this account. He laughed it off.

Steve's behavior was in stark contrast to Bill's. Both had a difficult time saying no to customers. Right after I got the OEM job, I watched Bill struggling through price negotiations though managing to stay involved, if obliquely. After we'd had a chance to chat at length about how to avoid putting him on the spot in the future, he changed his tune. From then on, whenever customers asked him directly for price concessions, he redirected them point-blank: "You'll have to work that one out with Joachim." A decision he hopefully never regretted, allowing him to remain firmly in the driver's seat while empowering me to obtain the best possible deals. *Auftragstaktik* at work!

SEARCHING FOR A SUCCESSOR

All these little incidents added up. Having to endure Steve's emboldened management style and observing the command philosophy drift away from Auftragstaktik with iceberg resolution, a tactical standard which had served MS so well, I longed, at last, to leave. My favorite successor was still Richard Fade, who was by now back in my group, working diligently and hard as ever despite the unresolved drama in his family. Steve agreed with my judgment, but as we seriously approached our preferred candidate, he remained unwilling to commit long-term. A concern to me and a much larger one to Steve, desiring to build a lasting team. In the fall of '00, after a visit to Europe, Steve had made up his mind, telling me he would like to see Richard Roy, the current German country manager, to succeed me. Another German, what was so great about them? I asked Steve to allow me an opportunity to contact my friends in Germany before giving an opinion. Their feedback left me unimpressed, ranging from "The guy is too political and not decisive enough" to "He's not well liked by a number of customers and employees alike." Respecting my well-placed informants, I advised Steve not to appoint him. Reluctantly, I offered to remain on board in the event we failed locating a suitable long-term replacement.

Steve wouldn't change his mind. I was unsure why. Not trusting my judgment any longer, being totally convinced of his own, honoring an already-made commitment, or was he just plain playing boss? The next step for me was to invite the chosen one to my—actually his—fiscal year '02 planning meeting. A good test for the candidate. We prepared well, and as he showed up, I told my direct reports what Steve's intent was. The meeting went well, but my guys were not impressed by the newcomer's attention to detail or by his general business sense, echoing my German sources. Back in Redmond, I reported our collective observations. There was still time to reconsider, but after a couple days, with rumors now flying, my boss announced him as the next leader of the OEM division. I promised a smooth transition. Richard Fade, sidelined, did not mind remaining steadfast and cooperative.

Steve then had made the decision to split the OEM personnel, selling to system builders off from the main group, and integrate them into the retail sales organization. To make the transition an easier one, he planned giving my successor control over a large part of that organization. In a follow-up meeting with all area managers—one of the fabled monster marathon sessions—the new organization was being hacked out from whole cloth. Introducing the new concept caused consternation. The major opposition was derived from the fact of a less proven man having oversight of such a diverse sales group and the reluctance of regional managers to release control of more sovereignty to the center. After Richard Roy returned to Germany, now fully comprehending the weighty and labyrinthine task, he flabbergasted us. He simply resigned. I found an enraged Steve and, after discussing the situation, convinced him to reconsider Richard Fade. So it happened; the most capable guy got the job. I felt providence.

The two of us immediately started working together super closely, preparing the transition. All decisions impacting the group beyond my departure we made jointly. We started visiting customers together. Fortunately, a lot of them knew him already, and as the clock was running out on my reign, I could relax and confidently detach myself from my job. My people would be in good hands, and the business should continue to prosper.

APPELLATE COURT RESPONSE

In June of '01, the appellate court had at last arrived at a conclusion. By then, I was just a scant few weeks away from leaving OEM. Like most employees, I was plenty happy with what I read. But the ruling was by no means the home run our legal team had been gunning for.

The good results first: Jackson, like Sporkin before him, got removed from the case. We were not guilty of unlawfully tying IE to Windows—Jackson lost his favorite argument a second time—and we did not illegally attempt to monopolize the Web-browser market. The court vacated all remedies of Jackson's final judgment—meaning no break-up of the company was imminent.

But here the good news ended. The appeals court upheld most of Jackson's finding of facts despite the "seriously tainted proceedings" and said it found "no evidence of bias," effectively opening the floodgates for a stampede of lawsuits. The court confirmed Jackson's opinion on all Sherman Act violations, except the ones mentioned above. The case was ordered to go into a liability phase in front of a newly appointed judge. After four years, the case would drag on. And on! Studying the appeals court's ruling, it looked like our legal team—in the hectic whirl of the moment—hoping Jackson's finding of facts, together with his verdict, would be kicked out, altogether had missed the mark. The old hand had beaten us handsomely. The magazine *Slate* described the outcome correctly: "If Microsoft won the day, the Justice Department won some moments." Real costly ones!

Dave Kopel from the Independence Institute called Jackson "a textbook example of a bad judge," who had "made an ass of the law, and of himself." Chief Justice Edwards of the appellate court called his behavior "beyond the pale," believing he had violated his judicial oath. Edwards had good reasons after reading the interviews Jackson had been volunteering and in which he labeled Bill a "drug trafficker," referencing US executives "gangland killers" and "stubborn mules who should be walloped upside the head with a two-by-four." Woody West, in *Insight on the News*, headlined his article, "Jackson's Actions: Betrayal of Trust," calling for his immediate resignation. The court found his conduct embarrassing and ruinously shy of his obligation as a federal judge opining his "insistence on secrecy"—his embargo in regard to the journalists he had been secretly talking to—"made matters worse." Ken Auletta's reportage in the *New Yorker* described his flaps succinctly, including Jackson saying Bill owned "a Napoleonic concept of himself." The laundry list of the judge's outrageous, rude, and unfair comments ran drearily on and on. After citing the ones above, there is no need to further pile on.

The Feds—now with a new team firmly in place—reworked their remedy proposal and, as MS had hoped, no longer insisted on a breakup. The new administration had finally made up and changed her mind. Yes, politics played a role before, and politics will do so again. In August of '01, Colleen Kollar-Kotelly took over as new judge. Under her tutelage, the parties agreed to a settlement resulting in yet another consent decree. She issued her final judgment in November of '02 just as I was leaving the company. MS took it in stride. The Feds were on board, and so were several, though not all, of the state AGs. Objecting to the compromise like Gateway and the Reback gang, they propelled the proceedings off onto yet another orbit, continuing the legal cage-fighting contest. By then I had left the company for good, and Richard Fade and his predecessor had to work through implementing the new decree. Not much fun. Eventually confirmed by the appellate court, it remains, with some small adjustments, in effect today.

Milton Friedman, the late Nobel economist, called the Feds' move against MS a dangerous precedent signaling extra government regulation in high-tech markets. Google is experiencing this right now. So why did MS lose the case segmentally, and why did MS become labeled a big bad monopolist, driving poor Bill to chase a new legacy in an easily pursued venue called philanthropy? First of all, I firmly believe Jackson was not only openly hostile to the appeals court, he was also covertly biased against us. The wealth and depth of comments made to journalists in the secret bowls of his chamber signaled huge and significant prejudice, which could not have popped up overnight. His grossly unfair proceedings in his courtroom underline this. Every single word of his should have been canned. Tossed, crossed, and banned.

The key victory for the Feds had been to convince both courts that we, in fact, possessed monopoly power within the whittled-down, narrowly defined market space. We certainly possessed limited pricing freedom and had earned an impressive share yet without ever charging monopoly prices or harming consumers.

Another reason for convicting us lay in our harsh negotiation tactics and the overly blunt manners our people employed in e-mails and various other documents. Written to impress internal recipients and rarely acted upon, they resulted—much to my surprise—in intent getting punished. Not upholding our copyright, as I had come to naively expect, was another injustice. The notion of allowing distributors to change the look and feel of a product without the copyright holder's consent was the hardest of all for me to swallow. Copyright law exists to promote the arts and not to restrict them. Infringing, without compensation, harms the interest of the rights holder. The court's failure to respect this sets a bad precedent for future litigation in the software industry.

What did the final judgment mean for MS's OEM business? The new conduct rules started with telling MS not to retaliate against OEMs for shipping competing products or services. Well, I failed to see how we ever did—frankly expressing and pointing out that what we did not like in a free speech society is OK with me. Long live the First Amendment! MS was forced to offer all OEMs identical contracts for Windows. The decree ordered her to publish an OEM price list but allowed reasonable volume discounts. Market-development allowances needed to be offered to all OEMs uniformly. Restricting the placement of icons on the Windows desktop and in the boot sequence got liberated.

The next section of the decree had to do with laying open Windows application programming interfaces and communication protocols. I never understood why we had used unpublished ones in the first place; it was unfair to the partners we otherwise supported well. MS was further ordered to make all her middleware products easily removable for end users (bravo!) and license her IP portfolio at a reasonable fee. The rest of the document addressed the selection of an enforcement committee, establishing a compliance officer inside MS. The same arrangement Sam Palmisano, now IBM's CEO, had warned me about.

The licensing of our IP had always been an option and should not hurt the company. It could only create an additional revenue source. And did! The OEM-related restrictions were minor. Strangest of all was to have a published price list and totally unified contracts, but a lot of other industries succeed nicely under exactly the same set of policies. For sales personnel, the decree made life less sporting, as selling became a "take it or leave it" exercise. The rule for sure established the level playing field I had always promoted. But as long as extra marketing funds could be used when undertaking joint promotions, MS was not hamstrung favoring cooperative and collaborative customers. All in all, the new consent decree meant OEM business as usual for the division, while having to endure increased scrutiny through the newly established compliance institutions. It for sure had no chance of curbing MS's dominance in the OS arena.

The European Union, spearheaded by economist Mario Monti, now Italy's premier minister, chimed in responding to a complaint filed by Sun and a second one by RealNetworks. Finding MS guilty of having violated European fair competition rules, he, as competition commissioner, imposed a huge fine. I paid little attention to that ill-founded condemnation and the following charade of a trial. The rules in Europe are bizarre and convoluted because anticapitalistic principles reign supreme. European bureaucrats, in essence feeding off the situation, germinated in the US courts, forced MS to provide a Windows version without its multimedia player. Rob Glaser from RealNetworks had finally gotten pseudorevenge. What the US justice system did not accomplish, the EU did—ironically benefitting an American company. Not so fast, my friends in Europe tell me they have yet to find a single PC in Europe loaded with this crippled version of Windows. Euro socialists versus common sense.

In summary, the harm done by the US regulatory measures was not directly lethal to MS's ongoing success. Losing a tremendous amount of talent, though, as a trial consequence, was far more damaging. The subsequent compliance scrutiny then instilled a cautionary conservatism and unheard-of hesitancy, slowly converting the company into an increasingly appeasing and less assertive competitor. The combination of all of what I just described left MS weakened, disabling the company from tackling emerging challenges as boldly as in the past.

Ironically, the Feds never proved their key point that the integration of IE functionality into Windows constituted a violation of the Sherman Act. The appeals court outright rejected Jackson's reasoning for a per se rule violation. In a surprising gesture of grandiose generosity, the judges offered the Feds another opportunity to seek a conviction using the rule of reason first used in 1918 in the famous *Chicago Board of Trade v. US* case. Not interested in prolonging the case any longer, the Feds dropped the tying complaint once and for all, making them, in this context, the biggest loser of the case. The infamous state AGs who stubbornly refused to come to their senses, loosely spending hard-earned monies from taxpayers' wallets long after the Feds and MS had settled the case—fared even worse when they lost yet another appeal. Not that it had any consequences for them!

Last but not least, let's talk about consumer harm. The Feds probably managed to do just that by imposing a firm Windows price list on MS. The net effect of no wiggle room in price negotiations: consumers will pay slightly extra for their PCs. Allow me, finally, to at least say thank you to the Feds and the many states who participated in our witch hunt for forcing OEMs into having to accept government-dictated, "take it or leave it," nonnegotiable contracts! The era of sending OEM customers a postcard containing the newest license agreement without legal review, as I had joked about during the Windows 95 reception, had finally arrived. The helping hand of the government made sure of it.

i

Parting

In the spring of '00, I found time to reflect on how my personal management style had contributed to my group's success story. I was engaged in a leadership class with my team, which I embellished by personally teaching lessons garnered while running OEM. My goal was to transfer as much knowledge and experience as possible and open my secret management toolbox. Let's look into it:

Successfully negotiating for me always required ironclad preparation. Nothing primed me better for the at-all-times astonishing and challenging give and take. Ever having to shoot from the hip meant I had not been meticulous enough with my groundwork. I further discovered that negotiating with open cards while showing flexibility within means made me a much tougher opponent. If threats were introduced, I ignored them politely, leaving the other side predictably speechless and often stranded. Any threat is hollow if not perceived or acknowledged. The seeds of pressures must find congenial soil to grow, and denying that hold strengthened my position. Consequentially, I avoided responding with a threat of my own; it could have only exposed my vulnerability. Tit for tat hardly ever works; I prefer logic. The Chinese have a proverb saying whoever introduces the first threat eventually loses. I subscribe to that even when deeply hurt or emotionally entangled. I learned to separate my inner self from the business at hand. At the same time, we are all human and not perfect, as my encounter with Theo Lieven showed.

When planning your next move in a competitive battle, do your homework much like for a university exam. Involve insiders and learn from them. Nobody can always be right. If I still made a mistake because my assumptions were obviously wrong, I was humble, corrected my misstep at once, and moved on. Do not let your ego get in the way; focus on the result and not on the eggs having landed in your face. That mess can be wiped off.

Planning is hard managerial work, and it's not humanly possible to foresee all possibilities at all times. Surprises are a certainty in life, and so is change. As leader, dealing with change is your primary task. Hanging on to policy principles for too long promises failure. To arrive at valid strategies means you need to participate in the legwork and evaluate a wealth of details in advance. Some you can delegate to qualified and trusted staff but not without verification. I made myself unpopular by asking the hard questions. Being hands on in planning a carefully drafted blueprint pays off and avoids micromanaging the execution of a mediocre one.

Be not ever happy with status quo; I never was! Nurture new ideas and help validate them rigorously—I know no better learning process. Be willing to experiment against better judgment as long as the risk is reasonable. As top change agent, you will have to accept risk; don't be afraid.

One of the tricks I used was going on a fishing trip once in a while, comparable to Bill disappearing for his annual think week. Reading up on urgent topics started the ball game. Down at the river, as I patiently waited for a bite to occur, my brain had lots of time to reflect and wander off, connecting the dots. I always came back with newly gained insights and shared them with people I trusted and respected. These were not messianic intuitions or improbable visions but pragmatic conclusions I had drawn about improvements I wanted to implement. To transform them into well-conceived and change-inducing strategies or organizational adjustments, they needed to be challenged and picked apart without me religiously safeguarding them. Only to-the-bone analysis validated them properly and made them operative. Each person I involved in that process felt empowered, having influenced my thinking and the eventual outcome.

The next step was to define a metric to measure success and build enough flexibility into a plan to accommodate the inevitability of change and surprise. Fleshing this out and assigning accountabilities finished the task. I stayed personally immersed to the end, making sure our chosen go-to-market plan or intended organizational change was as easy to explain as it was to execute—with Auftragstaktik in mind. Detailed enough to comprehend and loose enough to allow for initiatives and flexibility in the field. As a result, these innocent fishing trips of mine often resulted to bold changes for the organization and our sales and marketing policies. Fishing, as it was, for relevant, timely, and—in hindsight—vital changes. I was careful not to attempt them too often. Once every twelve to eighteen months was the most I deemed effective. Otherwise, I would have projected the image of a juggler having too many balls in the air at all times. Fearing uncanny moves, people would have questioned my motives or, worse, lost their trust in my ability to lead with levelheaded purpose and focus.

Running an organization successfully meant having my key business ratios in my head at all times. This cut straight through the fog and allowed me to stay in the driver's seat. Looking like magic to outsiders, it kept people on track and removed any clutter, immediately helping me in negotiations or in validating strategies.

When formulating a strategy, I tried to mimic the way the Prussian generals wrote their orders. Clarity and brevity needed to trump all. The order for the invasion of Normandy was communicated on 133 pages. No soldier could possibly remember all of its details or repeat the order in its entirety. It read like a cookbook—prescriptive, leaving hardly any room for independent thinking though plenty for bravery. The German attack plan for the Netherlands in WWII was contained on four pages. The American general Omar Bradley did even better when he formulated the instructions for Cobra the breakout out of the Normandy theater on just one and a half pages. The ones who needed to recall them verbatim could easily recite them anytime. My annual OEM business plans—developed by six to eight people at most—were presented on less than four pages. I preferred two! Crisp and clear. People knew exactly what to do, kept key objectives in their heads, and understood the elbow room they had.

When presenting annual plans to a larger audience, I nailed the core issues down onto PowerPoint slides containing only five objectives. I always told the attendees that if we achieved just the first three 100 percent, we would be heroes. Each objective was followed up with a slide containing five key implementation points. Again, the first three being sufficient to follow through and the last two, if completed, were the icing on the cake. I never fooled myself: most people can only keep three things in their heads at any one time.

I never forgot that the success of MS's OEM business, and therefore, my own depended on a successful team. Nothing motivates a team more than being empowered. I extended my trust generously—letting go. As a leader, you set the pace and the expectations. Ambitious challenges make people follow you. People get bored and lazy when the yardstick is set too low. I was always honest with them when evaluating their performances, understanding they hated conciliatory or mollifying judgments.

There is no need to know everything that goes on in an organization. First of all, you can't, and last but not least, you shouldn't; it distracts from your aim and makes you a spymaster instead of a leader. Trust and knowing all show no positive correlation. Putting people under a microscope does not help them grow! Auftragstaktik does. Order them to take the steepest hill without detailing the terrain. That is how they learn to follow you.

Once upon a time, one of my business managers sent me an e-mail while on the road, asking for a decision. To make my life easier, he listed two options. A and B. I was known to formulate even shorter e-mail answers than Steve's supertight telegram style provided. In this case, it was just a single word. "Yes." Unsure, he consulted with my admin. Nothing unusual—a lot of people did after receiving a handwritten note. Only she could decipher my physician-style writings. Now she was asked to interpret my cryptic and ambiguous response. Her prompt advice: "He means A. He did not bother with B after finding option A OK to implement." The oracle of Delphi had spoken, and her advice was abided by. The actual truth was quite different. I did indeed read both options proposed and decided to leave the decision to my business manager. No babysitting needed with neither option a bad suggestion. Smart guy he was—why would he posit his second choice first?

As fiscal year '01 ended and I prepared to depart the OEM division, I found a copy of an old PowerPoint presentation on my home PC, the one I had given to the board outlining my ten-year-old business projections. Reviewing my old forecast, I asked my controller to find out how much Windows business had been pulled in by the enterprise division, and as we added these numbers to ours, we had a nice surprise. The combined revenue figures exceeded my now-ancient predictions by approximately half a billion dollars. I felt super good about the astonishing result. Having produced best-case scenario results, I had not let the board down.

That June, I attended my last annual group meeting in Las Vegas, Nevada. We stayed in a hotel outside the main city center and, in good old OEM fashion, combined the nightlife and a visit to the fabulous Siegfried and Roy show with team-building exercises, including a challenging hike through the local mountains. An excursion to the Grand Canyon ended for me in an incredible and terrifying helicopter flight so I could catch the flight back to Las Vegas. The annual awards dinner the same evening promised to be special and, for the first time, even a black-tie event—just to honor me. I appreciated the gesture, but deep down, I resented the ensuing formality, and my crew knew that very well. Nevertheless, they had a terrific time roasting me. Bill and Steve were present by video; my friend Richard Fade had made sure of it. The session brought a warm wash of sentimental tears in my eyes. The remembrance was truly funny, rewarding, and a good snapshot of my time leading the team. Except I had the last word, and this was how I addressed the assembly of my battle-hardened cohorts with a quickly prepared handwritten farewell speech:

> My dear friends and comrades in arms, I would have loved
> to finish fiscal year 2001 way above budget. It just did not

happen. The economic downturn caused by the bursting dot-com bubble denied us this last feat.

After nearly fifteen years, I hereby pass the scepter of my beloved OEM division onto Richard Fade. You will be in good hands. Let me remind you what I said when announcing my retirement last December: "OEM has never been a one-man show—OEM is a well-fused and oiled team." With the incumbent fighter or better resident rebel moving on, he can help you no longer; good or bad, only the future will tell! Tears, regrets—not me. The only thing which bugs me is losing my trusted team.

While learning a lot in all those years and giving a lot at the same time, trust and empowerment have been the pillar of our working relationships. Paired with forward-looking thought leadership, they were the main forces helping us to achieve the incredible and unthinkable we daringly set out to confront and achieve. Trust, in particular, will remain the foundation for our customer relationships long after I have gone. Without, none of the bonds we created could have held together in challenging times and to the better of PC users and the entire IT industry. Just listen to some numbers:

During my time, we sold 60 million pure DOS licenses, over 520 million Windows copies, nearly 150 million little critters, brought in $40 billion in revenue, and made a lot of shareholders happy with nearly $25 billion in profit before taxes. In addition to the traditional OEM business, we built a by-now $150 million CE business and an astonishing $1.5 billion system-builder business the subsidiaries have envied and loathed us for. Unfortunately, most people who did the hard work and made that business a roaring success no longer reside in our group; nevertheless, let me salute them for their great effort and dedication to our cause and how vigorously they fought the pirates on the high seas.

Why did we win that impressively? We understood that striving for uncompromising product excellence and tenaciously building and defending market share could enact a self-accelerating economy of scale. If our competitors would have understood this better and attacked us earlier in the game, designed better OSs, and recruited a higher number of independent software vendors, we would have had a much harder time succeeding. We worked hard and played hard; with me around fun, was always part of the equation, and because of the lucky stars we were operating under, all of us eventually got rewarded. Seriously, where are our competitors now? They are all history, eating our dust as we left them behind on the various battle fields:

DRI / Novell / Caldera, Lotus, WordPerfect / Corel, IBM, etc., etc., etc.

In summary, let me say thank you for helping me to shape the spirit of this division. Give yourself a great applause for all your accomplishments. They are by no means mine alone. Let me finish with what General McArthur said about soldiers leaving the army: "Old soldiers never die, they just fade away." With all my heart, I wish you under Richard's leadership a lot of success for years to come, promising you two things in parting: I won't get bored and—and just in case you are afraid—I won't come back. Good-bye and good luck.

As I climbed down from the stage under roaring applause, Richard Fade, in his tuxedo, rushed toward me, shook my hand, and thanked me for what he called "a nice passing of the flag." Both a bit teary, we looked at each other, and before I could respond, he frankly admitted, "JK, you will be a damned tough act to follow!" With a big encouraging grin on my face, my prompt response was straightforward: "If anyone can do it, you can!" And just as I started to turn around, he addressed me one more time, smirking: "Hm, already retired?" He must have just now recognized that I was the only guy in the room wearing khaki to make a point. A tux has never and will never be a preferred outfit for me. Amen, be true to yourself to the last second!

When leaving my post after eighteen years of holding key executive positions, I took a treasure trove of experiences away with me. Something the company did not want to lose right away. So I found myself with a new assignment helping to redefine MS's Linux strategy. Steve referred to me as his Linux czar. The press got hold of this new moniker and leaped into reckless speculation over my ensuing evildoings, my personal life, and any other available gossip someone could dig up. My scant familiarity with Linux and the closeness of my final exit did not warrant any of this.

While running OEM, I had observed the progress and success of Linux astutely from a distance. For Steve, wanting to gain ground for our PC-server business, Linux was a much bigger thorn in the hide of MS's galloping ambitions. By '01, Linux had gained an undisputable foothold in the server arena as the new concept of *blade computing* took hold. These are stripped-down servers on a single board deployed in clusters and connected up through specially designed mounting racks to increase raw computing power for data centers. Linux, a lean, adaptable, and modular OS with a small footprint, was ideally suited for such application. By comparison, MS's monolithic NT servers were deemed too large, harder to maintain, and neither dependable nor secure enough. Our server group was deeply humiliated by these attributes and tried a variety of avenues to inhibit Linux's breakthrough. Immune to the usual tricks our marketers flourished, the volunteers fearlessly stood their ground—like minutemen under British attack.

After careful perusal of the complex Linux situation, I came to several conclusions: Linux will continue to be attractive for certain server applications for at least the next five to ten years. Hardest to predict was how genuine of a threat and at what time a move to the desktop might occur. There were several strategies available to combat its progress. My challenge in my new role as a consultant was how to effectively influence MS's actions. A tough assignment considering this was a company in transition. One that was drifting into some kind of migratory listlessness and questioning the fierce collective passions that once had catapulted her to exhilarating new frontiers.

We could not use a direct, head-on strategy to win. As mentioned before, we had a respectable offering but no means a cutting-edge solution. Convinced that we could eliminate our disadvantages over time, a pure containment strategy could not be pursued either. For that to succeed, we would have needed a reputation of hitting launch schedules dead-on and having the resolve to create a considerable leaner, meaner, and supermalleable NT version targeting blade applications at a reduced price. With price being the lesser of an issue, I discovered no appetite to do the needed development work. With inertia ruling our immediate culture, MS's development nerds just continued to add features. Lack of former competitive vigor, induced at least in part by our legal quagmire, was another hindrance. An indirect strategy would have worked only if we had an ace up our sleeve. Changing the ground rules on Netscape had allowed us to eventually win the game with higher integration and through buying time until other tech advancements became available. Not obtainable for Windows server, it seemed. A strategy of peaceful coexistence, another viable option, ran counter to our former cultural belief system but was what, at the end, roughly evolved. I came to the conclusion that we needed to apply several of these options over time to counter Linux on the server and prepare for a desktop assault. Complicated to execute, and hard to convince the MS's marketing and development folks of.

Starting my new assignment, I immediately felt a power gap. I had no direct reports, no marketing budget, zero staff, and hardly any admin support. I suddenly had to deal with dozens of uncontrollable cooks in different parts of the development and marketing kitchen. A lot of them pursued their own ideas and were not accustomed to listen to or follow an outsider. I had led an effective and well-oiled team who respected me for my judgment and joined accomplishments. In my new position, deference was in short supply, replaced by a surfeit of egos, each knowing more than the next and probably in need of a lesson in effective team play and discipline. With the company in legal jeopardy, the former speed-of-light velocity downshifted to one of cautious circumspection, not a terrific prerequisite for winning against courageous, inspired volunteers. Public image and avoiding questionable publicity were more important than winning with unpopular measures such as suing over patent infringements.

Having no track record with this new group, I came up short. Failing an assignment wasn't commonly found in my resume. I admit I early on lost the drive to sort through a labyrinth of diverting opinions and was tired of the political gamesman ship now required to succeed. Despite being offered a bonafide nonconsultant position, I left the company at the end of 2002 at last. This chapter of my life was closing, and after all I had accomplished in OEM, I knew I could no longer top my career in this by-now transformed company. I had better plans for continuing an active life. My professional phase was over, and a new cycle of the private Joachim Kempin was about to begin. I never regretted my decision and still very much enjoy my life after Microsoft.

A STATE OF DISORDER

"Fortitude is the marshal of thought, the armor of the will, and the fort of reason."   —Sir Francis Bacon

WHAT *L'AUDACE*?

Shortly after I left the company, I ran into Richard Fade and was surprised to hear that Steve Ballmer, in several OEM business reviews he had attended, continued to pose a simple question: "What would Joachim have done?" Being remembered that way felt rewarding, and maybe my old boss truly missed me, but I saw no reason to reverse course and go back to work. MS had changed dramatically compared to when I had joined up. She now employed a workforce of over fifty thousand people. I had been about the four hundredth when I joined. When I left, only ten others if I excluded Steve and Bill had served longer. My old comrades had voted with their feet following the exodus of others who had joined the company much later. At the end of my career, I felt like a relic from a bygone era, a dinosaur missed by the meteorite and its afterglow. Back when I arrived, the whole company was enthusiastically marching to what the French revolutionary Danton often quoted and what Frederick the Great, Napoleon, and General Patton appeared to have thematically adopted in abbreviated versions and acted upon in spirit:

"Pour les vaincre, messieurs, *il nous faut de l'audace, encore de l'audace, toujours de l'audace* et la Patrie sera sauvée!" (In italics is the shorter form; Patton used the condensed form of "l'audace, l'audace, toujours l'audace.") All of these versions express the same philosophy in a military sense, for which they were originally meant and applied for. Danton's original remark roughly translates into the following:

"In order to overcome them, gentlemen, you need audacity, extra audacity, and audacity forever, and the Fatherland will be saved." Danton used this powerful slogan to fire up the French population and its military during the revolutionary wars, as did French emperor Napoleon, hardcore Prussian king Frederick the Great, and my other hero General Patton.

When I joined MS, the above slogan characterized our shared understanding of how to be victorious—unspoken, but in spirit. "To win against competitors, employees, we need audacity, extra audacity, and audacity forever, and we will defeat them." The top two guys had instilled and drilled it into us. (After reading the above you might think that Steve Jobs was spot-on when he once said for MS it was all about winning and not about great products. Sorry, but without them there, what was there to win?) A modern interpretation of the meaning of Danton's words and the way I always thought they should be understood can be found on the Wasabi Venture website:

"A bold commander never gives up the aggressive initiative to his enemy. As a start-up, even when you have great success, you have to keep going. Great sales success is not an excuse not to do more marketing. Great new features in a product are not an excuse not to build the next great thing. This probably means extended time of 15 hour days, doing sales calls on weekends, answering support tickets at crazy hours, and never losing the audacity that got you success in the first place. It is in this effort that you will win. Your competitors are not your friends. They are the enemy, and you must step on their throats and cut off the very air they breathe.[54] Every customer/user they add is one you never are getting. Start-up life is a Blitzkrieg land grab, and you need to push every day to steal the next inch of ground. Winning is what this game is about, and that is what determines who gets to cash the big checks at the end of the day."

Organizing a relentless pursuit is the best way to decisively finish a battle and win the war as the Prussian general Carl von Clausewitz once wrote. This was what the company I knew practiced. The company I was leaving had her wings clipped. An important part of her business was now regulated. Blitzkrieg was no longer in management's vocabulary. In this context the result of the Feds action against MS can best be described with the famous words of Joseph Fouché, Napoleon's police minister when characterizing the unjustifiable murder of a political enemy ordered by Napoleon: "C'est plus qu'un crime, c'est une faute!" (It was worse than a crime, it was a blunder—unfairly diminishing America's number one IT Company!)

In the aftermath of the Feds strike, openly venal attorneys representing opposing parties encircled MS like scavengers ready to maul her bandaged carcass. They engaged the company in a host of class action lawsuits. Companies like Gateway and IBM posing as injured parties went for

---

54 Paul Maritz supposedly said this once and was chastised for it.

another part of the kill. The lion I had faced only partially accomplished what his extended pride eventually finished. The finding of facts by the discredited judge served as a scent trail for the blood-thirsty predator bunch. After the dust settled and the mop-up crews had shut the lights and left the building, the legal foragers had extracted approximately six billon US dollars, including the fines paid to the notoriously off-the-wall EU bureaucrats.

The company, wanting to get rid of these distractions and the resonating hell and damnation of the press, caved and settled most matters out of court. The total settlement sum, in perspective, was roughly the profit my group had generated during my last year at the helm of OEM. The conspiratorial pack of wolves and leeches certainly did have their pay-day nicely lining their pockets. MS dug into her cash hoard and paid up. Having lost her edge, she did so without mounting too vigorous of a defense. Neither the remaining management team nor our own lawyers were willing to slug it out at length. The historically combative mode was drifting toward political appeasement.

Having lost the legal battle partially and having paid up without depleting all cash allowed the company to take stock and regroup. Best of all her market position was unharmed. Windows and Office, the leading bread winners, were unharmed—by 2002—and so far unchallenged. The current versions Windows XP and Office 10 were solid products and had a strong following with enterprises and consumers alike. The profitable OEM business was still growing and piracy was slowly being reduced further. The enterprise server market—competitive as ever—was at least a break even business. Xbox and all online services, while not profitable, were steadily gaining momentum. MS researchers were creating more patents than ever, rivaling IBM in quality and quantity. In a nutshell, when it came to revenue, profits and product's health, breadth and depth, the US regulatory measures had been by no means lethal.

Was there any damage at all? Foremost the company lost a lot of talent, and second of all the public relations (PR) beating had given her a bad image. People leaving the company in droves had reasons. Not wanting to work for a PR damaged company was one, having made enough money to retire another. Salaries were now up to industry standards. But Steve had dropped stock option rewards. Correct or not it sent the wrong message. Employees interpreted his action as management having lost faith in the future of the company. In particular, developers felt that breakthrough contributions would no longer be rewarded like before. To improve pay one definitely needed to climb up in rank. With jobs on the top scarce, people believed their upside was slipping away.

Last but not least, plenty of other attractive opportunities were available to explore. The Google start-up was making a name for itself, revolutionizing Internet search and Web-advertising methods. Something MS had been talking about for some time but had dropped the ball on. Apple was gaining strength with new innovative notebook products and ever improving OS technology making working on leading edge technology tempting even for MS developers. The open software movement was stronger than ever, and all Linux distributors were vying for smart programmers. Social network services, still in their infancy, were rumored to be developed. All of these formative enterprises were still dishing out stock options, why not try and make a second run, add to one's fortune and once and for all secure a plentiful future beyond MS?

As a still growing company, MS responded and created more managerial opportunities for employees. More layers, more committees and more review cycles were the result. The speed of thought once eminent slowly disappeared. To increase visibility, political standings and future advancements, career minded employees now had to make sure that their peer reviews went well. Contributions became secondary for advancements. Political correctness counted as much and sometimes more.

I received several e-mails and phone calls from some of my old cohorts not understanding the company any longer. They complained in particular about the stack ranking method the company deployed during the review cycle. It had started when I was still around and I had ignored it, believing it was unjust and cruel for my in general well performing employees. Now it was followed to the letter and if you landed on the bottom of the stack, you had to fear for your employment. Like other employees, my ex-coworkers expressed their dissatisfaction in regularly conducted annual employee surveys. Nobody seemed to read them, get the message or appear to care. It surprised me knowing Steve's sensitivity. I concluded the bureaucracy which had bugged me years ago had by now cemented itself so deeply that it must have become a management blessed modus operandi. MS's wheels began to churn slower and slower. Products got delayed and influenced by designing them in committees. The company was losing her way, as Jim Alchin the head of the Windows group in '04 remarked.

Steve Ballmer was now fully in charge of the company. He tried to stem the tide introducing an enlarged leadership circle, promoting more people than ever to VP and making sure people got incentives in form of stock, not options—somehow restoring a little bit of faith. Before I left, I attended several of his rallies, listening to pep-talks he gave. They sounded repetitive and fell mostly on deaf ears. Even at his best the eternal fiercely gesticulating optimist was no longer the convincing demagogue he had once been. People just needed to follow the news cycle to understand reality.

Attacked in the aftermath of the trial, MS seemed to be giving up the noble fight, caving in everywhere by settling the left over matters. Management rained money on winning plaintiffs while at the same time tightening long established and generous freebees employees took for granted, causing more consternation. Clear signs of weakness, and as much as Steve tried to keep his posture in internal meetings, the air smelled of mollification. The subsequent compliance scrutiny fostered a cautionary conservatism inside. The stench manifested itself. An unheard-of hesitancy took hold, slowly converting the company into a less assertive competitor. Why stay put and become a political animal, when there were way more exciting opportunities waiting outside? So the exodus continued and enfeebled MS further when tackling emerging challenges.

The investment community for sure judged it that way. MS was throttling back her fighting spirit, dampening her ability to compete. The mass-departures of key contributors slowly but surely hampered her ability to respond fast enough and with sufficient innovations. Bill obviously crumbling was the last sign that the community needed to make MS's stock price head south. Fortunes vanished. Millionaire dreams evaporated. More brain trust left the company strongly doubting Bill's and Steve's renewed leadership efforts and MS's future business prospects.

Vista, the next version of Windows after XP, took five years to see the light of day. Not only was it two years late, it missed a plenty of new features like a long touted innovative file system. To Jim Alchin's dismay, competitors introduced that feature he had long envisioned first. Internet Explorer was staid and contained more bugs and security holes than usual. A tablet version was dropped from Vista's development schedule. A disgruntled Alchin who had supervised its development considered this release his last act in a company he no longer believed in. Three years earlier he had told Bill that he now preferred an Apple MAC over any IBM PC clone. Superior design and an advanced OS were the reasons for his frustrating opinion—sadly true and it showed up in Apple's ever increasing share of the total PC market.

Encouraged by Vista's deficiencies and bloat, Google released her own Linux-based OS called Android in '07. Having a small footprint it immediately gained a solid foothold in the rapidly evolving smart phone, tablet and netbook market, and is now poised to gain notebook and potentially desktop market share as well.

There were other indications for MS's falling behind. When Apple introduced the iPod in '01, an innovative music player, and complemented it two years later with an online music store (iTunes) for MAC and later Windows PCs,

MS was caught flat footed. ITunes allowed for music sales and downloads. Its launch clearly demonstrated who was ahead in exploring this new and highly profitable online service frontier. It took MS until 2006 to come up a competitive product called Zune, only to discontinue it a couple of years later. It simply crashed.

The year '04 brought us MySpace and Facebook. The popularity signs for social service networks had been on the wall since AOL introduced an instant messenger service in '01. While matching it somehow two years later, MS's management never got or accepted the real message of how much the younger generation truly enjoyed an always on chatting opportunity. The company for sure had the technical ability to show some leadership here but was seemingly out of touch because management was not convinced by the by then obvious. Over 900 million people using this service today shows how much one bad judgment can hurt a company.

During '04, stiff competition for Hotmail, MS's e-mail service,[55] arrived when Google entered this service segment with G-mail. One of the attributes and why people liked it was its ability to store unlimited user information in the cloud, something MS's chairman belittled when it was announced, but had to follow several month later as users left Hotmail in droves. It took Google some time to get her service right but she eventually cleverly cross linked it to her search engine pioneering the meanwhile infamous context advertising and accelerating her ad revenues dramatically. Today the e-mail market is pretty much split among three companies: MS, Google, and Yahoo! all retaining around 300 million customers. MS's recent announcement to revamp her service and make it look and feel more Outlook[56] like shows that the fight for member share is by no means over. While all these services are free, they produce revenues through advertising placements, which Google is the master of.

That same year Apple added insult to injury as she improved wireless streaming of audio, video, and photos with her Airplay online services. Internet Explorer finally became competition with a browser called Firefox. There were rumors that Google was working on her own version. MS's contribution for '05 was launching Xbox' second version, while she missed buying Skype cheaply from eBay. The next year another innovative service for exchanging short phone messages was born in Twitter. MS missed out on buying YouTube—Google beat her to that. Instead, MS contributed a mediocre version of a barely revamped browser IE 7, making her tremendously volatile to any new seriously contrived entry.

---

55 Acquired and operated since 1996 and earlier called MSN Hotmail
56 Outlook is MS's personal information manager contained in her office suite.

Painfully experiencing her own weakness and outfoxed by competitors offering highly innovative online services, MS in 2005 finally acted by acquiring Groove Networks. Her CEO and founder Ray Ozzie was a well-known and highly respected software pioneer. He is the inventor of Notes,[57] which Lotus acquired and which was the alleged reason why, in '95, IBM eventually bought that company lock stock and barrel. After a short stint with IBM, Ray founded Groove. Here with some seed money from MS he developed a new product allowing multiple users to work collaboratively on computer files simultaneously. For several years Bill had had his eye on him and through the Groove acquisition he finally succeeded luring him into MS. What bugged me a bit was the way he publicly expressed his feelings about working for my old company. Asked how he felt about being employed by MS, he answered it did not feel evil and was not inconsistent with his core beliefs! Coming from him the world sighed with relief.

Less than a year later Bill promoted him to Chief Software Architect, a mantle he had bestowed on himself six years earlier. Soon the industry was at buzz assuming that Bill's time had finally come. A year later he indeed left the company, retaining to this day his chairman title. Hiring Ray was his last act to prepare MS for a less PC centric future. Let's see if Ray was the resurrection and the light MS was hoping for or another fatal choice. Ray, who had been preaching for a long time the advantages of providing computing services in the cloud for a mobile society, seemed a great fit for turning MS around. Intellectually unmatched in a company relying on the continuance of Windows desktops, he did not disappoint. Shortly after his arrival in October of '05 he published a study headlined "The Internet Services Disruption." Inside I found a description of the historical aspect of the Internet revolution and the seamless and less complex user services MS needed to offer to stay in the game or better regain some magic. Bill had formulated a similar strategy years ago, yet the company had never acted with fortitude upon it. Who else than the ex-Chief Software Architect was to blame?

Looking at the '06 company organization chart one can easily spot one reason for what I call MS's failing grade. At least seven high powered people in the company were in charge of inventing and evaluating strategies. Too many cooks in a small kitchen? Some of them like David Vaskevitch, Craig Mundie and Eric Rudder were longer time employees and were either Steve's boys or Bill's boys. Part of the two inside mafias who for a longtime had nurtured disagreements about how to excel! As Ray Ozzie entered the fray in '05 he found a very difficult state of affairs to deal with. Remember he was a thinker and tinkerer with an acute sense for simple software novelties. At the time he considered himself a mercenary

---

57 A popular e-mail and collaborative workspace software package form the mid-1990s

fulfilling an engagement that had been part of selling Groove to MS. As an outsider he had been put in an awkward situation to work through and with every sub-culture MS possessed. He soon found out that his above mentioned memo did not cause immediate traction. (It was rumored to be written by Bill with Ray only editing it—who knows.) On the other hand, a newly created small sized development team led by him charged forward, advancing and realizing the propagated vision he wanted the rest of the company to believe and follow.

Before his efforts eventually bore fruit MS had to endure more pain. In '07 Apple introduced a nicely designed very capable cell phone—calling it iPhone. Far superior to any other phone of that time, it took the market by storm. Ten years earlier, when I was still around, MS had entered the same market licensing Windows CE to mobile phone manufacturers but still had nothing to show for. Her phone vendors had let her down and the development team lacked the resolve and perseverance to improve the product sufficiently to get it recognized and win design-ins.

At the end of '08 Google finally launched her own browser christened Chrome. It took MS four months to respond with IE 8. While vastly improved over its former version, the geeks nevertheless loved Google's speedier and leaner version much better and after a decade of nearly unchallenged leadership there was a lot of agony in store for MS. In its first year, Google's Chrome won nearly 20 percent usage market share and has increased it ever since.

Responding to Google's search engine crusade, MS tried three times before she got it somehow right. First an MSN search was offered, then with Ray Ozzie's influence Windows Live Search followed, and finally in '09 the Bing search engine appeared. While it lacked Google's finesse and bells and whistles in version one, it was good enough when combined with some cash payment to gain an alliance with floundering Yahoo. Today the product is feature- and speed wise pretty much on the same footing with Google's offering. But this has come at a price, sucking six billion dollars out of the company just to be more or less an at par contender.

Windows 7 followed the same year, signaling to the world MS was getting serious yet again by delivering quality goods like this leaner, faster and rock-solid OS release—yet it still missed a tablet version. Adapted for cell phone use one year after, it allowed MS to build an alliance with Nokia and reenter that market with quite an impressive product. Two other vendors including Samsung followed suite. Together they conquered only 4 percent of the smartphone market. Hardly making a dent! Four month later Windows 7 was followed by the solid and streamlined Internet Explorer version 9, putting it finally back on par with competing browsers. In 2010 Apple stunned the IT world by announcing a computing tablet christened iPad, powered by an ARM CPU. This was Apple's second entry in this category after her '93 tablet, called Newton, had miserably flopped. While MS and her OEMs had more or less given up on this category by 2005, Apple had seized the opportunity by taking advantage of vastly improved ARM technology. State of the art miniaturized semiconductor components and the availability of new plastic materials allowed her to create a very powerful, slick and visually appealing product accompanied by a sexy and cool looking software interface.

Despite judging Apple's tablet entrance a significant setback for MS, I conclude that after 2007 the company was seriously aiming to get her groves back. MS was still not considered cool enough for many but at least she was losing less ground. With Windows 8 *ante portas* there is even more hope. By 2010, after having been five years with the company, Ray Ozzie was still unconvinced that MS was on a path to recovery. He expressed this on stage, contradicting Steve in a widely publicized dialogue and left MS one month later. For him, Google's open source and free of charge Android and Chrome OSs guaranteed a better future. Today he is in the process of hiring programmers for an undisclosed product based on Google's flagships. He might regret it this time around.

Reasons may be found in the departure of several other key MS execs. One is Jeff Raikes and the others are David Vaskevitch and Robbie Bach. With no more Bill-leaning girls or boys around, Steve's authority was bolstered. Combine this with Bill's decreasing interference leaving MS's destiny mostly to her CEO, and you understand what has beyond doubt cleared up Steve's command center. No doubt he won the power struggle manifesting his control. People now follow his orders to the letter There is one further indication for this. When comparing the current org structure on the top to the one existing in '01, Steve has effectively cut his number of direct reports by two third. This could mean two things: more autonomy for the ones he appointed or more hierarchical command layers for the troops. The former might mean less micromanagement, the later could mean decision making will take longer. I, the ever hopeful, tend to believe it signals an attitude change and an expression of trust for his underlings because now he simply can—they are "his" guys.

Today the retreat of MS's aging chairman seems nearly complete. How did this ever happen? When I met Bill first he was an energetic, imaginative and inspiring young man. I adored his uncanny ability to cut through the chaos and to the chase, attracting—undoubtedly with Steve's help—excellent people while inspiring and directing them to get the right things done to propel the PC industry. With Bill cultivating the art of expressing paranoia over competitive threats, emphasized by his alter-ego and mouthpiece Steve, employees were kept on their toes in the early days. But as the company proceeded, the claims repeated themselves too often. Making a mountain out of a mole hill each time a paradigm shift could be vaguely spotted like a shooting star on the night sky signaled his intellectual pump was running dry—the pool shallow. Parading out the self-same scare tactics over and over became less credible and suggested he was in fact feeling genuinely threatened about losing power, visionary guru status or both.

The label "richest man in the world" had always counted less for him than being a recognized industry leader. The Feds stamping monopolist—the scarlet M—on his forehead was a monumental humiliation for him. His desire was to be honored as savior from proprietary IT chains and for causing a personal information appliance revolution for the masses. This was what was written on the revolutionary banner he held high. As he pursued his destiny with resolve and fortitude, unyielding counter forces awoke. Not willing to dance to his tune, the establishment felt threatened and pushed back hard. Struggling desperately and with perseverance to achieve his objectives and win the glory and the rewards he desired, he answered their fighting forces with extra lethal powers of his own, surrounding himself with acquaintances obsequiously applauding and buttressing his callous aim. They rubbed off on him and because failure was never an option for him personally, I spotted a megalomaniac trait. And so did the suspicious Feds who nailed him on the cross for relentlessly trying to achieve what they considered high-handed supremacy of the IT universe.

Jackson harshly opined he had developed a Napoleon complex. I disagree. Bill was neither short in stature nor did he have other handicaps to overcome. He was indeed shooting for the moon and having attained tremendous wealth and success—and therefore power—the humble and guilelessly ambitious person I had met years earlier transformed. Even the well-intending influence of his wife could not bring him back to earthbound reality.

Steve stayed humble. Unlike Bill, he was not an intellectual who would get drunk on his own ambitions and futuristic visions. Only numbers truly excited him. Customers summarily induced reality in him and so did harsh market responses. He knew and admitted to me the company should have demonstrated better industry leadership under him, sharply criticizing his own performance. Bill was less critical of himself, wanting to reign supreme while never admitting he had experienced his own Waterloo more than once. Only when handing the reins to Steve he indirectly confessed how much his reactor core had been impacted, and off he retreated into the intellectual realm of Chief Software Architect and philanthropist.

A man of simple genius or a man of character? While each of them exists, few possess both. Intellect does not breed conscience and conscience does not breed intellect. The men or women we admire for their intellect—we do not necessarily follow! Perhaps the company would have been better off if Bill would have left earlier rather than tangentially spending time saving his legacy and subsequently missing waves of new opportunities! I wish him all the luck in the world with his new endeavor focused on and supposedly solely filled with good deeds.

I met with Bill a last time in his office and then later in his house for a farewell party of all antitrust trial participants. While the latter was fun, talking to him in his office was intimate and rewarding. He thanked me for my hard work as we chatted about past episodes we had experienced and shared. He proudly mentioned the next version of Xbox would be able to function as a full blown PC. I could not believe what I heard and remembered what I had read in the WW II German army manual: "Great success requires boldness and daring, but good judgment must take precedence."

In July of '08 MS's burned-out chief software architect took the next step and left for good, retaining his title as chairman. As described, he had miserably failed in guiding the company through the Internet and social network revolution of the last twelve years of his reign. Journalists have written he and Steve re-booted the company after the settlement was reached with the Feds; I disagree. Where are the leading edge and breakthrough software products MS was supposed to deliver? Bill's focus was far less on MS's well-being but increasingly on his charitable foundation, distraughtly evading a tainted image. No longer limping home from combat bloody and bruised. Victorious! Inside the company he was still liked, but striving for renewed popularity had gotten in the way of making the tough choices and surrounding himself with first class advisors who drew strength out of disagreements. A simple glance at what the company missed tells the story:

Yahoo!'s and eBay's rises, Google's search engine popularity, being number one in Internet advertising and music/video download revenues, the birth of Facebook, Twitter and YouTube, not beating Linux on the server side, losing ground to Oracle and SAP, Apple's iPod, iPhone, and iPad emergence, the rise of Google's Android powered phones and tablets, and Firefox's and Chrome's browser popularity to name not just a few.

When I left MS, the two primary and profit-generating businesses were

Windows operating systems and Office productivity applications—sometimes labeled the Old Faithfuls. They are still the foundation MS is resting upon. Funds and talent necessary for entering new fields of competition were wasted as management failed to convert MS's staggering and established impetus and tremendous R&D spending[58] into the next tech success story.

Employees got mired in committees and were cautioned to not drive on as boldly and heedlessly as before. No longer stepping forcefully on competitors' toes! Let's prevent being put on notice again—controversy to be avoided! Deceleration meant losing decisive battles in the arena of the IT coliseum during Internet services and social network expansion time. Lush and fertile fields were left open and accessible for the Googles, Facebooks and the Twitters of this world. With MS culture and agility severely curtailed, a variety of start-ups enjoyed nimble and unhindered grazing.

I met with Steve in the spring of 2010 and offered him unabashed outsider's feedback. The man had tried hard and worked feverishly; I applauded him for his effort and the financial results he had produced. He is not *"Bad Boy Ballmer"* as a book with the same title wants people to believe. And don't get confused when he looks aggressive and mean making an announcement; he is just acting! He remains a decent human being and capable business executive with whom I continue to have spirited differences. Fair to me and the people who worked for me—we respected each other. (He unexpectedly and surprisingly called me recently to wish me happy birthday, my 70th. I am very thankful for that!) During our meeting I gave him what he, at the end, labeled "a friendly pep-talk." Not expecting anything else—I didn't disappoint. The company by now had bloated to 90,000 employees. (I still remember when Bill criticized IBM having 1000 people working on OS/2!) This was no longer the mean and lean tiger I knew.

The non-OEM sales and marketing groups had now been converted into a more centralized structure. Being a government regulated business, the OEM group was no longer reporting directly to him. Nevertheless her profit contribution seemed too large to be run—in true McKinsey style—as a small part of a product division after having been overseen by the COO for a long time. A sales group managed by an administrator? Hard for me to understand as brilliant as he might be! Or should I say no wonder the company failed to win the mobile phone and tablet wars? After Richard Fade's cautiously fore-shortened stint, four other executives have earned the honor of managing OEM. The frequent management changes left no room for consistency, bold maneuvers or lasting imprint. Take it or leave it

---

58 Nine billion US dollars in 2011 alone—the largest in the IT industry

is the choice customers now have according to the Feds' edict and MS's covenant. Theo Lieven would throw another of his famous fits if he was still in the PC business. On a separate note, how much fun is this for its sales personal and current management team?

The system-builder business, once wrongly separated from the main group, got reintegrated a couple years down the road. While it was robust enough to survive and show gains as long as it was run by capable ex-OEM managers, it nevertheless distorted the rest of the subsidiaries business. Its healthy revenues were used to justify overall headcount growth without helping market penetration for Office and server products. After reintegrating it four years later, the new OEM management team, having lost touch, showed less passion to nurture and grow it, resulting in a severe decline of revenue. I loved that business and lavished attention on it. People wondered why. Simply put, these customers were much closer to their clientele than the large manufacturers, and understanding their business well provided me with a great learning experience and beneficial feedback.

In other parts of the OEM business, major challenges have not been answered in developing countries, the People's Republic of China (PRC) and India in particular, because of failed OEM marketing activities, political tolerance of renewed piracy attacks made possible by hackers, and unaddressed security loopholes. For the first time, in 2005 an OEM security key for Windows Vista found its way into the hacker community and in 2009 the same happened with a Windows 7 key from Lenovo, opening the door for uncontrolled distribution which can only be fixed through altering security specific BIOS code through a new service pack. If MS would have the guts and threaten to impose huge fines for negligence of this kind, no damage might have ever occurred. Too radical considering current company policies and politics.

In concrete numbers: According to the research firm Gartner, PC sales grew by 3.8 percent in 2011 while MS's Windows revenue decreased by 1.4 percent. Taking into account that Intel reportedly scored the highest CPU sales gains in emerging countries, MS's failure, growing Windows revenue in lock step, can only have two main reasons. An approximate 10 percent price erosion—not true as my sources tell me—or most probably—a failure to control piracy in developing countries. The probable cause for a $600–$650 million revenue loss for the company![59]

A few of my old crew still work in the OEM group and a couple who had

---

59 With 8.4 billion shares outstanding, this translates into about seven cents of profit per share.

left actually came back. I hope the old timers do not suffer too much when they remember the old days full of unbridled energy, inspiring eventfulness, fun, volatility and tumult.

I encouraged Steve to listen carefully to IT world-shattering ideas his employees were concocting and spin off new businesses to sponsor a MS-led Silicon Valley like phenomenon of new entrepreneurs, right here in the state of Washington. Ray Ozzie, who by now had left the company, had once asked for the same. My message: Take full advantage of any mind blowing new opportunities they come up with instead of bogging entrepreneurial spirit down with internal bureaucracy, artificially-imposed standards, too many committees and management layers. He answered: "I hear you!" Two years later I am still waiting.

Management's success comes down to trusting others. Can MS's CEO trust anyone other than himself, and can he let go of micromanaging results? The recent developments might actually signal the latter. I remember Kurt Kolb, my super smart business manager, approaching me during the Java mess in 1996/97 and asking me if he needed to keep me in the loop of the ongoing investigation and negotiation. I answered: "Absolutely not, call on me only if you run into serious trouble causing us to eventually lose this potentially multimillion dollar lawsuit or believe we might get an injunction stopping Windows." I recognized he was puzzled and, reinforcing my opinion, I told him "You work for me because I trust your judgment." Kurt later considered it a management lesson learned.

Is Steve really the man to turn the ship around or is he too entrenched in the culture he needs to revive? Therefore his appointment to run the company may have been a mistake. As Richard Forster observed: "More companies fail because of strong cultures than weak strategies." Are insiders or outsiders better suited to cause long and lasting change and a successful break with old habits and management styles? Analyzing Lou Gerstner's, Steve Jobs' and Eric Schmidt's (Google's former CEO and now executive chairman) job performances suggest there are many examples where outsiders trump insiders.[60]

Steve B. has kept revenues and profits nicely growing. But so far he has not applied himself enough to be recognized as the turn-around product fanatic and perfectionist the company desperately needs. Without being a credited nerd and respected visionary, he continues to struggle with creating and inspiring a top-notch R&D team. MS needs to find her equivalent of a Steve Jobs! Sales, marketing and operational strength alone will not be

---

60   In this context, I consider Steve Jobs, after running NeXT and Pixar, an Apple outsider.

enough to effectively lead the overhauling of a company in need for product excellence.

I own zero shares in MS. I sold them when I left and never touched her stock again. I still admire the company and I am hoping she can rejuvenate herself. Talking to Steve offered a warm nostalgic taste of the good old days as we spiritedly and loudly rehashed the ancient and almost surreal stories. In remembrance I saw the spark in his blue eyes reappearing. We departed as we often had, as equals disagreeing in what I had contributed in our one hour long discussion.

In the summer and fall of 2010 I spotted rumblings in the press asking for his resignation. For sure the shareholders of the company who had made out like bandits in the early years had nothing to laugh about since he had taken the reins. MS's stock price has fluctuated between $21 and $36 over these years despite growing into a by now $75 billion plus revenue machine with most admirable profit margins. Wall Street and the investment community still seem disillusioned, not trusting the leadership qualities of the man at the helm.

There is one simple reason for this. MS's echelons didn't act decisively on their own judgment when considering how long the PC revolution would last and when the PC could be seriously challenged as number one information appliance. As early as the mid '90s her executives clearly understood the alternative trend to increasingly mobile devices. Notebooks and Netbooks casted a first light on it. Several efforts were made to engage in e-book readers, smart phones, pocket PCs and media tablets. Plagued by notorious product delays, mediocre product quality and insufficient features, the company failed to gain the coveted number one position in any of these categories as OEMs refused to invest and build these forward-looking devices.

Their reasons, first there was not enough passion or talent to get the software done right first time. Second, MS's R & D was hamstrung by management through committee and political infighting for resources. And third, most of these numerous and too ambitious projects were mainly undertaking to cover MS's flank and discourage competitors without a true belief in their validity and success.

Apple's new CEO Tim Cook recently said in an interview that Steve Jobs—Apple's former CEO—"grilled in all of us over so many years the company should revolve around great products and that we should stay extremely focused on few things rather than try to do so many that we did nothing well." This is very different from Bill's and Steve's aim to mingle in all categories at all times. Needless to say both philosophies offer a path to success but the one MS has chosen makes retaining focus much harder.

A lot of talented employees responded to this operating style like Rob Glaser, the founder of RealNetworks, by leaving the company. Their main motives, their egos, were being bruised and constrained by an ever-growing Windows centric view governed by bureaucracy. Robbie Bach's[61] departure falls into the same category. The top honchos never thought of setting some of their talent pool free, sponsoring new ventures with angel money and creating partners for mutual benefits and life! Instead, the ones who left were sometimes treated as enemies often accompanied by ugly legal wrangling. Never forgotten and making it ultimately harder for MS to obtain cooperation in emerging markets.

Spreading resources too thin or losing key ones hurt further and resulted in neglecting core products. Competitors were provided ample opportunities to catch up or even bypass MS's offerings. Internet Explorer is a premier example in this context. Having won the integration argument in court, IE and Windows are still joined at the hip. This fall it will be in version 10. But now serious competition has arrived. Neglected during Windows Vista reign, it spurred the release of two nimble browsers Firefox and Chrome. As a consequence, MS's browser usage share shrank to below 50 percent, forcing MS for the first time ever to advertise the usage of IE heavily on TV categorizing it as a *modern* browser. Worse, it motivated Independent Software Vendors to predominantly write browser applications for her competitors. Losing these ISVs—even if only partially—is not only agonizing but potentially life threatening. What federal regulators did not accomplish MS's negligence and the free market did. Proving once again competition never sleeps and will attack—even almighty MS—supposedly still protected by network effects.

---

61 SVP in charge of Xbox and mobile devices

PC DOWN SPIRAL

PC sale in industrialized countries has decelerated. Some of this can be attributed to the currently shaky economy but most is grounded in replacing PCs with ever advanced smart phones, tablets and doing computing in the cloud.

A phenomenon which has the industry stuck in a crisis potentially dooming IBM PC clone manufacturers. Gartner[62] estimates that by 2015 1.1 billion smart phones and 370 million media tablets will be sold annually compared to 400 million PCs. While total PC growth over the next three years will hover between 10–15 percent, smart phones sales will double and tablet sales will triple during the same time frame. The importance of the PC will continue to diminish. Cheaper, safer, faster and higher available connectivity and sophisticated cloud services could quicken this trend further.

MS's management team is at least partially responsible for the calamities of the PC clone industry. I always considered the PC manufacturers our partners as they listened to and depended on us. With MS now considered a follower, they had no one to look up to when their need was the greatest. The vibrant energies the company still possessed were constrained by regulations, self-imposed failures and her inability to create disruptive innovations. Management by consensus took its toll, by default, by choosing the staid, low risk path over leadership, pro-activeness and bold adventures on the wild tech frontiers.

The changes the IT industry has gone through since I left are nothing short of mind blowing. IBM needed no Ray Ozzie to understand the cloud and its services opportunities. Sam Palmisano, now CEO and chairman, used it to develop IBM into the most powerful IT consulting and services company helping customers to improve their competitiveness. The company is still selling mainframes but no longer produces PCs. In Sam's view, the cost-

---

62 A leading market research company

effective manufacturing techniques that the PC industry deploys often trump marketing and engineering. Not believing IBM could ever master them sufficiently and make enough bucks with PCs, he sold that business branch to Lenovo, the PRC-based company. I disagree. When compared to Dell's, IBM's cost of sales was the true reasons for her non-competitiveness. Streamlining manufacturing and R&D should have therefore played a much lesser role in considering bailout. I believe Sam divested that business just because he wanted IBM to focus where she could more easily excel.

IBM's failure to keep her PC lead is squarely grounded in her inability to market successfully to consumers. Her PCjr[63] experiment demonstrated this as early as 1984. She was and remains primarily focused on commercial enterprises. The main reason why OS/2 never appealed to consumers! Consumers voted for the elegant simplicity of Windows. Lou Gerstner and Sam Palmisano therefore lost the OS war and the PC battle while building a consulting empire and retaining IBM's mainframe grip.

When looking at Apple's rejuvenation and stellar rise under Steve Jobs you immediately understand Apple's consumer appeal. Neither Gerstner nor Palmisano were able to match his passion for perfecting products as both guided their company persistently through the metamorphosis of a different kind of phenomenal comeback.

There are now six dominant IBM PC clone companies in the world, representing around 60 percent of the market. The club includes Hewlett-Packard, Acer, Dell, Lenovo and Toshiba and as an emerging competitor the notebook- and tablet-focused ASUS. None of these companies has a valid smartphone or tablet offering of their own. Samsung, a meanwhile less important PC player, is the only one giving Apple true headaches in both fields. And true to form Apple is attacking her in court with patent infringement claims—landmines she probably cannot disregard any longer.

Ten years ago when I left MS, Apple's total PC market share hovered around 5 percent. Now it has increased to around 15 percent—making her the overall number two PC market contestant behind HP. She now lists as the most valuable company on the stock market trumping MS's evaluation by a factor of more than two. Talk about a reversal of fortunes! Perfecting the hottest information appliances paid off handsomely. With PCs, smart phones and media tablets all depending on software to succeed another sign has emerged as to why I believe that MS has let her OEM partners down.

---

63   PCjr, IBM's first PC launched to specifically attract consumers

Linux is vibrantly alive and went from a hacker advocated system to achieving consumer fame when Google released Android for smart phones, media tablets and notebooks. Google benefitted further when deploying millions of Linux server clusters as part of her cloud infrastructure. Well ahead of her competitors, she used them first to dominate the search engine world and later to provide all kinds of Internet services including e-mail and context advertising.

Red Hat attacked Windows with a Linux-based consumer version of her own in '05 only to abandon it one year later scared of a patent challenge MS mounted and the company is now focused mostly on workstations. Dell began offering a Linux version called Ubuntu for her consumer PCs in '07. Other derivatives are trying to gain as well, notably Mandriva and Linspire/Freespire—the danger to Windows' reign is growing. The competitive moves were no doubt encouraged by the shakiness of Vista. The Linux and hacker community relentlessly demonstrated its vulnerabilities exploiting its security flaws. Dirty tricks? Not really, MS deserved to be caught flat footed. Complacency had snuck in! Released in '09, Windows 7 put a temporary halt to these types of attacks, but I would not count them out forever—the volunteer army knows how embarrassing they can be—experiencing plenty of their own.

MS's response to the appearance of so-called netbooks was to sell Windows Vista at a reduced price for use on these slow performing and stripped to the bone notebooks. I hope I am still well informed enough to say the OEM royalty for these machines was hovering in the neighborhood of $20. Windows 7, with a smaller and efficient kernel, fixed most of Vista's performance issues. The OEMs saw their Windows prices for netbooks—a meanwhile dead category—go up as a result. Looking at MS's total business, a Linux PC desktop breakthrough with Android for example remains the most immediate threat to her profitability, endangering Office also. Finally the high efficiency of competitive server farms and the availability of in the cloud residing and superior Office-like applications—free ones—could be MS's true and final death knell. All these attempts to attack MS's crown jewels will take time to play out and I believe effective counter measures are already being planned and will be launched with the appearance of Windows 8 and Office 365.

Before I try to predict MS's long-term prospects, let's look at the importance of the Internet one more time. In 2011 a UN report called the right to access it a basic human right and was urging all countries not to limit its use even in a political crisis. Who would have ever imagined such a classification in 1995?

A COOL RENAISSANCE TO BE RECKONED WITH!

"We always overestimate the change that will occur in the next two years and underestimate the change that will occur in the next ten years. Don't let yourself be lulled into inaction." —**Bill Gates**

In the past, MS's successes have often been achieved by coming from behind to later dominate a product category through clever marketing actions, sales tactics and most importantly perseverance. Being the underdog, the press enjoyed the uphill battles the company had to win, cheered her on and sympathized with her numerous times. MS's intense campaigns mounted with resolve and fortitude bought time to perfect a product, successfully retaining end-users. In hanging on, they bet that their patience and loyalty would be rewarded with value added releases eventually addressing all of their needs. Sometimes they had to wait for a long time. MS seldom let them down. Executed with passion and Auftragstaktik by highly motivated employees, a relentless pursuit taking advantage of competitors' mistakes always accompanied MS's crusades as she won customers' hearts and minds. In short, her management team perfected the art of wrestling customers away from rivals, producing both staggering results and envy.

This was then. Having eventually achieved a kind of top dog status and having been condemned of being a monopoly increased the odds against her. Today the company is tightly watched and intensely scrutinized by the press, her competitors and the Feds, making it harder to win with the same-self tactics. Her competitors have gotten better at releasing nearly perfect version one products and responding to her marketing tactics. While still at her disposal, rough and tumble sales and marketing maneuvers are being used way more cautiously. Fearing the Fed's wrath or a public outcry nearly rule them out. Today the company has to rely on top-notch performance or severe competitive failures as the only means to improve her image and win a prosperous future. Indeed a tough spot to be in!

As expressed before, after a string of failures gloom and doom about her future is continuing to pile up. I see this differently. A radically new strategy to win is already being deployed and a couple of factors favor a turn-around. First and foremost the immense strength of creating profits with existing products hand over fist. Considering the immense war chest of her competitors, this will help less to win decisively, but for sure means management can take a larger calculated risk. Second, she is the only company engaged in all key IT fields except mainframes: Game consoles, smart phones, PC operating software, tablet software, PC server software, office productivity applications, cloud, e-mail and search services and related development tools, and she possesses an impressive consulting arm. Most companies she competes with do not have the same type of depth or breadth. Like Apple they do a few things right and win in a narrow number of categories. A lot of analysts consider MS's broad engagement activities a disadvantage. I disagree and believe MS's management will use this in the short term as an advantage, in particular when I look at the hype surrounding Windows version 8. Longer term she will have to switch gear and restructure through keeping the strategically important parts and divesting the less important ones which will shine brighter when pursuing an independent and less doctrinarian path.

A long time ago Bill Gates talked to me about the concept of "Win-tone" as a way to propel the company to the sought-after number one spot in the IT universe. He envisioned that when customers used information technology they would always encounter a unique Windows signature. Embossed like a trade mark, highly visible and accepted by businesses and consumers warranting quality, value and perfection. (I can hear the critics already: "God help us!") What Bill once outlined in a memo in '98 and what he discussed much earlier inside the company went way beyond what journalists reported. He wanted to create and own the ultimate dial tone of the IT universe. The means to achieve this was to supply seamless services via mega-servers. A vision emphasizing ease-of-use, less complexity and cross-platform connectivity! Seven years before Ray Ozzie published "his" memo on the same subject. The only thing Bill did not do was label his idea continuous and ever accessible cloud computing. Do you really always need a nickname?

Has MS ever abandoned this ancient yet brilliant idea? I think the dream is still being pursued with all possible vigor and resolve. Ray Ozzie revived it and MS's cherished principles of "being in for the long haul" and "never giving up," as often publically expressed by Steve, guarantees its renaissance. What took so long is simple to explain. First of all, Bill's old crew had to leave. Second, Steve had to whip a new team into shape so it would be loyal to him and act in unity following his command. Then, helped by his team, he had to personally embrace the general concept of how to own this imaginary dial tone and put his personal imprint on the ensuing crusade. This is the way this gutsy, proud and self-centered man motivates himself, works best and acts with fortitude. He loathes following others—even his own chairman. His newly formulated and meanwhile company-wide adhered philosophy took at least three years to be worked out and take hold, recently leading to respectable product offerings. Office 2010 comes to mind and so does Windows 7 and IE 9, all solid pieces of work in regard to value, security and stability. Cloud computing finally got addressed as Ray Ozzie left his footprint behind with Azure, which MS now offers at a cloud development tool and an extensive fee based service[64] option. Having closed that gaping loophole successfully means trouble for her competitors.

Steve has always believed that Microsoft equals Windows and Windows equals Microsoft. How often did he tell me? For him this product represents the soul and heart of the company and any return to magic, and with him at the helm, will have to start there. He considers not following him on that path heresy—Ray Ozzie's disbelief in that expressed in his presence, on stage and in public undoubtedly led to his fall. When I met with Steve two years ago and told him all of my family's PCs were running Windows his eyes lit up. He is the ultimate Windows-forever man. I believe he is right on and he now has a chance to eventually conquer the mobile and chat eccentric generation F with its next version.

Windows 7 spearheaded a comparably small rejuvenation. I predict Windows 8 is readied as a much deadlier assault weapon. It reminds me a lot of the hoopla surrounding Windows 95. How we positioned and launched it. Every week MS is again in the spotlight. The blogger scene is ablaze. The product is poised to address the smart phone and media tablet market well and is readied to fit on yet to be invented cool form factors. In a nutshell, Windows is finally coming into the 21st century. The company has the technology and the will to provide the seamless services a mobile workforce and a social network crazy crowd demand—some of it even for free. She for sure can afford it. A *modern* Win-tone to the rescue!

Pre-announced at the CES[65] in January of '11 in Las Vegas, 8 flabbergasted the IT world by running on a tablet powered by NVidia's ARM-based CPU. I consider this move to ARM a scale 9 earthquake and wake-up call for MS's longtime allies Intel and AMD. A long overdue Windows variant not only for buyers of media tablets but potentially signaling that notebooks and even desktop PCs may be prone to a future ARM attack as well.

---

64   Called SkyDrive
65   Consumer Electronic Show

Over the last five years these types of CPUs have through increased performance found their way from modern graphic cards into smart phones, media tablets and netbooks. They are characterized by a very small footprint like the size of your finger tip,[66] a limited instruction set compared to typical CPUs found in today's PCs and extremely low power consumption. It makes them an irresistible proposition to be used in handheld devices. They need no cooling fans and nobody needs to fear getting burned when touching devices powered by them. Extended battery lives are their other obvious benefits.

Over time advanced silicon manufacturing technologies have not only shrunk ARM CPUs but enabled the expansion of their instruction sets. The leading brands like TI's OMAP, Qualcomm's SnapDragon and NVidia's Tegra 2 are well qualified to be price and performance competitive with Intel's and AMD's mid-range spectrum. NVidia's chip for example is capable of running 3D graphics without requiring an additional graphic processor, quite an accomplishment considering its size. With Windows 8 supporting several of them, MS will now be in a position to challenge Google's, Apple's and Amazon's ARM based mobile phone or tablet offerings.

To mount the attack, Windows 8 software will arrive with a modern and sexy looking tile interface.[67] A huge deviation and fantastic improvement from the by now 20 years old design![68] Its eye appeal is very pleasing and should help its acceptance as users learn how to navigate with it. Despite of some pundits already expressing skepticism about user's willingness to relearn, trust me, they will! Best of all, Windows 8 information appliances, be it a phone, a tablet, a notebook or a PC promise to provide users with the same-self modus of operandi. No need to bother with the annoyance of having to remember different key strokes or gestures when switching between devices or operating them with a mouse or a touch screen. Neither Apple nor Google have ever accomplished such uniformity. There are more than ten different Android/Chrome versions out there and switching between Apple devices is no easy feat either.

I am convinced *the cool ones* among us will eventually embrace this modern version of Windows wholeheartedly, master it and enjoy a slew of innovative ease-of-use features—like switching between foreign languages on-the-fly. Embossing METRO uniformly on all information appliances enables end-users to move up and down the device chain and its attached cloud services nearly effortlessly. Has the Win-tone of the 21st century lastly arrived?

---

66 Capable of functioning as a complete computer system on a chip (SoC)
67 Called Metro, which debuted in Windows 7–powered phones in 2010
68 Easing the transition for the nostalgic crowd, the familiar interface will still be available.

In general, the move to ARM will mitigate the trend to smaller mobile devices profoundly and give PC manufacturers opportunities to develop a variety of spiffy tablets. To make this even more attractive MS will include her Office productivity suite with the ARM version of Windows.[69] Storage for documents and spreadsheets will be provided in the cloud. A very gutsy move reminding me again of the Windows 95 playbook when IE got integrated into Windows. The difference this time: MS does not start with a mediocre product. Google's similar applications being already free to use in the cloud for sure inspired this move and should warrant no court challenge. Less need for constant connectivity for 8-powered tablets when running MS-Office applications means a further leg up over Google's solution. Short term this move might sacrifice some revenue, which I expect to be made up through service, storage or subscription fees.

Using both her crown jewels and selling them cheaply to get a late foothold in the mobile segments proves how serious MS is to recover lost ground and most importantly lead again. What we tried at the end of the '90s, offering Office through subscription, will be revived as well. The market might have matured enough to make this successful this time. Calling it *modern* Office and having it bundled with cloud storage four times larger than what competitors are offering and allowing for installation across multiple devices might just be the ticket.

Another trump card I expect MS to play is addressing end-user privacy rights issues. The contentious concerns surrounding Google and Facebook in this regard are serious enough to give her an opening. As a policy, I expect her to let end-users choose how messages and their content are being tracked and linked to advertising and how the "fair use" rule will get applied in respect to copyright issues. This is not strictly an OS issue. A proactive move by MS signaling a retreat from practices her main competitors are employing will create a lot of good will, if not admiration. I am sure the Facebook generation—sometimes accused of "digital narcissism"—will eventually demand this. Sympathy shall follow!

Two recent announcements point to this. Hotmail being worked over—finally! Advertising being suppressed in end-user to end-user e-mails in the improved Hotmail is a promising first step in the right direction. Internet Explorer 10 having a default setting that will disallow anybody to spy on user preference is yet another. Building a privacy engine into Windows that ISVs will have to adhere to should follow and could police and eventually foreclose privacy abuse once and for all. A faster adaption rate for such a version should easily mitigate potential losses of advertising revenue for MS for sure—but not necessarily for any of her competitors. A scenario well

---

69 Called Windows 8 RT—standing for Runtime

suited to cut deeply into Google's profit if not mitigated by Google paying ISVs to bypass such an engine. Always possible because it is just software! Will such integration get the courts busy again?

Understanding that nearly 50 million e-book readers will probably be sold in 2015 alone, MS's move to obtain a stake in Barnes & Noble's NOOK[70] business—even as she overpaid for it—needs to be understood an assault on Amazon. This move represents MS's second entry into e-book reader category. A sad issue to highlight for me knowing that back in '98 MS made a dedicated effort to invent this product category. But the developers involved in this effort were told to shut down because their solution was not Windows centric enough. True on its face value, but impossible to realize at that point in time with then available technology. You do not need Windows to read a book—MS-DOS would have sufficed and could have easily been replaced with more advanced technology later.

Naturally any media tablet can eventually access book publisher's libraries and download titles. Yet the more dedicated e-book readers pioneered by Amazon and perfected in her Kindle Fire versions should continue to be cheaper to obtain because of their limited feature sets, and if not Amazon has the financial muscle to subsidize them and/or attract buyers by bundling other enticing goodies with them. Together with Barnes & Noble's size and reach, MS bringing Windows 8 to NOOK has nevertheless the potential to reduce Amazon's share and get rid of another Android infested device. MS has deep pockets and Steve loves to get a foothold in consumer segments and win against open-source-based devices. Be prepared for a fierce battle to progress and expect Amazon's Kindle sales to be hurt.

As I have explained throughout the book, to gain a coveted top spot in the IT industry a company needs to be recognized as the coolest and most groundbreaking innovator. In today's landscape this means you have to know how to excite and energize the Facebook and Twitter generation. If you succeed, the progressive media will hail you and promote your product through plenty of PR—for free. MS urgently needs to relearn this and do it pronto. She once owned the tennis shoe/sneaker and jeans generation. Will a solidly executed engineering strategy for Windows 8 be enough to earn accolades from this audience? Market challenges have changed tremendously since Windows 95 was launched. The appearances of Google's, Facebook's, Twitter's and Apple's products have profoundly influenced user experiences and expectations. Today's excitement is created by hip-hop minded entrepreneurs who appeal to a wide audience projecting revolutionary sentiments of change and progress. Their groovy—or should I say Ray Ozzie—like messages fascinate and

---

70 Barnes & Noble's e-book reader based on Android

reverberate in particular with the mobile information addicted younger generation.

MS's image over the last decade seemed stuck with buttoned up business suits or ex-nerds who have become less popular as they moved through the ranks and obtained middle manager status. When announcing Windows 95, we brought Jay Leno up on stage and the PC played "Start me up"—the famous Mick Jaeger song every time I booted it. Gimmicks? The launching of 8 will need to create a special atmosphere demonstrating a brighter future, one the young generation can believe in, trust and talk about as being utmost cool (or as MS would say, *modern*). For this an image remake of MS needs to occur. MS needs to think Facebook generation, not business community, and leave the booming voice and piercing blue eyes cliché behind. Creating a new social forum for the current generation expressing outmost exhilaration is in my mind the ticket to make inroads and cause a wave of excitement!

While the former paragraph may sound too tactical for some, let's return to the more strategic issues. Developers now engaged with Apple and Google need to be converted and ensured that resulting opportunities and profits will be theirs to keep without MS turning from friend to foe rather earlier than later. There is too much fear in the development community today that MS might not keep her promises. Her current CEO might have a trust deficit in this community, and if so he will need to fix this or leave. Without loyal partners MS is nothing!

To underline my point, Windows 8 has already caused a stir with ISVs. Making it difficult for competitors to port their browsers into the new ARM/METRO environment is not the way to succeed. Returning to the secrecy of the 90s or pursue a walled-garden approach in regard to content as some already claim might awaken the Feds again. Users want choice, and ISVs need a maximum of design freedom for modern Windows to succeed.

My old OEM customers are venting about MS developing its own proprietary tablet following Apple's lead. This has happened before and was code named "Courier" early in this century and shut down to avoid hurting OEM partners. No doubt most dropped the tablet ball by not seriously supporting and promoting earlier MS efforts and as such pursuing a proprietary route now is understandable. What is troubling is the secrecy it has been shrouded in. During my time this would never had happened—I would have made sure of it. I am aware that several OEMs have already made substantial investments to develop their own tablets. They feel like being punched in the face. Further perturbing is that MS's plans to offer a proprietary notebook. It will compete head on with partner offerings and endanger current OEM relationships. If both actions backfire, I envision chances for Linux and Google's Chrome OS to make inroads with OEMs and take significant market share away from Windows.

I further believe the P&L[71] math does not work for either device. Not considering any losses to Linux and Google as discussed before, the new tablet and notebook should increase revenues but I remain very skeptical about the profit picture. MS does not own a factory and has a track record of having trouble with sourcing hardware components and producing devices as cheaply as her competitors. Hardware margins are super thin. To obtain any, the accompanying software will have to be nearly given away. If competitors respond with launching loss leaders, this picture will get even uglier and compare most unfavorably to MS's traditional royalty model. The situation will worsen if as a result OEM market share for Windows will get lost through defection. Revenue deterioration will for sure follow and red ink will almost certainly be spilled. I do not know who did the math on this project. The slim revenue gain with not much hope for real profits combined with losing partner's trust and loyalties seems not worth that risk.

A much smarter move was and still is available. Understanding the frustration of Apple winning the tablet and notebook category and OEMs' failure to produce them at all or not making the most attractive design choices, MS should have nevertheless shared her intent to explore both segments early on. In parallel she should have spun off the people eager to design either device, given them some seed money may be combined with a loan and management would have basically created another OEM competing with her current customers. An arm's length relationship would have been easy to impose and execute. Doing so enables MS to either limit her financial exposure in case of failure or to increase her chance for a nice profit when selling her stake in the future. Should I say back to the drawing board there is no need to play with fire?

Having a stake in such company would raise few eyebrows and probably just stirred up other OEMs' desire to finally get tablet and notebook designs right. There is no doubt that MS did exactly that—hats off. The device which sticks out the most is her proposed media tablet. Adding an innovative wireless keyboard makes it a hybrid located between today's notebooks and tablets. When combined the slick design promises to totally obsolete notebooks in a few years when solid state drives[72] will become cheap and small enough to replace traditional hard drive storage units. Its glossy plastics and sexy eye appeal beat Apple's, and if the flat keyboard finds acceptance, MS's design may be a real hit. Let's see how Apple will respond!

---

71   Profit and loss
72   Storage units based on computer memory with no movable parts acting as hard drive

Despite getting the tablet design right first time I have yet to spot a Steve Jobs like product fanatic in today's MS. I hope Steve Sinofsky, president of MS's Windows division, will step up to play this role. Like others I always wait for a service pack to be released before trusting a new OS version. He will need to correct this notion with product excellence right out of the chute to gain vital momentum. This is in particular important for changing MS's fortune in the media tablet market where Apple, Google and Amazon are seen as leaders.

The cooperation with Nokia on the Windows 7 smart phones is off to a good start. A jointly developed Windows 8 phone promises to build on this early success. Yet only a few contenders, in particular weak ones, are neither enough for an early splash nor for enduring success. With Apple having a nearly insurmountable image and perceived quality lead and Google's Android systems outselling Apple's two to one, time to market is more than ever of the essence. Only a broad-stroke attack with a multitude of eager and healthy partners promises to regain the lead. The company needs to go back to the future and cut super attractive flat-fee deals to gain market share and combine it with going totally for broke by massively paying for partner advertising to propel this new breed of phones on a steep trajectory to success. Especially after Google bought Motorola's phone business! MS acquiring a similar supplier would not be an effective counter move either. Neither MS nor Google have enough expertise to be a successful low margin component or device manufacturer.

Other way more effective weapons exist to stimulate market share for Windows 8–powered mobile devices. The integration of slick video communication means—using Skype features restricted to Windows—seems to be one promising idea. The reduction of communication costs is another. Amazon is showing the way here by offering limited free cloud storage and an annual $50 data plan—less than half of what the large carriers charge today—for the just announced new Kindle e-reader. Deep-pocket MS needs to respond and beat this!

Finishing the device discussion, let me comment on Apple's rumored plan to market her own TV sets and Google having similar intentions with her so-called Fiber TV services. Obviously both companies are working on integrating the Internet, mobile devices and PC experiences into a revolutionary TV realm comparable and way beyond to what Apple currently offers with Airplay. A slick and advanced proprietary Apple or Google inspired TV set will find plenty of buyers—even for an initially high price. Google's closest partners Samsung and LG will make sure of that.

The difference, Apple's move in particular will create more enemies in the industry and I predict any ingenuity advantage shall be short-lived. The TV industry, already plagued with super low margins, will fight tooth and nails to stop this provenly dangerous competitor right away. First, patent challenges like the ones Apple and Samsung are experiencing with smartphones and media tablets can be expected. Second, Apple has no distribution depth beyond her own stores, a handicap her competitors will explore to their advantage.

Third, over the last two to three years efforts to connect TV sets to the Internet have been quite successful and constantly improved. Whatever novel idea Apple comes up with will therefore cause a storm of imitations in a very short time span. A lot of traditional TV manufacturers are PC, phone and tablet producers. They are more than capable of finding ways to hook up their devices to TV screens. Let's not forget the TV market is profoundly different from the phone or tablet market. The average replacement time is seven years and not the two to three years Apple is used to with her current product offerings. The giant Sony has lost money in this market for years, and companies like Panasonic and Sharp are struggling equally. Why should Apple be the exception? The best way to make profit here might be to dominate the movie download industry or sell annual software updates, which take advantages of hidden hardware options. The latter has never been tried before, and MS—the software DNA driven company—might be much better positioned to take advantage of such an opportunity when she enters this field in particular if she is willing to share some revenues with her TV manufacturers.

"SmartGlass" technology will provide MS with sufficient technology to at least equalize Apple and Google. It will let users control a TV, an Xbox or other entertainment devices from a tablet or phone, using Windows 8. Add to this the availability of Xbox's music services[73], which my old friend Yusuf Mehdi, now VP of MS's interactive entertainment division just announced. It makes me believe that MS is finally right on the ball and ready for a counterstrike. The technology will work across all competitors' devices as well and more trouble for competitors might be ahead if MS's Wallpaper[74] efforts will bear fruit.

Nothing hinders MS to license the SmartGlass software and a TV adapted version of Windows 8 to Apple's competitors. This allows them to quickly integrate a low cost ARM CPU into their TV set or the remote—whatever works best. Creating a great opportunity for these OEMs to beat Apple to the punch! Add MS's Kinect interface or the even more advanced "Leap

---

73 An easy to use service allowing music and streaming video downloads and subscriptions
74 A MS research project using a thin film TV screen like wallpaper

Motion" controller[75] to the equation and the relative passive way to operate a TV, as my generation knew it, will soon be forgotten. I predict generation F will immediately embrace and enjoy such a dramatic and cool—sorry, *modern*—feeling providing changeover.

At the end, the possibility to easily fit Windows technology into TV sets might be the most unforeseen yet overwhelming reason why that OS finally had to meet ARM. The battle for dominance in the living room will be renewed and this time it will not be fought over game console's market share. It will go way beyond and create a chance for every end-user to have the most encompassing and enjoyable digital entertainment experience the world has experienced so far. Consumers and businesses will be allowed to simultaneously access what they desire—at their fingertips—in the living room or on an office wall. As always, all this naturally comes down to just the right software and synergistically emerging digital devices creating opportunities galore for all market participants.

With this in mind, let me paint a broader picture for MS and explain how she needs to maneuver most effectively in this fast changing digital landscape. The need for her to be a nimbler company comes to mind first. This is along the lines of what Ray Ozzie recommended five years ago and what I have personally mentioned to Steve. Bring the company back into start-up mode. Easily achievable through spinning off some business units! Being smaller and doing a few things right has a much higher chance to conjure some immediate magic. Let only software DNA guide the company. Hardware will be produced cheaper by others. Consequently, selling intellectual property and services are the way to go to immensely reward shareholders. They have been waiting for this for over a decade.

Even Apple's luck will run out one day. It has only lasted this long because the company filled a void and was able to keep her prices high. This is coming to an end and her perceived high quality image is nothing other than a marketing gimmick—her competitors are already at her heels. Let's see what happens when she only incrementally improves her products like she just did with iPhone 5, when her novelty image peels off and when Apple users cannot be retained as much as today and the user pool from other vendors dries up further. Good–bye, Apple outlets, and good-bye to her current darling status. All the more reasons for MS not to venture into making tablets and notebooks! She needs to put her skin in the game differently by using her corporate cash hoard and seed some desired developments. Reap in some rewards when they succeed, thus limiting exposure and OEM defections as she goes along.

---

75 These type of controllers let you interact with devices through motions like hand signals

Steve Ballmer's recently declaring that Microsoft is from now on a "devices and services company" is the wrong direction for the company. I agree with Paul Thurrott opining in *Windows IT Pro* that somebody in MS might suffer from a "bout of reality distortion" suggesting that "Microsoft might end up jettisoning the one important differentiator that it has against monoculture competitors such as Apple and Google: the diversity of its ecosystem." If carried out to the detriment of the company—it should lead to the CEO's dismissal.

As tempting as it looks today, copying the Apple model does not guarantee MS's future success—only following software DNA does. Freeing avant-gardists from the chains of the company's bureaucracy and committee politics will lead to pioneer novel go-to-market strategies and foment enduring partnerships. Can the CEO overcome his lust and thrust of empire building and value a leaner and meaner new culture and spearhead such transition? As English historian Edward Gibbon once said: "We improve ourselves by victory over our self. There must be contests, and you must win."

There is no reason why this could not be achieved in two steps. First spin off the Xbox and device division including all software and services currently contained within. The resulting entity should be big enough to stand on its own feet, prosper and shortly thereafter launch a successful supplemental IPO[76] to gain some operating cash and fund new development efforts. Such move should help to defend Xbox's market share against the emerging and well-funded Ouya game console as well. A $99 device based on the Android OS promising to deliver the gaming software through the cloud for free—we will see! I further recommend to go one step further and incubate the promising new Windows 8 devices there as well. OEMs might regain trust.

Second, improve productivity of the distributed work place using Office, Windows, and foremost the cloud and Skype. Observe the car manufacturers as they come up with more and more tricks to save fuel. Do the same with software DNA to save things like battery lives for any mobile device and connectivity cost for enterprises and consumers alike. Last but not least, perfect voice and handwriting recognition to increase security, input accuracy and speed.

Ideally, MS's board will embrace a divide and conquer strategy and in the long haul restructure the company into a holding company of tech innovations causing one cool market revolution after another to happen. Instead of coordinating it all with the current bureaucracy, cut the rest of the software divisions loose and allow them be increasingly entrepreneurial

---

76 Initial Public Offering of stock

when doing the hard work. As a result, I would expect huge savings in coordination cost, faster responses to interruptive technologies and a flood of innovations coming to market earlier. In the process the company needs to get rid of all distractions like her doomed retail stores. (Fortunately I don't need to ask for the same in regard to the MSNBC partnership, it just got severed, hallelujah—what took so long?) These are relicts from a bygone area. Next to address would be ineffective R&D investments and discontinue them. Billions of dollars with not too much to show for in pure research is not healthy for shareholders. Long-term these changes will lead to a new management team. For sure initiating this much change constitutes a challenge to MS's board which has been pretty lame duck in the past. It will need to start demanding—with fortitude—not only revenues and profit gains but leadership from the executive suite following a change in chairmanship. In the best interest of the company and her shareholders!

A Last Word

"A heart to resolve, a head to contrive, and a hand to execute."
—*Edward Gibbon*

As mentioned before, MS in 2001 abandoned her stock option plan, significantly altering how employees from then on received long lasting rewards. When I joined, stock prices were nowhere on the horizon. After MS went public in '86, people had substantial incentivized rewards to look forward to, inspiring them to push the company to the loftiest heights of success. In May of '05 Bill gave an interview declaring the largest mistake he ever made was to shower employees with stock options. Had he gone nuts? I sent him a flame-mail, never receiving a response. Without concrete anticipation of being thusly rewarded, much of the wealth of talent he and Steve had accrued would have marched down the road to any number of competitors at a dizzying and disturbing rate, far faster than anything he could have ever stomached. He is right in one regard. His closest personal friends for sure got way too many.

I know from people inside the company how tough it was for them to sit back and watch the world rocket past under Bill's stewardship, only to then watch Steve's cultivation of MS into a civil service like asset management company. After government intervention, the once cutting-edge tech racehorse stopped winning, and was locked into a barn with neither Bill or Steve nor the board ever daring to break its lock. Taking risks was suddenly deemed uncomfortable and hanging on to outdated principles was favored instead—creating a cozy, however highly profitable, comfort zone as a result. In that context I agree with current press articles, the challenge for MS board members is not to replace her current asset management team but to attract a race horse Wall Street is willing to bet on.

Change I can believe in means returning to Auftragstaktik leadership principles. Intuitively and instinctively adopting them in the early years created not only a successful start-up but a cadre of empowered and motivated employees willing to sacrifice and go several extra miles. The challenge facing the board today is to renew that winning spirit in a much larger company. A herculean task because most of the key proponents and practitioners of this leadership style moved on—lured away by too much wealth, other lofty opportunities or lack of belief in MS's future. If it can no longer be passed from one management and employee generation to the next, formal training will be needed to reverse course. Splitting the company into loosely coupled and more autonomous portions should make this an easier task to achieve. After mostly dealing well with operational complexity, healthy transformative change will inevitably follow!

This newly bred generation of hungry and ambitious warriors need to be able to empower themselves to lead the next social networking and mobile device revolution beyond what we have experienced so far. Cut them loose and it will happen! Monetizing all applications through the cloud and offering its usage for a pittance will probably be one consequence and should spur a tremendous innovation tsunami. Only as recognized innovation leader will MS win Independent Software Vendors (ISVs) back who today eagerly and often with priority add value to competitors' products. ISVs remain the backbone of all operating and middleware vendors' success, and if their defection cannot be stemmed and reversed, mighty Microsoft will unravel fast. With competitors like Apple and Google loaded with cash, MS's money has lost its luster when buying favors. Keeping promises and not upsetting partners will therefore be even more important in the future than it has been in the past.

In the early years, money in the form of stock options indirectly bought a lot of talent. Employees accepting these thought about a brighter future and directed their passion and employed their skills to move a noble cause forward. To rekindle this I believe a return to employee co-ownership is badly needed for genuinely enduring change and long-term prosperity, bringing me to my final point.

Compensation and recognition are not necessarily correlated and are very delicately balanced as we have seen with the Linux volunteers. They have contained their venal spirit and gotten their kicks out of fighting the good fight. In moving their cause forward, admired and honored in their own hall of fame they indeed beat the establishment against all odds. With a hired workforce the game changes and when I joined, stock option incentives did play a significant role in how people performed and how they wanted the company to succeed. Before the dot-com bubble burst, MS created a lot of wealthy people and a lot of them stayed around to foster her cause. I was one of them. Monetarily I could have retired years earlier; I had fun, pure and simple, gaining contentedness by helping to win competitive battles and beating the odds while leading my knights. I had the greatest job in the world and I was thankful for the freedom granted so I could perform at my best. No regrets whatsoever.

But without hope for reward and a possibility of securing my future, I would have left much sooner. Opportunities were aplenty. Fame and pride in your accomplishments present only one side of the coin. Eventually you need to eat and want to enjoy your life after working hard and heal the scars after spearheading your last crusade. Today I am able to do just that. Following the Roman soldiers as they graciously accepted their land grants when retiring from their beloved army. Without that security blanket, this book would have never been completed.

My book would be incomplete without analyzing Microsoft's future potential. So let me consider her competitors' aims, if her products are threatened by substitutes, and what type of bargaining power her suppliers and customers possess.

In doing such an analysis product costs always come to mind. In regards to Apple, this has never been an issue for Microsoft. Microsoft's software products have definitely low, per unit cost. But they are not the lowest in the industry. Linux and its derivatives benefit from a volunteer army and a free and open source platform. In this context Microsoft has to compete by offering superior value and has done so successfully for years.[77]

Nevertheless devices and services powered by free software always incur ancillary costs. This levels the playing field. There is no free lunch. Customers will be asked to pay a different kind of penalty when using what they obtain for free like having to endure a barrage of advertising and/or an invasion of their privacy. Security concerns fall into the same category. Most customers rely on a trusted company (like Microsoft) to stand behind her software. In a corporate environment, costs get shifted: what you don't spend up front (and usually more!) gets moved over to technical support and development costs. So instead of thinking just about price buyers are forced to consider trade-offs. This serves to Microsoft's advantage. Not being the lowest cost provider therefore becomes less of an issue.

Now let's look at product differentiation where I believe MS holds an edge. First of all the company has huge brand recognition in general and with Windows in particular. Second Windows 8 will uniquely unify how a variety of computing and mobile devices will be operated and seamlessly communicate with each other and in the cloud. MS's solutions encompass the living room, the office, and the mobile space. Its characteristics and

---

[77] And if this principle is violated, competitors like Google march in and take market share away as we have seen with the Chrome browser

potential reach are unsurpassed, and her services and applications are attractive—exceeding most competitive offerings at least in the short term.

Microsoft's competitive aim is clearly defined as I see it: Applications, tools and services for mobile and stationary computing, and communication devices interconnecting users to the rest of the IT world. No other company covers that as broadly and in-depth as MS. Not being too narrow in her aim and not being a monoculture works to MS's advantage.

We can expect that current competitors will vigorously answer MS's challenge and even new competitive forces might emerge. MS is used to deal with all of them by moving her products forward—not resting on her laurels. Management's revived resolve and fortitude should suffice to regain and keep the company's lead for the foreseeable future.

Product substitutes for her top revenue producing products already exist. So far this has not endangered the company's ecosystem. As long as she can hold onto a critical mass of independent software and hardware vendors she will do well. With Windows 8 she is threatening her competitors with a superior concept, which will take time to reverse engineer. If the Facebook generation buys into this new paradigm as the new standard for coolness, MS will break through.

When it comes to bargaining power of her suppliers Microsoft has to fear if they will get upset by her ambitious plans to produce competing hardware devices and eventually defect. Producing these devices cheap enough is less of an issue. Not endangering the ecosystem the company depends on needs to be avoided at all cost! In regard to customers, the company has always pursued a low pricing policy, and I am convinced management knows how to make sure that price will never stand in the way when deciding to purchase any Microsoft product.

Winning over the Facebook and Twitter generation—while no easy task for Microsoft—will be central to success, and I am more than ever convinced that Microsoft will pull through with revived audacity and new magic. Replacing her twenty five year old logo and adapting it to the Windows 8 look further underlines how serious the company is taking this launch. As an optimist—while critical when needed—I believe that *resolve and fortitude* will not only keep her prosperous—but against all odds and pundit opinions—facilitate her conquering of unheard-of heights! Google's and Apple's glory days are numbered.

APPENDIX:
HOW MICROSOFT GOT HER STRIPES

For less familiar readers, let me put the nascent mechanics and the growing pains MS endured into context as I background the meteoric evolution of the PC market and showcase its competitive forces. In '83 when I started working for MS, the company was just 8 years old and had only 400 employees. Co-founded and led from inception by a young and charismatic CEO and chairman, Bill Gates, she gave the impression of a risk taking entrepreneurial software pioneer. She came alive at the cradle of the PC revolution when Bill's dream to deliver meaningful and affordable computing systems for personal use became a tangible probability.

The deliverance of such a system in the mid-70s depended on the availability of an affordable and powerful enough semiconductor chip called a microcomputer—the heart, soul, and work horse of early Personal Computers (PCs). In a modern PC, that component has been replaced by a much more complex and powerful Central Processor Unit (CPU). But like a microcomputer, todays CPUs need additional hardware components, peripherals, and software to become a fully usable computing platform. As the hardware components were invented and became available, the focus of the Information Technology (IT) industry shifted to a kind of magic glue, called software. It was needed to make the different parts working together and enable the common man to effortlessly perform meaningful computational tasks.

Software is nothing other than a collection of instructions a computer can understand and execute. Cobbled together in a meaningful way and able to perform a computing task, they represent an application program. To execute one on a computer system, you typically need an intermediary, which experts call an Operating System (OS). A mystic piece of software code which talks directly to its CPU, manages all hardware components, and enables the utilization of the application program you want to deploy.

Writing a software programs requires a tool known as a programming language. This is an artificial language that links the instructions a programmer writes to an OS so that the computer system can perform the desired operations. Depending upon the type of application programmers want to write, a number of specialized programming languages are available to them. One of the simplest is called BASIC.[78] Bill Gates started MS by adapting a version of it to a variety of microcomputers. Its availability enabled the programmer community to quickly write useful applications for a new platform. With retail distribution channels for software unavailable, he sold his tailored product directly to the early manufacturers, who in turn installed it on and sold it with their microsystems.

Utilizing the primitive micros of that time required way more tech knowledge than using today's PCs. No wonder merely a smattering of unsophisticated end-users bought them. Its momentum increased when Tandy and Commodore jumped in the fray, delivering easier to use systems and inspiring more usable and sophisticated application programs to appear. A fundamental breakthrough occurred in '77 when Apple Computer introduced a PC called Apple II, leapfrogging competitors' designs and usefulness. Soon overpowering its contenders by attracting a number of quality applications programs, Apple rapidly became the market's trendsetter. Two years later her position was further manifested with the appearance of a spreadsheet program[79] called VisiCalc, running exclusively on the Apple II and turning her flagship into a formidable analytical business tool. The press grandiosely labeled it a "killer application," postulating how hardware and software progression depended on each other for eternal success. This number crunching program alone, unbelievable as it sounds, motivated people to buy an Apple II—killing its competitors! Software to win—hardware to function!

The next formidable entry happened in '81. I can still remember the ads Apple launched that fall headlined "Welcome IBM. Seriously!" The largest computer company in the world legitimizing PC technology motivated her to arrogantly embrace the threatening giant publicly—a gutsy bit of ad copy. Having beaten IBM by four years to the market, nobody at Apple, my then employer, genuinely expected IBM's new baby could or would endanger her lead. Apple fans counted in the millions and were quite loyal, and with an abundance of slick application programs readily available, the incumbent seemed clearly advantaged.

Over the years Apple's strength had been manifested by an army of

---

78 Beginner's All-purpose Symbolic Instruction Code
79 A software application simulating an accounting worksheet

Independent Software Vendors (ISVs)[80] and the huge number of dedicated applications they had created for the Apple II. Their faithfulness was a prized and powerful defensive asset, and Apple's management knew darned well how essential they were for her success. As soon as IBM showed up, Apple increased support for this software community. I was part of mounting a paramount defense effort in Europe. My management firmly believed we were protected from IBM's onslaught for now by hiding behind a much richer application environment. Way too cocky as history shows. Within less than two years the imaginary protective shield crumbled. The supposedly iron-clad application barrier to entry[81] eroded when ISV's entrepreneurial spirit risked supporting mighty IBM following alternative and less expensive consumer's choices and enterprise sentiments. It for sure crossed Apple's plan to hold IBM in check long enough until she could ambush and beat her with the advanced systems that Steve Jobs, the company's co-founder and CEO, was working on.

The two contenders varied tremendously in how they had arrived at their products. Apple mirrored the traditional proprietary go-to-market approach, which the computer industry had deployed for decades. Excluding her CPU, bought from Motorola, all key parts of the Apple II were designed and manufactured solely by or for her. IBM on the other hand had, for the first time ever, deviated from this tactic. Spinning her newly formed PC division off, she freed her from the exclusivity dogma to meet time to market objectives.

Not making the operating system an exception from this newly crafted policy surprised the industry the most. Yet a bit more cautious, IBM offered a choice of two functionally equivalent ones from two different vendors: Digital Research Institute Inc. (DRI) and Microsoft (MS). Hedging her bets, she avoided picking a winner in the ensuing race for popularity—leaving it to be ultimately decided at the point of purchase.

Hard to imagine, but the original IBM PC came without any OS, making installing it an afterthought—either performed in the purchaser's home or on the premises of his or her friendly retailer. The reason for separating hardware from software can be found in an antitrust case IBM settled years earlier and now applied, without being legally obliged, to her newest creation.

The two vendors IBM had picked as OS suppliers had vastly different backgrounds and expertise. DRI had established herself as a widely recognized leader in early microcomputer OS technology while MS had

---

80 A company in business to develop and sell software for a variety computer systems
81 Defined as a hard or impossible to overcome hurdle for a competitor

made a name supplying programming languages for the same ecosystem. Both knew each other from an earlier business relationship which had MS compelled to license an early version of DRI's OS, CP/M,[82] to adapt and sell it together with an add-on card to the Apple II community. A prosperous and profitable business as it turned out. Now the rapport between them was poised to turn ugly, as DRI considered MS's alternative offering an invasion of her turf.

The method MS had used to develop her newest darling added insult to injury. To deliver in time for IBM's launch, she obtained a clone of one of DRI's CP/M versions called QDOS[83] from a small software company in Seattle, Washington.[84] It served as a starting point to develop an IBM PC tailored derivate christened MS-DOS.[85] After a court challenge about its legality launched by DRI got defeated, the first OS war was about to begin.

MS's cloner image and DRI's market leader reputation should have given the latter a huge advantage over the newcomer. Not so. The first mistake made was how DRI dealt with IBM. While MS licensed[86] her version at an aggressive price, DRI's CEO believed that his company's reputation as market leader warranted excessive royalties. IBM therefore offered DRI's version at $495 at retail while asking only $39.95 for MS's version, which she sold under the name of PC-DOS. The low price instantly resonated with early IBM PC buyers who veered to the MS's solution en masse. Round one for MS!

To make inroads into Apple's empire, IBM desperately needed to attract and recruit a sufficient mass of ISVs. Only rich application availability drives PC sales—just the existence of hardware combined with an operating system doesn't get that job done. IBM smartly enlisted both companies for the challenging recruiting task. DRI did a lack luster job helping IBM out. If she would have made an honest effort and reconsidered here excessive pricing at the same time, history might have turned out differently. Instead serious assistance for IBM came mainly from fastidious MS. She engaged vigorously willing to help other ISVs, even her competitors. It turned the tide and soon the development community followed the most supportive and mass attracting vendor. Round two for MS, without deciding the battle for dominance once and for all!

---

82   Control Program/Monitor
83   Quick and Dirty Operating System—released as 86-DOS
84   The designator 86 indicated both OSs were to run on the Intel family of 8086 CPUs.
85   Microsoft Disk Operating System
86   A software license is a contract governing the usage or redistribution of software. In general, it does not transfer the ownership of the software from the licensor to the licensee.

IBM's early decision to buy freely available and generic components from third parties made cloning her hardware quite feasible. A high temptation for start-ups and IBM's existing hardware competitors to enter the PC fray developed. Early on Bill had bet that nobody would tolerate that IBM could have and eat the PC cake alone. Following his belief, he engaged in an early crusade lobbying for and preaching the cloning of IBM's architecture.

The largest remaining obstacle to a successful cloning of the architecture was a piece of software code called the IBM BIOS. It contained the Basic Input Output System an IBM PC needed to manage peripherals like printers, hard disks, etc. These programming instructions should have been an integral part of two the OSs. Instead, IBM had written the essential software herself. Without it, no one could replicate her PC architecture. Worse, IBM was showing no intent to or interest in licensing it to lurking competitors. To protect it from unauthorized use, she had copyrighted its code. An insurance policy and a company called Eagle Computer soon felt IBM's legal wrath after copying it without permission. But as with all software, clever programmers can legally overcome such obstacles.

A small entrepreneurial ISV in Massachusetts called Phoenix Computer Systems was first in line attempting to clone IBM's code. Reinventing software, even if copyrighted, is perfectly legal as long as the imitator writes a functional equivalent piece of code and refrains from identically replicating the original instructions. This was understood as being one of the vulnerabilities in the software industry and easier to accomplish when the code's volume is relatively small. To pull this off, developers use only the functional specifications of the targeted software as a starting point. As long as this principle—called a dark-room approach by attorneys—is strictly followed, the newly from scratch created program is deemed legal and will survive court challenges. Phoenix encountered none.

Once available, companies such as Compaq, Acer, Nippon Electric Company (NEC), Hewlett-Packard (HP), Zenith, Dell, Siemens, Victor, Olivetti, Tandy[87] and Osborn took up the challenge and designed IBM PC clones. Under Bill's directions, MS overtly promoted them with valuable tech insight and advice, which did little to engender warm and abiding friendships within IBM. Energized to compete, the avalanche of new entries accelerated the affordability of PC technology and presented MS with the opportunity to win new customers for MS-DOS. Not to be outdone and with a highly compatible but not identical OS at hand, DRI immediately engaged as well, drawing MS into a fiercely and long lasting battle to win over emerging PC manufacturers.

---

87 Now called RadioShack

All were joined by risk taking ISVs drawn to potentially high volume sales opportunities and, compared to Apple's, a promising and more homogenous application environment. Business and home users alike quickly recognized the groundbreaking benefits of choice, speed, and efficiency provided by multiple vendors. Accessible and affordable PC power and software had arrived once and for all! On their own desks. In their own homes. As Bill predicted, truly profound pioneering advances in productivity and ease-of-use freed consumers from timeshare, minicomputer, and mainframe dependencies, gutting IBM's terminal sales, to her detriment. For existing computer paradigms, the wave created by the desire to buy affordable PC clones had just turned disruptive.

Coinciding with the launch of her PC model AT/339[88] in '86, IBM began preinstalling MS's OS exclusively on all her PCs. The crucial change of heart signaled the beginning of the end for DRI's product—and in the swift trajectory of tech history eventually doomed the company herself. The market had spoken and IBM confirmed customer sentiments.

Not so fast, changing to a pre-installation routine was not caused by an overwhelming preference for MS-DOS alone. Like always, in the IT industry competitive pressures were solely to blame! The PC world had just gotten more hassle free and user-friendly. IBM's competitors, influenced and encouraged by MS sales reps, forced her hand to deliver out of the box ready to use systems. Costlier or not, the industry leader bit its tongue and gruntingly—after five years—lastly followed competition.

Apple, the early pioneer, was hardest hit by the onslaught. By '83, IBM PC clones outsold Apple's PCs three to one. The coveted number one spot was taken over by IBM and never regained. Adding to Apple's misery was collective reluctance among the business community to buy IT technology from a dwarf like her. Wasn't an IBM approved technology a much safer bet for the future?

PC technology took off like an Atlas V rocket. Before it got introduced, most computing tasks were accomplished by using terminals connected to either mainframes or minicomputers. Their central computing power was shared between several users simultaneously. This often led to bottle necks determined by computer capacity or connectivity bandwidth. End-users at home had no access to computer power at all. With the arrival of the PC later accompanied by network connectivity via regular phone lines, the stone age of computing had finally come to an end.

---

88   AT means Advanced Technology

Usability and access were not the only obstacles plaguing the IT industry. Applications specifically written for one computer system in native assembler programming language would not run on any other without costly adaptations. The invention of high-end programming languages like FORTRAN[89] and COBOL[90] improved that. A program written with such programming instructions promised to be written once and then run on any computer system as long as the respective language translator—typically a compiler—was hosted on the targeted system. Its role was to transfer the high-end instructions into system specific code the computer could execute. Recompiling a program was fast and cheap! Problem solved.

Unfortunately, the availability of these high-end and supposedly standardized programming languages did not mark the end of the incompatibility story. Leave it to the restless human mind to innovate, improve and embellish them with non-standard extensions. Proprietary in nature and therefore reintroducing incompatibilities, programmers used them anyway to optimize programs and gain execution speed. Trade-offs such as these, while well-intended, played right into the hands of computer manufacturers who assumed that any incongruities between systems impaired the possibilities of customers switching vendors. Investments, in particular in databases and other commercial applications, often exceeded customer's hardware outlays. Having to throw these away when changing hardware vendors constituted, in the eyes of computer manufacturers, a formidable barrier of entry for competitors, casually described as a choke point or legally characterized as a lock-in situation. IBM had created one of the largest with her 360 mainframe system, which was followed by every minicomputer vendor under the sun with its own unique and proprietary approach.

Customers, though, did change computer vendors. The lock-in chains were always weaker than imagined and feared by advocates and politicians. Entrepreneurial spirit combined with brave risk taking always found a way to break them. When I worked for Digital Equipment Corporation (DEC), I brought the thinking of a considerable number of clients around, switching them from IBM's to DEC's gear, guiding them through the process. The highly competitive IT industry contrived multiple disrupting technologies over decades. They offered so many advantages to potential buyers that switching of vendor platforms every so often became an absolutely necessity, if one wished to remain competitive and cost conscious. PC technology was dead on one of them.

Bill's personal vision and goal was to fundamentally change the status quo of the nascent PC industry way beyond what the early micros and their

---

89   A formula translation language suited for scientific and engineering computations
90   COmmon Business-Oriented Language suited for financial computations

successors had already achieved. First, he promoted and supported the cloning of IBM's PC architecture. Second, he helped establish MS-DOS on most PC systems. (DRI scored on the rest with her relative MS-DOS compatible CP/M-86) The caveat? Hardware brand became less important than software marque and changeover inconveniences disappeared overnight! A properly installed MS-DOS or CP/M-86 bridged any vendor gap perfectly well, serving as the magic glue to make the normally tricky application compatibility issue stick. This allowed the industry to address a set of novel of performance challenges and provide for healthier vendor differentiation. As a result, application programs sold in higher volumes than ever, and to the joy of consumers, they sold at progressively cheaper prices.

As that notion caught on, Apple's attractiveness shrunk. I dubbed the resulting and overwhelming creation process a 24/7 phenomenon. The lights at Apple went out each and every day when engineers left heading for home. Not so in the PC clone industry. Around the globe PC vendors, their component manufacturers and ISVs never stopped implementing innovative ideas and producing compelling products. Opportunities popped up nearly overnight and attracted groundbreaking entrepreneurs who took their chances. Absolutely amazing to witness, creating not only legions of satisfied fans but more importantly attracting additional buyers. Prices came crashing down as the competition was heating up and everybody was striving for the number one spot.

Becoming a volume leader in one of the lucrative segments was a coveted rung on the ladder. The exceedingly competitive and never-ending struggle to achieve and defend this was nevertheless not for the faint of heart. Considering the rewards, the enduring *mêlée* produced frothy and protracted competitive wars. The winners took no prisoners. Characteristically, they took most or all of a certain market segment, something as stressful to defend as to obtain. MS, always under pressure, performed accordingly.

The economic consequences in regard to the emerging PC business model would later be described as being fueled by network effects. The more people used PCs, the more economical the model turned out. Leading us to the 64 billion dollar question: Would Bill's groundbreaking realization lead to a computing paradise or a choke point known as MS-DOS compatibility? Answers differ without challenging the following: Achieving and policing MS-DOS and later MS-Windows compatibility created a pro-consumer marketplace. The issue yet to be settled was who would or should ultimately be allowed to control this rising pyro-software empire? MS, the PC manufacturers, or the Feds? This keystone topic, volcanically controversial and conceptually riveting, eventually evolved into the most media-exploited and judiciously contorted battle in recent IT history. Dissecting how unfairly this bloody war was fought and eventually concluded was one of my penultimate inspirations for writing this book!

Early on, most PCs were sold through retailers consisting of specialized mom and pop shops, large chains such as Best Buy, or anything in between. In 1983/84, at the end of President Reagans first term, the economy finally climbed out of a 16 month recessionary period. Jubilant and hopeful, the industry bet on sky high expectations for accelerating PC demand. Component manufacturers wildly expanded production capacity. New competitive entries further distorted the picture. While PC usage and sales did expand tremendously, the entrepreneurial industry nevertheless overestimated sell-through. The till-then rocketing component producers throttled back—too late. Their over-expansion left lots of goods unsold in volume, piling up in warehouses. Innovation accelerating, the inventoried components had a shelf life of six or nine months at best. By the end of '84, producers were forced to find a home for their surplus. Either that or begin fearing for their business lives. With their traditional direct customers, the larger PC manufacturers, saturated and at full capacity, they experimented with pushing overstock through a fledgling component distributor network.

Selling below or close to costs jump-started a whole new breed of start-ups, whose entrepreneurs understood PC demand could further be stimulated by lowering prices. In particular, small business owners were no longer willing to accept the premiums that brand name PCs demanded. Not the only concessions they were looking for—like advanced users, they were on the hunt for customized PCs in configurations most retail stores did not offer. The juicy wave of bargain basement PC components allowed responses to both. Soon, small system-builder shops—less respectfully called screwdriver outfits—sprung up all over the world, specializing in customized PCs at reasonable cost. Most of them found their clientele in their local communities and gained their trust by supporting them well.

Other emerging stars from these frugal beginnings, notably Gateway and Dell Computer, expanded their horizons in '85/86 beyond local business opportunities. They envisioned a nationwide direct sales burst, accurately predicting tech-savvy users could be attracted by the ability to order customized PCs by phone. By no means a slam-dunk, however. There were inherent investment risks in launching such a pioneering high-tech marketing venture over the impersonal land-line frontier. Hiring technically knowledgeable sales personnel and using cleverly crafted print and TV advertising, their phones started ringing. And ringing. Buying PCs through mail-order took off and rapidly established itself as a valid procurement alternative. Manning phone banks and streamlining operations were soon the biggest challenges for these newcomers who were coping with unexpected and explosive demand. A new way of acquiring PCs, allowing tech-savvy customers to circumvent the middle men, had passed the test and was there to stay! With increasing technical knowledge this buyer segment continued to expand, eroding the benchmark standard of PC prices. Unburdened by high overhead costs, manufacturers conquering the new terrain were reasonably profitable from the get-go. Established retailers and old-school marketers sat up and took notice. The tides were turning. Business was being swept away from under them.

The reigning brand-names grudgingly acknowledged the emerging system-builder competition and fast expanding mail-order rivals. Their empires had blossomed into well-established and financially sound business institutions. Goliaths versus the Davids of the new world order! With the old guard deluded by a short history of success and already hobbled by inertia, the new players experienced no meaningful resistance, gaining a substantial foothold. With giddy start-ups nipping at their ankles, the stubbornly entrenched establishment missed a glaring opportunity to aggressively plunge into the fray and counterpunch the new competitors when they were still vulnerable. Helped by Reagan's tax policies and increased deficit spending, the accelerating economy increased PC demand across the board. Flourishing upstarts discovered plenty of room for unfettered growth. The quantifiers, pundits and hired-gun market-mavens missed their rise. They either disregarded the emerging phenomenon altogether or analyzed it incorrectly for quite a long time, underestimating world-wide PC demand and unit sales. Professional analysts were alarmed to discover that by 1987/88 nearly 25 percent of WW PC volume was sold by system-builders or direct marketers.

Instead of paying attention foremost to changing trends in PC purchase behavior, retail-driven brand name OEMs[91] were pre-occupied with building larger and highly efficient production facilities locally and streamlining their operations. To whoever wanted to hear it, they crowed on at length and in detail, exhorting how efficient their manufacturing process had become and proudly offering tours to inspect their grand, stalled kingdoms. In hindsight, the trend of investing in high-wage burdened local manufacturing plants in the '80s—dubious at best—diverted management's time and focus from the increasingly threatening sales channel transformation. A first sign for shake-ups to happen!

The technology improvements and related performance advances in PC software and hardware from '81 to '86 had been truly astonishing. Intel's Central Processor Units (CPUs), the main performance engines of the IBM PCs, were now at least eight times faster than six years earlier without

---

91 Original equipment manufacturers stands for PC manufacturers.

experiencing substantial price increases. Moore's law[92] was holding up nicely. Mainstream PCs had gained increased memory size and speed, and hard drives of vastly greater capacity and shorter access times had become common. Data access had gained velocity from wider internal bus architectures. They had widened from 8 bit to 16 bit internal computing highways with 32 bit designs knocking at the door. These hardware engineering advances led to new usages. A PC was no longer merely a desktop device. The most powerful ones worked as office servers, effectively competing with low end minicomputers and signaling an end to this approximately 25 year old technology.

In the expanding universe of Central Processor (CPU) technology, Intel was the undisputed leader followed by AMD,[93] which made her mark selling Intel clones at lower prices. Intel had achieved at least 80 percent market share and defended her turf with shrewd marketing tactics, often stopping AMD from making inroads into the top brand name manufacturers. We observed with astonishment how Intel consistently got away with seemingly anti-competitive actions. AMD complained to the Department of Justice (DOJ) and the Federal Trade Commission (FTC) and so did customers, but so far no government action was forthcoming. AMD enjoyed most of her success with European, Asian, Latin American and lower volume US manufacturers. The cheapest PCs in 1986 were still Intel 8086, 8088 or 80186[94] based. The bulk of the market had moved to systems powered by advanced 80286 CPUs. In '86, Compaq stunned IBM by delivering the first 80386 based PC, using an expanded Compaq designed memory bus system including memory protection and 32-bit data transfers. The race to 32-bit computing was on! Constrained by production shortages, Intel played a masterful game—or a dirty one depending on what side you were on—allocating production capacity to her most loyal customers. The industry was at buzz.

When Intel soon thereafter developed and released a Reduced Instruction Set Computer (RISC) CPU, named i860, as a response to a similar Sun Microcomputer offering, the industry was immediately rife with rumors. How secure was the future of Intel's current architecture built on Complex Instruction Set Computer (CISC) technology? Would RISC technology prove disruptive enough and hinder Intel to produce a successor to her flagship 80386 CPU? As a consequence, most software, and in particular OSs, would need rewriting from scratch. Fortunately the rumors were short-lived, and as insiders we quickly understood Intel had already started

---

92 The numbers of transistors on a CPU will double every eighteen months. Read more at http://www.answers.com/topic/moore-s-law#ixzz1hrV9tsp5.
93 Advanced Micro Devices
94 The larger the number, the more advanced is the CPU

working on an upward compatible successor called the 80486 eventually to be released in '89. Lucky us!

In 1986 we released version 3.2 of MS-DOS. Our intense sales and marketing efforts had by now restricted DRI, our main competitor for the IBM clone business, to primarily serve the low end of the PC market. Her prices were now half of ours and counting. Her management had learned from earlier mistakes. Like AMD with CPU sales, DRI had gained ground in developing countries and with a few larger European manufacturers. Most dominating brand name OEMs were for now firmly in our camp. An OEM buying primarily AMD's CPUs was always a danger sign for me, as they were drawn to DRI like flies to the honey pot.

History had reversed itself—CP/M-86 was now considered an MS-DOS clone. It did sound strange knowing that MS-DOS had been derived from an early CP/M-86 clone. To DRI's dismay, MS was for now unarguably in the lead. She found herself in the role of a scrambling and tenacious runner-up who never—in great competitive spirit—gave up correcting the imbalance. Her unrelenting attitude kept MS's OS development on its toes and guaranteed product improvements from both companies.

In '88 DRI changed the name of CP/M-86 to DR-DOS. Signaling compatibility, she continued to use MS-DOS identical version numbers. Not leaping ahead of us further fostered her clone image. With both OSs comparable in performance and features, DRI—for the world at large—appeared inept at charting her own destiny—the MS-DOS benchmark therefore remained the undeniably yard stick. A few journalists and advanced users disagreed with this entrenched perception, though imparting little effect on MS's image as the standard bearer.

As a result, people wondered how compatible DR-DOS and MS-DOS truly were. Their code had never been identical. Users therefore experienced well documented and small functional differences despite DRI's strident denials. MS's support organization, which I managed for several years, diligently kept track of them. Any public mention of apparent incompatibilities made DRI management outright furious! Allegations were always instantly rebutted. OEM customers who were licensing from both companies concurred. Our answer to them: Switch to the "original" and your customers won't suffer. DRI's public relations teams tried hard but never mastered the daunting job of changing public misconceptions.

With the appearance of 80286/386 CPU-powered PCs, UNIX had become a viable OS alternative for PCs. It was derived from AT&T's original code, developed in the 70s and written in a non-system specific programming language called C, developed between '69 and '73 by Dennis Ritchie at the then AT&T owned Bell Telephone Laboratories. A novelty at a time when OSs were normally written in assembler language uniquely tied to the computer systems they were designed for. The brilliance of writing UNIX in C eased its portability to other systems. As long as a C-compiler existed for the targeted computing platform, a port could be accomplished in a relatively short time frame. AT&T spread UNIX's popularity further when licensing it with few restrictions to a variety of manufacturers. Porting it to several minicomputer systems was the first serious attack aimed at the proprietary OS empires the IT world was built upon long before Bill succeeded with the MS-DOS standard in PC land. UNIX running on minicomputers had the potential of bringing down switching cost for customers. Not wanting to believe in or promote that type of progress, Digital Equipment Corporation's CEO Ken Olson called UNIX a form of "snake oil." As such, UNIX was a disruptive technology and contributed heavily to the downfall of the leading minicomputer companies of that time before PC servers became feasible and finished that job. Why didn't UNIX conquer the PC world as convincingly?

Simply put, all companies who developed UNIX for the IBM PC were thinking and planning within the cloistered, proprietary walls of sovereign systems unwilling to answer the call for unification. IBM called hers AIX, Sun distributed SOLARIS, Hewlett-Packard called her version HP-UX, and many others left the name UNIX unchanged. MS and SCO (Santa Cruz Operation) used the name Xenix. Introducing vendor specific incompatibilities robed UNIX of a golden all-encompassing standard setting opportunity. The other most important ingredient for a successful run truly existed. Available for nearly 20 years, the UNIX environment was rich in application programs. On the other hand, to port them to the variety of incompatible UNIX systems was costly and time-consuming. The larger memory footprint and the higher-end CPUs UNIX needed to function well added to its disadvantage. Last but not least and as important, UNIX was hard to operate by computer novices. Lacking their crucial acceptance, it nevertheless successfully veered off into the high-end PC workstation realm used by computer experts for graphical designs and engineering tasks. The availability of superior graphic libraries and design tools were the main reasons for its popularity in that user segment—again software availability drove hardware use. All in all, the fragmentation of the UNIX standard and its consumer unfriendly operating interface served to MS's advantage and made—for now—for a less dangerous and only niche occupying competitor.

In the early '80s, controlling OSs with command line user interfaces—like MS-DOS, DR-DOS and UNIX—finally came of age. The trend toward using a Graphical User Interface (GUI) instead can be traced as far back as

1945. Several experiments then and thereafter led to a handful of scientific papers substantiating the advantages of operating a computer with a mouse-driven cursor and graphical window technologies. The competition for developing modern GUI shells—as we use them today—eventually heated up. Earlier systems like the Xerox Alto and the Three Rivers PERQ failed to gain market acceptance. Apple was slightly more successful with the experimental Lisa[95] system she released in '83, and gained tremendous customer acceptance and industry accolades with her Macintosh (MAC) PCs one year later.

Steve Jobs had developed the latter to regain Apple's lead and eventually wipe out PCs powered by MS-DOS, potentially hurting us and the IBM PC–clone industry alike. The full blown GUI operated MAC was years ahead of competition. Its commercial success nevertheless lagged Job's revolutionary vision. Being priced too high dampened customer acceptance and led to Steve Jobs ouster in '85. A huge blunder for Apple's shareholders and fans! His role as product guru was taken over by Jean-Louis Gassée, my ex-colleague, who eventually met the same fate. The PC industry including MS continued to prosper while Apple went into a tail spin.

The PC industry knew that Apple was on the right track. So the GUI race continued. In '84 Tandy ambitiously ported her semi-graphical user shell from the archaic TRS-80 to her MS-DOS based PCs. DESQview[96] appeared as yet another none-command line interface for IBM PCs. Commodore fitted her high volume 8-bit Commodore 64 model with a proprietary GEOS[97] shell, later porting it to the Apple II and MS-DOS. Written for 8-bit computer systems, it was instantly obsolete once 16-bit computing took hold. At about the same time, UNIX was enhanced by the graphical X WINDOW system, while Sun's SOLARIS got a PostScript-based[98] GUI. Where was MS?

In '82, VisON showed a prototype GUI for MS-DOS at Fall COMDEX[99] in Las Vegas. (The COMDEX fair is the Mecca of PC Tech where careers, products, and entire start-ups rise and fall each year.) Launched a year later, its commercial success was hampered by requiring too much memory to perform satisfactorily. (Computer memory prices were 100 times more expensive than today) That same year, IBM jumped in with both feet by announcing TopView, her secretly self-developed text-based shell for MS-DOS, shipping it a year later with the AT model, her first Intel 80286

---

95 Apparently named after Steve Jobs's daughter
96 Developed by Quarterdeck Office Systems
97 Graphic Environment Operating System
98 A programming language normally used for desktop publishing applications
99 Stands for Computer Dealers' Exhibition

powered PC. For the startled MS, her breakaway drive for independence classified as augury.

The embedded revelation: We had fallen behind the curve by not replacing our command-line-oriented MS-DOS interface with a sexy GUI. True to form, we decided to compete, prematurely. Hastily announcing our own development—eventually named Windows—with a flourish of trumpets and drumrolls, an avalanche of promises and fanfare, innovative info-speak, and a sprinkling of smoke and mirrors. Attending the announcement in NYC and flying for the first time on Air France's Concord to get there in time, I watched our mock-up being demonstrated by Bill Gates personally. Partly written in BASIC, the new GUI prototype ran on a variety of PCs. Slow and failing a couple of times, without dampening the simmering excitement in the glorious Helmsley Palace Hotel ballroom. Limping, we had at last arrived in GUI Land!

In true competitive spirit, DRI made her own effort to literally upset the Apple Macintosh cart and MS altogether. The second MS announced her intent, DRI disclosed her workings on a GUI named Graphical Environment Manager (GEM). The early and already well-functioning prototype DRI demonstrated at fall COMDEX '84 created extra urgency. Fueled by leapfrogging competitive entries, the growing GUI acceptance was threatening MS's fragile house of cards, rapidly unveiling the end game. Bill had no choice but to reveal our still nascent cobbled-together code earlier than he was comfortable. The NYC event, supported by a lavishly well-orchestrated public relations campaign, served as a stake in the ground, upped the ante and put us at least back on the map. Bill regained his top industry guru status. We were again the darling of a press, casting the spotlight most vividly on the ever-more fascinating dark-horse.

Our faithful OEM customers understood the competitive dangers generated by Apple's MAC and Commodore's Workbench[100] driven Amiga platform. Therefore they supported our Windows' efforts enthusiastically. Without endorsing GUI wholeheartedly, the IBM PC–clone industry would have lost its edge. The responsibility was not necessarily MS's alone but, with Apple owning a predominance of patents in this field and us having licensed them, who else could have securely and successfully pursued that user-friendly path? Our situation seemed to worsen after DRI delivered her version unexpectedly in February of '85, well ahead of our Windows release. Luckily we got an unanticipated boost when Apple marched DRI into court for patent infringement—defending her Apple cart for now and helping ours. DRI relented and later redesigned the product to avoid further

---

100 A GUI operating system for Commodore's proprietary and PC-like Amiga system

patent disputes, resulting in a much-crippled and long-delayed version. Re-released in '86, it never gained popularity.

No one foresaw how long it would take for MS-Windows to show up or how underwhelming it would perform. The OEMs who had signed on early were growing justifiably upset with the interminable delays. Bill echoed my customer's sentiments. I heard of several loud and emotional meetings where his temper boiled, calling people out by name and hammering on them to work harder and to hasten its release. Meanwhile I had to mollify and coddle unhappy licensees. My response to their restless discontent was the repeated mantra: "The wait will be worth your while"—oh my, oh my! Bill, deeply concerned about Windows' performance, painfully lost that argument against actually shipping a barely acceptable version in late '85. Fully functional and in many ways innovative, but slower than a duck on ice with hardly any none-MS applications taking advantage of its innovative features. Failure on the horizon?

The product, like others of its kind was hosted on top of MS-DOS, remained consumptively resource-hungry and was less elegant than Apple's solution with its overlapping window frames. Using up a ton of memory, it needed lightning-quick storage devices and top-end CPUs to work reasonably well. Memory was now a bit cheaper than six years earlier, when the original IBM PC had been introduced with just 64 KB,[101] yet still significantly costlier compared to today. Most OEMs, while hailing Windows a step in the right direction, were reluctant to burden their PCs with sufficient hardware to make it shine. Licensing Windows in addition to MS-DOS and adding all the hardware bells and whistles sharply increased costs. In the eyes of most users Windows lacked applications and was therefore simply not enough of a must have. Our bottom-line and price point-wary PC manufacturers consequently casted Windows as a dim afterthought. All through '86/'87 MS struggled mightily to keep that business alive and growing.

ISVs were still learning how to design smooth and well performing Windows applications and were not releasing them as fast as expected, stalling its acceptance. Those who had released Apple MAC versions successfully jumped out ahead. The others discovered GUI applications consumed extra development time compared to text or character-based ones. Windows' breakthrough was by no means guaranteed.

With manufacturers not biting, maybe pushing a relatively inexpensive $99 MS-Windows package through the retail channel could revive its image and drum up business. Just before Christmas '86, MS ran a monstrous print and public relations blitz, a real doozey in retrospect—SVP Steve Ballmer,

---

101   Hard to imagine anymore. Apple's iPod, her media player, today has millions more.

Bill's right hand man, the main force behind it. The twelve minute video with him heralding Windows can still be found on the Internet, a provocative enactment in the harsh light of over two decades' hindsight. Not too many PCs in use could actually run Windows reasonably well without upgrading raw hardware performance. Our marketing ploy nevertheless worked, enhancing overall product visibility and keeping the Windows application sales dream alive with struggling ISVs. Windows developer conferences where we induced them to write better ones filled up. Steve, personally functioning as the chief agitator, convincingly indoctrinated excitement and belief in Windows future. Developers working on non-Windows apps began sensing powerful headwinds.

This renewed push to mobilize ever-greater legions of ISVs supporting Windows along with an eye dazzling public spotlight on the product tremendously pleased the few OEMs who had licensed it. Steve's propaganda machine kept them in our camp. For OEMs shipping just MS-DOS PCs, like Compaq and IBM, there was good news as well. Customers buying Windows at retail often needed additional hardware to beef up their systems. The ripple effect, OEMs began including MS-Windows testing in their quality control efforts, yet another bit of beneficial synergy much appreciated by us. The smoldering volcano was becoming visible.

MS made her ongoing Windows push effective by doubling personnel devoted to helping other software companies write Windows apps. We had realized early on that cooperation, innovation, and success were interchangeable and mutually re-enforcing concepts. People working in the Windows support group were called evangelists. Over time—as they succeeded—they had a huge impact on the richness of its application environment. They were the unsung, true, and only heroes who eventually made the platform relevant and most popular with end-users and OEMs alike.

Igniting Windows with marketing money and adding the evangelists felt like the magic of exponential muscles visibly at work. Steve truly nailed this one, neatly clicking the final cogs in place, engineering a critical and decisive move to lay a strong and unshakeable foundation for providing quality Windows applications to consumers and businesses alike.

Not resting on their laurels, MS and IBM had already started working jointly on a successor to MS-DOS and Windows called OS/2. Its design called for adding security and networking improvements to its combined feature set while making sure that existing MS-DOS and Windows applications ran unaltered on the new platform. A key for easing the transition for end-users and guaranteeing financial health for hard at work ISV partners. With its release date set for late '86, the world was anxiously waiting to put OS/2 through its paces.

The huge hardware performance achievements and incremental improvements in OS technologies had helped producing high performance PC software applications. Some of them had gained popularity beyond belief. Lotus had bypassed VisiCalc as the leading spreadsheet company with her Lotus 1-2-3 product in the US, trailed by MS-Multiplan for MS-DOS and later MS-Excel for Windows and the MAC. The success of her spreadsheet was based on graphics capabilities she had integrated into her flagship product. In Europe MS attacked her successfully with localized versions of Multiplan and Excel and denied her the lead. Ashton Tate was the unchallenged database powerhouse with her then superior dBase product, followed by Symantec's Sybase. In word processing, WordStar was still alive in '86 although it was losing its luster. WordPerfect was on its way to becoming the new text processing star, rivaled by MS-Word for MS-DOS, Windows and the MAC. Buying dedicated word processors or god-forbid typewriters were now a thing of the past, and spreadsheets had taken over as most popular analyst's and accounting tools. Brimming with energies, the software industry was well and alive and her innovation span guaranteed PC technology to strive and become an indispensable tool for society.

And how I earned mine

Born in Hannover, Germany, I finished high school before joining the German army for two years instead of the mandatory eighteen months, as determined by the draft system. This move allowed me to obtain the rank of lieutenant while laying money aside. Young men despising the draft employed all possible means to avoid it. Hardy and hale and with retreating to Canada not an option, hesitantly at first I served. Joining early, I avoided interrupting my later studies. The army complemented much of what I had learned so far in life and prepared me well for a far larger role. Long a history buff, I was deeply impressed by her leadership principles which originated from the early 19th century. Contrary to public opinion, the German army was not looking for blind submission of complicit sheep unquestioningly following its leaders. Even in peacetime, overshadowed by the cold war, she wanted confident young men willing to marshal their intelligence and capabilities to accomplish distinct and complex tasks. I soon discovered that within well-specified parameters, you had considerable freedom to achieve defined objectives. In stark contrast to the blind obedience drilled into me—often with the aid of a bamboo stick—during the strictest of upbringings. Resourcefulness was not only desired but rewarded! I found myself accorded ever-higher levels of autonomy and trust by the chain of command. Allowing me, for instance, to organize and lead an officer class of 56 trainees during the last three months of my stay. For the first time in my life, I gained a solid sense of what it meant to manage and be boss. I relished it, gave leading the cadre my best effort, and never looked back.

The savings from my army stint came in handy as I worked my way through university and obtained a diploma[102] in mathematics. Yet, my primary source of income soon became developing software programs for agricultural and pharmaceutical companies, applying my ever-growing statistical knowledge. I learned to write them in two different native assembler languages specific for International Business Machines

---

102   Comparable to a master degree in the USA

Corporation (IBM) mainframe[103] computers and in several higher level programming languages. When writing software programs was not enough to support me, I started teaching mathematics and physics classes at a local high school. Germany had a teacher deficit and welcomed non-degreed workers like myself after I had passed a half time math exam. I had a terrific time learning how to teach the ten to eighteen years old high school students. Young as I was, I strived to make learning fun for them. The teaching experience together with my programming background helped me to later land my first employment in the computer industry. As I made my way through university without being supported by my parents, I learned to appreciate my independence and how to be responsible for my own well-being. The invaluable lessons provided by working hard to gain and sustain my individual freedom and financial health influenced my work ethics for the rest of my life.

My professional career was officially launched in the fall of '72, when I joined Digital Equipment Corporation (DEC) in Munich. DEC's business was selling minicomputer[104] systems, and the PDP 11/20 was her newest baby. My first job was teaching computer programming classes to customers, and learning two additional assembler programming languages and the internals of several OSs in the process. The toughest teaching assignments given to me happened during so-called onsite classes. Taught at customer facilities where I had to educate a very clever and super knowledgeable audience. The demanding subjects the students desired to be instructed about were sometimes unknown to me ahead of time. Improvising I was sometimes only hours ahead of my students by reading up over lunch and teaching the self-same topic in the afternoon. Talk about having fun learning something new and being stressed teaching it to a challenging student body minutes later! My second job as sales trainer ended in a promotion to manage DEC's Munich-based training center, which I turned around, restoring its high profitability. Two and a half years later I accepted a marketing manager position in DEC's training division headquarters in Bedford, Massachusetts. Before departing DEC in 1981, I spent a year as marketing manager, this time in the banking and insurance product group in Merrimac, New Hampshire. Due to office politics I became bored and frustrated. Back in Munich, a short stint with National Semiconductors followed, where I learned how to sell semiconductor components to German computer manufactures.

Having missed the first wave of the PC revolution, I was easily lured away from struggling Nat-semi by Mike Spindler, soon to be Apple's European SVP located in Paris. In '81 he offered me a marketing manager position to

---

103 Large computers used in scientific and commercial applications
104 Downsized computers compared to IBM mainframes

guide Apple's European software ambitions and promised to promote me to head a to-be-founded European subsidiary of Claris, an Apple-owned software company. After that plan was dropped by management I saw no reason to stay. I relished the two years I worked for the company, learned a ton about the PC industry in general and how essential software is for her success in particular. Not wanting to play second fiddle in a mainly hardware focused enterprise, I moved on. Before leaving I attended a last meeting with Steve Jobs, who showed a portable Apple II and gave a progress report in regard to the Macintosh. Despite his prima donna aura I peppered him with questions, annoying Mike, who took me aside during recess saying "You do not cross this man if you want to have a career in this company." He was right on, I quit one week later…

A software guy at heart, a small company called Microsoft (MS) had caught my eyes. She was founded in '75 at the dawn of the PC industry. I got involved with her international division when brokering a deal to localize her Multiplan spreadsheet into several European languages for the Apple IIe introduction in March, '83. Helping me to get that job done was Jean-Louis Gassée, then Apple's French country manager and an avid admirer of MS and founder Bill Gates.

When Mike Spindler learned I had accepted a subsidiary manager position in Germany working for MS, he roundly congratulated me, expressing his desire for local cooperation. He seemed proud to have an ex-Apple guy run MS's new sub.[105]

My dream had long been running my own software company. Though technically never achieved, I had so far been consistently lucky to align myself with cutting-edge American IT entities. Each challenged me to make a difference, implicitly entrusting me with vaster operating autonomy than a typical German company would have done—tremendously helping my career. Off I went into another adventure, one that would last twenty years and would be filled with turbulence, hard work, and tons of fun.

---

105   We will meet him again as Apple's chief executive officer.

Acknowledgments

Working on this book for nearly two years, I received a lot of help from and encouragement from my friends, ex-coworkers, and my family. They taught me how to improve my writing style and bettered the accuracy of my telling. They spent countless hours reviewing several drafts and returned them with useful comments, which at the end made a huge difference in how this story came together. Thank you sincerely for all your help:

Peter Maul, my writing coach, teacher, and friend
Karl Schlagenhof, my friend and ex-business partner
Kurt Kolb, my ex-coworker and friend
Molly MacDonald, my beloved wife
Carl Gulledge, my ex-coworker and friend
Ron Hosogi, my ex-coworker and friend
Rich Ellings, president, National Bureau of Asian Research, and my friend
Ben Hsu, my ex-coworker and friend
Tyler Bramlet-Kempin, my talented son and immensely helpful editor
Karolos Karnikis, chairman and CEO, Vistula, and business partner
The Internet—I could not have written this without you!

GLOSSARY

A

Acer	PC manufacturer based in Taiwan
ActiveX	Frame work designed by MS for reusable software components independent of any programming language
AG	Attorney General
AIM	Alliance formed by Apple, IBM and Motorola to create a PC standard based on the IBM Power Pc architecture
AIX	Advanced Interactive eXecutive, Name of IBM's UNIX version
Alagem, Beny	Founder and CEO of PB
Alchin, Jim	SVP of MS responsible for the Windows development group
AMD	Advanced Micro Devices, Manufacturer of Intel like CPUs and graphic cards
Amiga	Family of computers marketed by Commodore
Amstrad	British manufacturer of consumer devices
Anglo-American	Countries governed by English common law
AOL	American Online, Internet services and media company
Apache	Distributor and developer of Linux software
API	Application programming interface, set of specifications for software programs

Apple	Manufacturer of PCs and consumer electronics
Apple II	8-bit home computer built by Apple
Application	Short for *software application*, set of computer instructions for a specific task
apps	Abbreviation for application
ARM	Advanced RISC Machine, a CPU containing a reduced instruction set for a computer
Ashton-Tate	ISV for database products, most famous one was called dBase
assembler	Low-level programming language specific to a computer system
AST	PC clone manufacturer from CA founded by Albert Wong, Safi Qureshey, and Thomas Yuen
AT	Advanced Technology, IBM designator for one of her PC models
AT&T	American Telephone and Telegraph, telephone and Internet services provider
Atari	Home computer and game-console manufacturer
Atlas V rocket	Powerful rocket delivering payload into space
Auftragstaktik	German for mission-oriented command philosophy
Austro	German, for Austrian

B

Ballmer, Steve	CEO of MS
BASIC	Beginners All-purpose Symbolic Instruction Code, high-level programming language
Bavaria	State located in the southeastern part of Germany
Beijing	Capital of PRC
Bell Telephone Laboratories	Former research arm of AT&T
BeOS	Multimedia OS for Mac, IBM, and PowerPCs by Be Inc.
Big Blue	Nickname for IBM
binary code	CPU instructions using only the digits 0 and 1

Bingaman, Anne	Assistant AG, head of the DOJ's antitrust division from '93–'97
BIOS	Basic Input/Output System, CPU instructions to manage peripheral devices
bit	Smallest unit of information for a computer system
blitz	German, lightning-fast action
blitzkrieg	German, speedy military action to obliterate an enemy
Boies, David	Lead prosecutor in MS's antitrust trial
boot sequence	Process of launching an OS
Borland	ISV producing software-development tools and programming languages
Bull	Groupe Bull, French electronic and communications company
Bush administration	Management of the US government during George W. Bush's presidency
Butler, Jeremy	SVP of MS responsible for territories outside the USA
byte	Digital information unit consisting of eight bits

C

C	Programming language developed by Dennis Ritchie
CA	California, state of the USA
Caldera	ISV for OS and application software
Canon	Japanese manufacturer of imaging and optical products
carpe diem	Seize the day
CD	Compact disk, optical storage disk for digital data
CE	Windows CE, MS OS for mobile devices and industrial controllers
CeBIT	Largest trade fair for IT and telecom solutions, held every spring in Hannover, Germany
CEC	Corporate Executive Committee, IBM's top management group

CEO	Chief executive officer
CES	Consumer Electronic Show, held each January in Las Vegas, Nevada
CFO	Chief financial officer
Chase, Brad	SVP of MS's Windows marketing group
Chicago	Code name for MS Windows 95 OS
Chicago school of antitrust	Institution of thought in regard to economics, law, and national policy
chip	Basic electronic component in form of an integrated circuit
Chongqing	Major city in the southwestern part of PRC
Chrome	Google's Internet browser and OS
Churchills	Cigars of a large size as preferred by England's former prime minister
CISC	Complex instruction set computer, computer architecture able to execute more than one instruction simultaneously
Claflin, Bruce	GM and VP of IBM's PC company, now chairman of AMD
Claris	Apple-owned ISV
Clausewitz, Carl von	Nineteenth century Prussian general, military theorist, and author
clone	Hardware or software system designed to mimic another
cloud	Grid or network-delivering computing services
cloud computing	Execution of software programs via the cloud
COBOL	Programming language used for business, financial, and administrative tasks
code	Computer programming instructions
COMDEX	IT expo held every year in Las Vegas, Nevada
Commodore	Home computer and PC manufacturer
Commodore 64	Best-selling 8-bit home computer from Commodore
Compaq	PC manufacturer now owned by HP

compiler	Program translating source code from high-level programming languages into CPU executable code
computer	Programmable machine carrying out sequences of arithmetic and logical instructions
Con gusto	With pleasure
COO	Chief operating officer
Corel	Utah-based ISV, engaged in word-processing software
C++	Intermediate programming language developed by Bjarne Stroustrup
CP/M	Control program for microprocessors, OS developed by DRI
CP/M-86	DRI's disk OS for the Intel's family of CPUs
CPU	Central processor unit, electronic component capable of executing the binary instructions for a computer system
Cro-Magnon	Early *Homo sapiens* of the European Upper Paleolithic

D

Danton, George	Leading figure of the French Revolution
database	Application program organizing a collection of data
DataQuest	Provider of research and analysis for the IT industry
dBase	Popular database produced by Ashton-Tate
DEC	Digital Equipment Corporation, manufacturer of minicomputers and PCs
Dell	IBM PC clone manufacturer
Dell, Michael	Founder and CEO of Dell
Deloitte & Touche	Professional services firm
deposition	Witness testimony given under oath
desktop	Screen background for a user interface operating a computer
desktop PC	Stationary PC

DESQview	User interface for MS-DOS designed by Quarterdeck
developer	Person designing or writing software programs
DOJ	Department of Justice, USA's top law-enforcement agency
DOS	Disk Operating System, OS needing a hard disk to function
sot-com bubble	Stock market rise and fall caused by enterprises engaged in Internet-related commerce between 1995 and 2001
DR-DOS	DRI's disk operating system for IBM PCs
Dreamcast	Game console manufactured by Sega
DRI	Digital Research Institute, ISV producing OSs and software development tools
driver	Program guiding the interactions between the CPU and its peripheral devices
DVD	Digital versatile disk, high-capacity optical storage disk exceeding that of a CD

E

e-book reader	Text for a book stored on an electronic device
eMachines	Korean investor-backed PC manufacturer, later bought by GW
e-mail	Messaging system allowing computer users to exchange data via a network
Ernst & Young	Professional services firm
Escom	German PC manufacturer
étude	Musical composition of considerable difficulty
evangelist	Software specialist helping other developers
Excel	MS's spreadsheet version for the Mac- and Windows-powered PCs

F

Facebook	Popular brand of social network services
Fade, Richard	SVP of MS, led the OEM group after I left

FE	Far East, all Asian counties
Feds	Federal officials or agents
Firefox	Open-source Web browser
Fischer, Franklin	DOJ's economic expert during MS's antitrust trial
FLP	Front Line Partnership, marketing pact between Compaq and MS
FORTRAN	Programming language for numeric and scientific computations
Franco	French
Frederick the Great	Eighteenth century king of Prussia from the Hohenzollern dynasty
Free Software Foundation	Nonprofit organization for the for free-software movement
FTC	Federal Trade Commission, regulates business practices supposedly to protect consumers and enterprises
Führungsstil	German, leadership style
FY	Fiscal year, period used for calculating annual financial statements

G

GameCube	Game console manufactured by Nintendo
Gassée, Jean-Louis	Former Apple GM and founder and CEO of Be
Gates, Bill	Cofounder and chairman of MS
Gateway	PC manufacturer, now part of Acer
GB	Gigabyte, one equals million bytes of 8-bit information
GE	General Electric, large conglomerate
GEM	Graphical Environment Manager, DRI's GUI for MS-DOS and DR-DOS
GEOS	Graphic Environment OS for the Commodore 64 home computer
Gerstner, Lou	IBM's CEO and chairman from 1994 to 2001

GM	General manager
Gneisenau, Neidhardt von	Nineteenth century count and Prussian field marshal
GNU GPL	License for open-source software, originated by Richard Stallman
Google	ISV and Internet services provider
Great Wall company	PC manufacturer in PRC
GUI	Graphical user interface, user shell to operate a computer with mouse clicks and graphical icons
GW	Gateway

H

hacker	Person accessing a computer system by circumventing security measures
Hades	Greek, the abode of the dead
Hallman, Michael	President of MS from 1990 to 1992
Hard drive	Information storage device containing a spinning disk holding large quantities of data
hardware	Physical components of a computer system
Harris, Jim	VP of MS and predecessor in my OEM job
HCL	Hindustan Computer Limited, India-based IT company
Heiner, Dave	MS's antitrust attorney
Hewlett-Packard	IT conglomerate
Hitachi	Japanese manufacturer of computer systems
HK	Hong Kong, city in Asia located in special administrative region of the PRC, formerly under British rule
Holley, Steven	Senior attorney, guided me through the antitrust maze
holographic images	Images stored in a three-dimensional format
Holstein	A region in Northern Germany
house of cards	Structure in danger of collapsing
HP	Hewlett-Packard

| HTML | HyperText Markup Language, program language for designing Web pages |

I

IBM	International Business Machines, computer manufacturer and IT services company
IBM PC	Personal computer designed by IBM
IBM PC clone	PC functioning like an IBM PC
IBM PC company	IBM's business unit in charge of her PC business
IBM PC compatible	IBM PC clone
icon	Pictogram displayed on a computer screen
IE	Internet Explorer
in camera	Latin, in private—not open to the public
injunction	Court order, requiring a party to refrain from certain acts
Intel	Semiconductor manufacturer
IP	Intellectual property, product of human intellect that can be protected by law
Internet Explorer	MS's Web browser integrated in Windows
Interpreter	Software executing programming instructions
iPhone	Apple's Internet-enabled smartphone
ISP	Internet services provider, company connecting users to the Internet
ISV	Independent software developer, company producing commercial software programs
IT	Information technology

J

Jackson, Thomas P.	Judge presiding over MS's antitrust trial from 1998 to 2001
Java	Object programming language and partial OS
Julius Caesar	Roman general and statesman

K

KB	Kilobyte, one equals one thousand bytes
Kindle	Amazon's brand of e-book readers
Klein, Joel	Assistant AG, head of DOJ's antitrust division from 1997 to 2002

L

LAN-Manager	MS's first network operating system
laptop PC	Portable PC or notebook
l'audace	French, audacity
Legend	PRC-based PC manufacturer, now called Lenovo
Lenovo	Same as Legend, bought IBM's PC business
LG	Lucky Goldstar, Korean-based electronic conglomerate
license	Legal instrument governing the usage, redistribution, and replication of software
Lieven, Theo	CEO and cofounder of Vobis
Linux	Linus Torvalds's UNIX version
Lisa	Apple's first workstation computer
Logitech	Swiss-based PC peripheral manufacturer
Lotus	ISV producing office applications, now part of IBM
Lotus 1-2-3	Lotus's spreadsheet product

M

MA	The state of Massachusetts
Mac	Apple's Macintosh PC
Mac OS	Operating system for Apple's Macintosh PC
mainframe	Largest multiuser computer system for scientific and commercial use
Mao	Communist leader of the PRC from 1949 to 1976
marines	One of the four branches of the US military

Maritz, Paul	Former group VP of MS's system-development group
MBA	Master of business administration, college degree
McKinsey	Management consulting firm
MDA	Market development agreement, contract regulating marketing incentives
Me	Millennium
media tablet	See *tablet*
melee	French, confused struggle
micro	Microcomputer, early PC
Microsoft	Largest ISV and PC-device manufacturer
middleware	Software components needed to connect to the Internet
minicomputer	Class of midrange computer systems located between PCs and mainframes
minimum commitment	Purchase guarantee in a contract
MINIX	Andrew Tannenbaum's version of UNIX
MIT	Massachusetts Institute of Technology
MLB	Master League Baseball
Moltke the Elder	Helmuth von, nineteenth century German field marshal
Morpheus	Greek mythology, the god of dreams
Mosaic	First popular Web browser
Mostek	Integrated circuit manufacturer
motherboard	Circuit board containing crucial electronic components like the CPU for a computer system
Motorola	Semiconductor manufacturer
MS	Microsoft
MS-DOS	Microsoft's disk operating system for IBM-compatible PCs
MT	State of Montana
Multiplan	MS's early spreadsheet version for MS-DOS

Mustek	South African PC manufacturer

N

NAP	Non-Assertion of Patent, clause restricting the use of patents lawsuits
Napoleon Bonaparte	Emperor of France from 1804 to 1815
Napoleon complex	Alleged type of inferiority syndrome supposedly affecting men short in statue
Nat Semi	National Semiconductor, electronic-component manufacturer
Navigator	Name for Netscape's Web browser
NBR	National Bureau of Asian Research, Seattle-based think tank
NCSA	National Center for Supercomputing Applications
NEC	Nippon Electric Company, Japanese IT conglomerate
netbooks	Stripped-down, lightweight, and inexpensive notebook PCs
Netscape	Web browser ISV
Netware	NOS produced by Novell
network	Interconnected computer web allowing sharing of resources and information
networking effects	How the number of users of goods and services impact their values
Neukom, Bill	MS's general council from 1985 to 2001
NFL	National Football League
NH	State of New Hampshire
Nintendo	Japanese game-console manufacturer
Nixdorf	German computer manufacturer
nom de plume	French, pseudonym or pen name
NonStop computing	Name for Tandem's server
NOS	Network operating system
NOOK	Barnes & Noble's brand of e-book readers

notebook PC	Portable PC
Notes	Program allowing communication and collaboration between computer users
Novell	ISV specializing in NOS and its development tools
North, Oliver	Retired Marine Corps officer, center of attention during Iran-Contra affair
NT	New Technology, MS's OS for enterprises
NVidia	Semiconductor manufacturer for graphic subsystems and ARM processors
NYC	New York City

O

object oriented	Programming paradigm using data structures and fields and their interactions to develop software
OEM	Original equipment manufacturer, exchangeable with PC manufacturer
Office	MS's suite of office productivity software
Oki, Scott	SVP of MS my first boss
Olsen, Ken	Cofounder and former CEO and chairman of DEC
open standard	Publically available specifications
Oracle	ISV producing databases and commercial applications
OS	Operating system, software managing computer resources and providing services for application programs
OS/2	Name for an OS jointly designed by MS and IBM
Osborne	Early PC pioneer

P

PAC	Political action committee, organization advancing political outcomes
Palmisano, Sam	IBM's chairman
Panthera leo	Latin, lion

Pascal	Procedural programming language
path-dependent outcomes	Results influenced exclusively by decisions made in the past
Patton, George S.	WWII US general
PB	Packard Bell
peripheral	Device attached and managed by a host computer
PC	Personal computer, range spanning from netbooks over notebooks to desktops and workstations
PCjr	PC junior, IBM's first consumer PC (1984)
PDP-11/20	Programmed Data Processor, one of DEC's 16-bit minicomputer models
Pentium	Intel's successor to the 80486 CPU
per-copy license	Pay-as-you-go software license
per se rule violation	Illegal act by statue, constitution, or case law
per-system license	Licensing on a per-model basis
Pfeiffer, Eckhard	Former CEO of Compaq
PGL	Price guideline, comparable to a price list
Phillips	Netherlands-based consumer electronics company
Phoenix	ISV engaged in PC software tools
PlayStation	Game console created by Sony
Pogo principle	Shooting yourself in the foot
Posner, Richard	Judge at the Chicago appeals court, legal theorist, teacher, and book author
PostScript	Programming language for electronic and desktop publishing
PowerPoint	MS's version of a slide-presentation program
PowerPC	IBM's workstation computer
PPC	PowerPC
PR	Public relations
PRC	People's Republic of China

prima donna	Italian, someone who behaves in a temperamental fashion revealing an inflated image of him or herself
Printaform	Mexican enterprise engaged in paper and PC manufacturing
processor	Short for CPU
program	Instructions written in a programming language to complete a computing task
programming language	Artificial language designed to communicate instructions to a computer system
Prussia	Kingdom and historical military state located in the northeastern part of Germany
Psion	UK manufacturer of rugged mobile handheld computers
Pyrrhic victory	Inflicts more long-term losses than immediate gains

Q

quid for pro	A favor for a favor
QDOS	Quick and Dirty OS, written by Seattle Computer Products Company

R

RadioShack	Tandy Corporation
RealNetworks	Provider of media software and services
Reback, Garry	Attorney specialized in antitrust law, hostile to MS
Red Hat	ISV, Linux distributor and support agent for Linux
Redmond	Town near Seattle, WA, where MS's headquarter is located
rep	Representative
RISC	Reduced Instruction Set Computer, computer architecture containing only simple instructions for speedy execution and lower heat consumption
Rockefeller Standard Oil	American oil company in the early twentieth century
ROM	Read-only memory
ROMable OS	OS residing in ROM

Rose, John	Former DEC and Compaq SVP
royalty	Payment made by a licensee to a licensor

S

SA	South Africa
Sacramento	Seat of the CA state government
Salt Lake City	Capital of the state of Utah
Santelli, Tony	GM IBM power PC and VP IBM PC company
Samsung	Electronics conglomerate located in South Korea
sans lumière	French, without light
SAP	German ISV engaged in database software
Scharnhorst, Gerhard von	Prussian General and chief of staff (1755–1813)
Schneider	Once a German PC manufacturer
SCO	Santa Cruz Operation, ISV specialized in a UNIX version called Xenix and its development tools
SCP	Seattle Computer Company
Sega	Japanese game-console manufacturer
Seoul	Capital of the Republic of South Korea
server	Computer linking others together and providing storage and communication services across a network
server farm	A collection or cluster of connected computer systems
Shanghai	Large city in the eastern part of PRC
Sharp	Japanese manufacturer of electronic products
shell	User interface for operating a computer
Shirley, Jon	President of MS from 1983 to 1990 and board member until 2008
Siemens	German electro conglomerate
Silicon Valley	Southern part of the San Francisco Bay Area in Northern CA, home to many high-tech companies
sitz Fleisch	German, the flesh one sits on

SmartSuite	Lotus's office productivity suite
SoCal	Southern California
software	Entities of programs, procedures, and algorithm providing computer instructions
software piracy	Illegal reproduction of computer software
Solaris	Sun Microsystem's UNIX version
solid state drive	SSD, storage unit containing computer memory and no movable parts, acting like a hard drive
Sony	Japanese consumer electronics conglomerate
source code	Text written in a computer programming language
Soviet Union	Union of Soviet Socialistic Republics, from 1921 to 1991
spreadsheet	Application program simulating an accounting worksheet
Spindler, Mike	Former CEO of Apple
SR	Sales representative
Stac Electronics	SoCal ISV selling hard disk compression programs
stock options	Right to buy stock at a fixed price, corporate grants to employees
Sugar, Allan	Founder and former CEO of Amstrad
Sun Microsystems	Workstation manufacturer and ISV, now part of Oracle
SVP	Senior vice president
Sybase	ISV for relational database software, now part of SAP
Symantec	ISV engaged in security software
system builder	Tiny PC-assembly outfit

T

tablet	Lightweight, touch-operated computing devices not containing a hard drive
Taligent	ISV engaged in OSs and development tools, now part of IBM

Tandy	Reseller and producer of electronics goods
TCP/IP	Transmission Control Protocol/Internet Protocol, rules Internet messaging systems
Technology Achievement Medal	National Medal of Technology, granted by the president of the USA to American inventors and innovators for significant tech contributions
Texas Instruments	Semiconductor manufacturer
The Research Board	International think tank headquartered in NYC
thin client	Slimmed-down PC relying heavily on cloud computing
TI	Texas Instruments
TLC	Tender loving and care
Tokyo	Capital of Japan
TopView	IBM's text-mode interface for MS-DOS powered PCs
Toshiba	Japanese IT conglomerate
touché	French, expression for acknowledging a telling remark
Triumph Adler	German company producing a variety of office systems
TRS-80	Tandy's microcomputer model 80
Twitter	Phone-messaging network for short messages

U

UNIX	OS originated by AT&T and ported on several computer architectures
US/USA	United States of America
USB	Universal Serial Bus, cable, connector, and protocol standard for computer peripherals

V

Viglen	UK PC manufacturer later bought by Amstrad
virtual machine	Isolated guest OS working independently from the main computer OS

VisiCalc	Early spreadsheet realization by VisiCorp
VisON	Early GUI for MS-DOS powered PCs by VisiCorp
Visual BASIC	Advanced BASIC version from MS
VP	Vice president
VTech	Video technology, HK based manufacturer of consumer electronics

W

WA	State of Washington
Waitt, Ted	Cofounder and former CEO and chairman of Gateway
Warp	Name for late versions of OS/2
WikiLeaks	Organization publishing submissions from anonymous sources
Wind River	ISV specialized in tools and OS for industrial devices
Windows	MS GUI for PCs
Windows CE	Mini version of Windows for use in consumer electronics and industrial controllers
Windows NT	Third-generation Windows OS, since 2001 the tech underpinning of all MS Windows versions
Windows X	GUI for UNIX systems
Word	MS word processor for MS-DOS, Windows, and the Mac
word processor	Application program used for composing, editing, formatting, and printing text material
WordPerfect	Word processor for PC OS from Corel
WordStar	Word processor produced by MicroPro International
Workbench	GUI for Amiga home computers
workstation	Stationary computer used for graphical design or heavy engineering work
Wright brothers	Two Americans credited for building the first self-powered airplane and making a successful flight in 1903
WTO	World Trade Organization

WW	Worldwide
WWI	World War I
WWII	World War II
WWW	World Wide Web, the Internet

X

Xbox	MS game console
Xenix	MS's and later SCO's version of UNIX
Xerox	Global document-management company

Y

Yahoo!	Internet portal provider for search and e-mail services
YouTube	Video-sharing website owned by Google

Z

Zenith	Consumer electronics manufacturer
Zilog	Z (the last word of) Integrated Logic, microcontroller manufacturer

REFERENCES

Auletta, Ken, *World War 3.0: Microsoft and its enemies*, Random House, 2001

Banks, David, *Breaking Windows: How Bill Gates fumbled the future of Microsoft*, Simon and Schuster, 2001

Clausewitz, Carl von, *Vom Kriege (About the war)*, Ullstein, 1980 (Dümmler 1832/1853)

Condell, Bruce and Zabecki, David, *On the German Art of War: Truppenführung, German army manual for unit command in WWII*, Stackpole Books, 2009

Eisenach, Jeffrey and Lenard, Thomas, *Completion, Innovation and the Microsoft Monopoly: Antitrust in the Digital Marketplace*, Kluwer Academic, 2001

Ellig, Jerry, *Dynamic Competition and Public Policy: Technology, Innovation, and Antitrust Issues,* Books LLC, 2010

Evans, David and Fisher, Franklin and Rubinfeld, Daniels and Schmalensee, Richard, *Did Microsoft harm Consumers: Two Opposing Views,* The AEI Press, 2000

Gerstner, Louis, *Who Says Elephants Can't Dance: Leading a Great Enterprise through Dramatic Change,* Harper Collins, 2003

Gordon, Allan, *Antitrust Abuse in the New Economy: The Microsoft Case,* Edward Elgar 2003

Isaacson, Walter, *Steve Jobs,* Simon & Schuster, 2011

Jordan, Jonathan W., Brother Rivals Victors, Eisenhower, Patton, Bradley and the Partnership that Drove the Allied Conquest in Europe, NAL Caliber, 2012

Kotter, John P. et al., "On Leadership", Harvard Business Review, 2012

Levy, Robert, *Shakedown: How Corporations, Government, and Trial Lawyers Abuse the Judicial process,* Cato Institute, 2004

Levy, Stephen, *In the Plex: How Google Thinks, Works and Shapes Our Lives,* Simon & Schuster, 2011

Liebowitz, Stan and Margolis, Stephen, *Winners, Losers and Microsoft: Competition and Antitrust in High Technology*, The Independent Institute, 1999

Maxwell, Fredric, *Bad Boy Ballmer: The Man Who Rules Microsoft*, Harper Collins, 2003

McKenzie, Fredric, Trust on Trial: *How the Microsoft Case Is Reframing the Rules of Competition*, Perseus, 2000

Miller, Fredric and Vandome, Agnes, and McBrewster, John, *Bill Gates*, Alphascript, 2009

Page William and Lopatka, John, The Microsoft Case: Antitrust, High Technology and Consumer Welfare, University of Chicago Press, 2010

Phelps, Marshal and Kline David, *Burning Ships: Intellectual Property and the Transformation of Microsoft*, John Wiley & Sons, 2010

Posner, Richard, *Antitrust Law*, University of Chicago press, 2002

Rand, Ayn, *The Voice of Reason: Essays in Objectivist Thought*, Penguin Group, 1991

Reback, Gary, *Free the Market*: Why Only Government Can Keep the Market Place Competitive, Penguin Group, 2010

Simson von, Ernest, *The Limits of Strategy: Lessons in Leadership from the Computer Industry*, iUniverse, 2009

Sugar, Allan, *What You See Is What You Get: My Autobiography*, Macmillan, 2010

INDEX

A

Acer, 44, 86, 128, 205, 264, 306, 331, 351, 357
Active Desktop, 157
Advanced Micro Devices (AMD), 87, 250, 264, 311–12, 337–38, 351, 354
Advanced RISC Machine (ARM) processors, 144
AIM (Apple, IBM, Motorola alliance), 71–73, 141, 351
Akers, John, 107
Alagem, Beny, 27, 202–3, 351
Alchin, Jim, 173, 292–93
Allen, Paul, 10
American Online (AOL), 76, 135, 151–52, 158, 200, 241, 294, 351
Amiga, 205, 341, 351, 369
Amstrad, 39–40, 58, 203, 351, 367–68
Andreessen, Mark, 173
Android, 191, 293, 297, 299, 307, 312, 314, 317
Antitrust Law (Posner), 218, 223, 250, 372
antitrust regulations, 95–96, 159, 213–14
antitrust trial, ix, 32, 35, 47, 89, 95, 159, 218, 223, 250, 354, 371–72
Apple Computer, 13, 41, 71–73, 87–89, 156–57, 183–84, 219–20, 245–46, 292–94, 296–97, 302–3, 306, 314–20, 328–30, 332
applets. *See* Java
application programming interface (API), 141, 174, 277, 351
AST Research, 196

Auftragstaktik, 3–4, 16, 27, 35, 48, 203, 258, 271, 273, 282–83, 309, 323, 352
Auletta, Ken, 276
Austro-Prussian War, 4
Azure, 311

B

Baber, Mark, 121, 242
Bach, Robbie, 297, 304
baijiu, 46
Ballmer, Steve, 12, 47, 49, 80, 185, 199, 259, 289, 292, 320, 342
Basic Input/Output System (BIOS), 24, 154, 301, 331, 353
Beard, Tim, 20
Beginner's All-purpose Symbolic Instruction Code (BASIC), 10, 140, 328, 341, 352, 369
Beijing, 44–46, 195, 197–98, 267, 352
Be Inc., 73
Belluzzo, Richard, 30
BeOS, 73, 245, 352
Berners-Lee, Tim, 123
Big Blue. *See* International Business Machines
Bingaman, Anne, 93, 96, 99–100, 213
Blattner, Jeffrey, 213
Boies, David, 213–14, 218, 222, 224–27, 229–41, 256, 353
Bork, Robert, 168
Borland, 73, 85, 353
Bosch, 144, 146
Boyd, Vig, 90
Brown, Mike, 104
browser, 123–25, 127, 132–35, 141, 148, 157–58, 167–68, 173, 183, 189, 221, 231–36, 239, 296–97, 304

373

Buffalo Bill, 201
Burning the Ships (Phelps and Kline), 105
Butler, Jeremy, 16, 35, 57, 134

C

Canion, Rod, 25, 67
Cannavino, James, 32–33
Canon, 30, 59, 353
Capellas, Michael, 207
CeBIT, 51, 53, 55, 353
Cecil, Jim, 28
central processor unit (CPU), 71–72, 98, 142, 185, 194, 250, 264, 297, 301, 311, 327, 337–38, 352–53, 355–56, 364–65
Chase, Brad, 85
Chicago, vi, 111–13, 115, 121–22, 124–25, 127, 161, 164, 254–55, 278, 354, 364, 372
Chowdhry, Jai, 267
Chrome, 191, 296–97, 299, 304, 312, 316, 354
Chuanzhi, Lui, 195
Claesson, Jan, 238
Claflin, Bruce, 115, 120–21, 129–30
Claris, 347, 354
Clark, Jim, 124–25
Clark, Mike, 66–68, 238
Clarkson, Larry, 197
Clauson, Roy, 115
Clinton, Bill, 79–80, 139, 197, 214–15, 252, 267
Clinton, Hillary, 197
Clow, Garry, 104
Cole, David, 172
Collas, Jim, 204
COmmon Business-Oriented Language (COBOL), 354
Compaq, 25–26, 47–49, 58, 66–68, 85–88, 107, 115–17, 143, 151–52, 206–7, 238, 252, 270, 337
Complex Instruction Set Computer (CISC), 337, 354
COMputer Dealers' EXhibition (COMDEX), 116, 120, 125, 136, 340–41, 354
Consumer Electronics (CE), 144–46, 263, 285, 296, 353, 369
Consumer Electronics Company, 193, 364
Creighton, Susan, 168

Cutler, Dave, 32, 141–42

D

Danton, Georges, 289
Daohan, Wang, 197–98
Dataquest, 355
Dell, 41, 65–66, 86, 107, 128, 144, 158–60, 191, 196, 207, 237, 270, 306–7, 335
Dell, Michael, 66, 201, 209, 270
Deloitte & Touche, 120, 355
DESQview, 340, 356
Digital Equipment Corporation (DEC), 26–27, 33, 86, 108, 128, 206, 333, 339, 346, 355, 363–64, 366
Digital Research Inc. (DRI), 2, 9, 26–27, 37, 39–44, 53, 55–57, 60, 75–78, 139, 144, 329–32, 338, 341, 355–57
Duers, Grant, 62
Dunn, Celeste, 151
Dynamo. *See* Mike Spindler

E

Easterbrook, Frank, 168
Edwards (chief justice), 275
Eisenberg, David, 188
Ellings, Richard J., 197
eMachines, 203, 205, 356
e-mail addiction, 6
Ernest & Young, 120
Escom, 41, 205, 356
Exley, Charles, Jr., 68

F

Facebook, 168, 294, 299, 313–15, 356
Fade, Richard, 45, 47, 49, 260, 273–74, 276, 284–86, 289, 300
Federal Trade Commission (FTC), 52, 93–94, 99–100, 155–56, 165, 171–76, 199–201, 204–5, 213–15, 217–23, 241–42, 250, 276, 278–79, 298–99
Fiber TV, 317
1512, 39
Fiorina, Carly, 207
Firefox, 294, 299, 304, 357
Fisher, Franklin, 218, 222, 224, 241, 246
FORmula TRANslation (FORTRAN) language, 333, 357
Forster, Richard, 302
Fouché, Joseph, 290

Franco-German War, 4
Friedman, Milton, 276
Fries, Ed, 264
Front Line Partnership (FLP), 86, 115
Furukawa, Susumu "Sam," 42

G

Gartner, 301, 305
Gassée, Jean-Louis, 41, 73, 340, 347
Gates, Bill, 9–13, 48–50, 59–61, 72–73, 94, 111, 159–62, 164, 201, 233, 295, 298–300, 310, 331–33, 341–43
   Road Ahead, The, 136
   think week, 56, 147
Gateway (GW), 65–66, 204–5, 230, 237, 356, 358
Gaudette, Frank, 80
General Public License (GPL), 186, 358
Germany, ix–x, 2–3, 7, 9–10, 16, 20, 35, 39–41, 51–52, 61, 128, 144, 273–74, 345–47, 352–53
Gerstner, Louis, 109, 130, 132–33, 162–64, 302, 306
   Who Says Elephants Can't Dance?, 112
Gibbon, Edward, 320, 323
Glaser, Rob, 47, 187, 278, 304
"Gold Rush" campaign, 127
Google, 105, 168, 191, 276, 292–94, 296–97, 299, 302, 307, 312–17, 324, 354, 358, 370
Gort-Allen, Brigitte, 197
Graphical Environment Manager (GEM), 341
Graphical User Interface (GUI), 31, 339–42, 357–58, 369
Graphic Environment Operating System (GEOS), 340, 357
Great Wall (company), 44–46, 358
Groove Networks, 295
Groupe Bull, 41, 203, 353
Grove, Andy, 250
Grupo Printaform, 55–56, 365
Guest, Kelly, 270
Gulledge, Carl, 87, 349

H

Hallman, Michael, 57, 80
"Halloween document," 188
Halo, 264
Hard Drive (Wallace), 100

Harris, Jim, 15–16, 24–25
HCL, 267, 358
Heiner, Dave, 93, 105, 156, 225
Hewlett, Walter, 207
Hewlett-Packard (HP), 29–31, 86–87, 107, 135, 143–44, 187, 189, 191, 203, 207, 217, 306, 331, 339, 358
Hindustan Computer, 267
Ho, Bosco, 43–44
Holley, Steven, 225–26
Honeywell, 144
Hong Kong, 43, 60, 195–96, 358
Hosogi, Ron, 42, 349
Hotmail, 294, 313
Hsu, Ben, 44
Hurd, Mike, 207
Hyper Text Markup Language (HTML), 183, 359

I

IBM. *See* International Business Machines Corporation (IBM)
"IBM First" campaign, 117, 121, 242–43
IBM-Works, 133
Idei (president of Sony), 59
Immelt, Jeff, 200
independent software vendors (ISVs), 9, 11, 33, 183, 189, 247, 263–64, 286, 304, 313–14, 324, 330, 332, 334, 342–43
"Information at Your Fingertips," 125
Intel, 15, 19, 26, 31, 55, 71, 87, 141–42, 185–87, 191, 250, 301, 311, 337
International Business Machines Corporation (IBM), 12–15, 29–34, 49–50, 71–73, 106–9, 111–13, 115–17, 119–21, 128–34, 161–65, 245–53, 328–34, 336–43, 351–57, 359–60
Internet Explorer 3.0, 134–35, 141, 148–49, 151, 157–58, 168, 171–74, 179, 183, 214, 218, 221, 233–34, 236–39, 255–56
Internet Service Provider (ISP), 230, 359
iPhone, 296, 299, 319, 359

J

Jackson, Henry M., 197
Jackson, Thomas Penfield, 101, 220
Japan, 42, 66, 134, 203, 263, 368
Java, 140–41, 165, 237–38, 302, 359

Jobs, Steve, 13, 33, 73, 157, 168, 184, 207, 220, 290, 302–3, 306, 317, 329, 340, 347

K

Kan, David, 268–69
Kazandjian, Diran, 90
Kempin, Joachim, 288
   life story, 345
   as subsidiary manager, 1–2, 5–7, 16
Kinect, 318
Klein, Joel, 168, 174, 213
Kleiner Perkins Caufield & Byers, 124
Kline, David
   Burning the Ships, 105
Kolb, Kurt, 302, 349
Kollar-Kotelly, Colleen, 276
Kontron, 144
Kopel, Dave, 275
Kotter, John P., 4
Koviac, William, 223
Kuehler, Jack, 50, 71, 73, 109
Kutaragi, Ken, 263

L

LAN-Manager, 47–48, 77, 142, 243, 360
l'audace, 289
Lautenbach, Ned, 162
leadership, 13
Leap Motion controller, 318
Lenovo, 195–97, 301, 306, 360
licenses, 2, 17–19, 29, 62, 82, 86, 127, 246, 285, 342
Lieven, Theo, 40, 51, 53, 128, 281, 301
Linux, 142, 165, 168, 185–91, 245–46, 257, 286–87, 292–93, 299, 307, 316, 324–25, 351, 360, 365
Lisa, 13, 89, 340, 360
Logitech, 89–91, 360
Lotus, 2, 7, 33–34, 50, 58, 73, 108–9, 132–33, 161, 242, 252–53, 295, 344, 360, 367
Lotus 1-2-3, 2, 132, 344, 360
Lotus SmartSuite, 109
Lucky Goldstar (LG), 43, 317, 360

M

Macintosh (Mac), 13, 31, 41, 78, 87, 89, 97, 124, 184
Malone (chief prosecutor), 173, 216–17
management, 16, 95, 293, 302, 305, 353, 361

Mann, Thomas, 238
Maples, Mike, 80
Maritz, Paul, 32, 49, 72, 148, 161–62, 210, 215, 236, 290
market development agreement (MDA), 88, 128–29, 252, 361
Massachusetts Institute of Technology (MIT), 186, 361
Maynard, Massachusetts, 26
McNealy, Scott, 141
Medhi, Jusef, 157
Metro, 312
Michels, 24
Microsoft, 11–13, 15–17, 24–26, 28–35, 39–45, 71–73, 130–36, 167–69, 171–76, 275–78, 286–97, 299–307, 309–21, 327–32, 338–44
MINIX, 185–86, 361
minutemen, 189, 287
Mirales, Jorge Espinosa, 55
Moltke, 4–5, 361
monopoly, 50, 72, 96–98, 100, 159–60, 171, 173, 214, 217–18, 220, 222, 245, 247–49, 254, 256
Monti, Mario, 278
Moore's law, 337
Mosaic, 123–24, 361
Mossberg, Walt, 112
Mostek, 206, 361
Motorola, 71, 73, 317, 329, 351, 361
MS Disk Operating System (MS-DOS), 2, 356, 361
MS-DOS 5.0, 65
MSN (Microsoft Network), 135
Multiplan, 2, 73, 89, 344, 347, 362
Mundie, Craig, 295
Myhrvold, Nathan, 264
MySpace, 294

N

Napoleon complex, 289–90, 298, 362
National Bureau of Asian Research (NBR), 197
National Cash Register Company (NCR), 65, 68–69
National Center for Supercomputing Applications (NSCA), 123–24, 362
Navigator, 125, 173, 183, 233, 253, 362
Nell Miller, 119
netbooks, 303

Netscape, 124–25, 134–35, 149, 151–52, 158–59, 167–68, 173, 183, 200, 229–30, 232–36, 238–39, 241, 253, 362
Netware, 47, 76–77, 139, 362
network operating system (NOS), 47
Neukom, Bill, 93, 104, 155, 172, 215, 222, 225
NeXTSTEP, 33
Nintendo, 263–64, 357, 362
Nippon Electric Company (NEC), 30, 42, 202, 264, 331
nonassertion of patent (NAP) clause, 105–6
Nook, 314, 363
Noorda, Raymond "Ray," 76, 78, 139, 259
Norris (IBM rebuttal witness), 133, 226, 241–43, 248, 252
Notes, 132, 148, 295, 363
Novell, 47, 73, 76–78, 103, 139, 159, 286, 362–63
NVidia, 311–12, 363

O

official price guideline (PGL), 15–16, 20, 364
Oki, Scott, 1, 7, 36, 39, 57, 68, 141, 363
Oksenberg, Mike, 197–98
Old Faithfuls, 300
Olsen, Kenneth Harry, 26–27, 108, 206, 263, 276, 339, 363, 371
Opel, John Roberts, 108
operating system (OS), 29–35, 47–50, 56–59, 71–73, 85–87, 107–9, 111–13, 116–17, 124–25, 129–30, 140–43, 145, 245–49, 251–52, 327–32
original equipment manufacturers (OEM), 7, 10, 15, 20, 25, 35–37, 47, 63, 67, 79–82, 89, 142, 216–17, 261, 301
OS/2, 13, 24, 29–35, 37, 49–50, 57–58, 71–73, 78, 85–86, 97, 101, 108–9, 111–12, 247–48, 251–52
Ozzie, Ray, 132, 295–97, 302, 305, 310–11, 314, 319

P

Packard Bell (PB), 27, 65–66, 129, 135, 202–3, 351, 364
Palmer, Robert, 206

Palmisano, Sam, 162, 164, 171, 237, 277, 305–6
Park, June, 43
PC junior (PCjr), 306
"Pearl Harbor" speech, 137
Pentium Pro, 142
People's Republic of China (PRC), 43–46, 61, 175, 195–98, 267, 301, 306, 352, 354, 358, 360, 364, 366
Pfeiffer, Eckhard, 67–68, 206
Phelps, Mike
    Burning the Ships, 105
Philips, 193–94
Phoenix Computer Systems, 331
PlayStation, 263–64, 364
Political Action Committee (PAC), 139, 202
POSIX, 141
Posner, Richard
    Antitrust Law, 218, 223
PowerPC, 71–72, 115, 161, 352, 364
Prussian army, 3–4
Psion, 145, 365

Q

Quick and Dirty Operating System (QDOS), 24, 330

R

Raikes, Jeff, 259, 297
Rand, Ayn, 95
read-only memory (ROM), 24, 77, 365
RealNetworks, 47, 159, 278, 304, 365
Reback, Garry, 95, 100, 156, 169
Reback, Gary
    Free the Market!, 167
Red Hat, 188, 307, 365
Reduced Instruction Set Computer (RISC), 71–73, 337, 365
Reiswig, Lee, 112, 130
Reno, Janet, 93, 96, 100, 139, 174, 256
RJR Nabisco, 109
Road Ahead, The (Gates), 136
Rollins, Kevin, 209, 270
Rose, John, 27, 206, 235
Rosen, Ben, 25, 66, 68, 207
Rubinfeld, Daniel, 218
Russell, George F., Jr., 197

S

sales team, 15, 33, 44, 56, 60, 179, 257, 302, 367

Saloner, G., 219
Samsung, 43, 105, 296, 306, 317–18, 366
Santa Cruz Operation (SCO), 24, 142, 339, 366, 370
Santelli, Tony, 115, 130, 134, 252
Schmalensee, Richard, 223
Schmitt, Manfred, 41
Schneider, 39, 366
Scott, George C., 36
scripts. *See* Java
Seattle, ix, xi, 9, 24, 44, 60, 68, 84, 197, 330, 365–66
Seattle Computer Products (SCP), 24, 365–66
Sharp, 30, 42
Sherman Act, 174, 214, 255, 275, 278
Shih, Stan, 44
Shirley, Jon, 9–12, 16, 23, 29–30, 35, 47–48, 57, 162, 366
Siemens, 2, 40, 331, 366
Silverberg, Brad, 85
Sinofsky, Steve, 317
Skype, 294, 317, 320
SmartSuite, 109, 132, 367
Smith, Chris, 134
Solaris, 72, 367
Sony, 59–60, 105
South Korea, 43, 366
Soyring, John, 108
Spindler, Mike, 13, 71, 73, 184, 346–47
Sporkin, Stanley, 100
Stac Electronics, 103, 367
Stallman, Richard, 186, 358
Stevenson, Bob, 161–62
Stimac, Gary, 25, 206, 209
Sugar, Alan, 39
Sullivan & Cromwell, 225
Sun, 72, 88, 140–41, 159
Swan, Philippe, 146
Swavely, Mike, 25, 67
Sybase, 344, 367

T

Taiwan, 44, 60, 90, 205, 268, 351
Taligent, 72, 367
Tandem, 206, 362
Tanenbaum, Andrew, 185
Tate, Ashton, 344

Technology Achievement Medal, 94, 368
Thoman, Richard, 116, 133–34
Thomson, Jon M., 162
Thomson, Rick, 89, 91, 265
3Com, 49, 130
Torvalds, Linus, 185–88, 190, 360
Tosaka, Kaoro, 42
Toshiba, 42, 306, 368
Triumph-Adler, 2
Twitter, 168, 294, 299, 314, 368

U

Ubuntu, 191, 307
UNIX, 2, 24, 30, 33, 72–73, 77–78, 97, 113, 124, 141–43, 183–86, 189, 338–40, 360–61, 366–70
Urowsky, Richard, 223, 225

V

Vaskevitch, David, 295, 297
Verges, Bernard, 2
Viglen, 90, 368
virtual machine, 140, 368
VisiCalc, 2, 328, 344, 368
Vobis, 40–41, 43, 51–53, 360
VTech, 43–45, 369

W

Waitt, Ted, 65, 201, 204, 241
warp, 112, 116, 119, 127, 132, 161, 248, 369
Wedell, Christian, 51
Weisfield, Kathy, 134
Weitzen, Jeffry, 205
Welch, Jack, 200
West, Woody, 276
Who Says Elephants Can't Dance? (Gerstner), 112, 248
WikiLeaks, 270, 369
Williams, Dick, 55, 75
Windows, 30–34, 49–53, 85–90, 105–9, 134–36, 140–45, 148–49, 153–54, 171–74, 187–91, 217–23, 234–38, 246–53, 310–20, 341–44
Windows 3.0, *33, 34, 49, 51, 52, 53, 58, 65, 95, 115*
Windows 7, *296, 301, 307, 311, 317*
Windows 8, *307, 311, 312,* 314–15, 317–18

Windows 95, 112, 127, 131, 134, 176, 179, 194, 203, 220, 229, 238, 250–52, 279, 311, 313–15
Windows 98, 165, 174, 199, 248–49
Windows Millennium (Me), 257
Windows XP, 257
Wind River, 144, 369
WordPerfect, 33, 58, 108, 159, 286, 344, 369
WordStar, 2, 9, 14, 58, 344, 369
Workbench, 341, 369
Workplace Shell, 108
World Books encyclopedia, 132
World Trade Organization (WTO), 198, 369

X

Xbox, 264–65, 291, 294, 299, 320, 370
Xenix, 2, 24, 40, 339, 366, 370

Y

Yahoo! 294, 296, 299, 370
York (IBM's chief financial officer), 131
YouTube, 294, 299, 370
Yuanqing, Yang, 195

Z

Zappacosta, Pierluigi, 89
Zee (CEO of Great Wall company), 44
Zemin, Jiang, 197
Zenith, 203, 331, 370
Zune, 294

CPSIA information can be obtained at www.ICGtesting.com
Printed in the USA
LVOW050723090213

319249LV00002B/43/P